AF389496

Potentially Toxic Elements Pollution in Urban and Suburban Environments

Potentially Toxic Elements Pollution in Urban and Suburban Environments

Editor

Ilaria Guagliardi

MDPI • Basel • Beijing • Wuhan • Barcelona • Belgrade • Manchester • Tokyo • Cluj • Tianjin

Editor
Ilaria Guagliardi
National Research Council of Italy (CNR-ISAFOM)
Italy

Editorial Office
MDPI
St. Alban-Anlage 66
4052 Basel, Switzerland

This is a reprint of articles from the Special Issue published online in the open access journal *Toxics* (ISSN 2305-6304) (available at: https://www.mdpi.com/journal/toxics/special_issues/Toxic_Element).

For citation purposes, cite each article independently as indicated on the article page online and as indicated below:

LastName, A.A.; LastName, B.B.; LastName, C.C. Article Title. *Journal Name* **Year**, *Volume Number*, Page Range.

ISBN 978-3-0365-6237-7 (Hbk)
ISBN 978-3-0365-6238-4 (PDF)

Contents

About the Editor

Ilaria Guagliardi

Ilaria Guagliardi is a permanent researcher at the National Research Council of Italy, Institute for Agricultural and Forestry Systems in the Mediterranean (ISAFOM) in Rende (CS), Italy. Her research interests mainly include environmental geochemistry; pollution phenomena in agricultural and urban soils, waters and sediments; geostatistical and multivariate analysis of data with the support of compositional and multifractal data analysis methods; the development of geospatial geochemical risk analysis methods for the optimized management of environmental impacts on a regional scale. In addition, she has also carried out water-related research, such as research on hydrological modeling, watershed management and landscape ecology. Contextually, she participates in intense editorial and peer-reviewing activities, being the associate editor and reviewer of more than 50 international scientific journals. She is the author of more than 50 scientific publications and an expert panel member for evaluating international research projects. She currently collaborates with several Italian and foreign universities and research institutes.

Editorial

Editorial for the Special Issue "Potentially Toxic Elements Pollution in Urban and Suburban Environments"

Ilaria Guagliardi

National Research Council of Italy-Institute for Agricultural and Forest Systems in Mediterranean (CNR-ISAFOM), Via Cavour 4/6, 87036 Rende, Italy; ilaria.guagliardi@isafom.cnr.it; Tel.: +39-0984841427

Citation: Guagliardi, I. Editorial for the Special Issue "Potentially Toxic Elements Pollution in Urban and Suburban Environments". *Toxics* **2022**, *10*, 775. https://doi.org/10.3390/toxics10120775

Received: 30 November 2022
Accepted: 9 December 2022
Published: 11 December 2022

Publisher's Note: MDPI stays neutral with regard to jurisdictional claims in published maps and institutional affiliations.

Pollution by potentially toxic elements (PTEs) is becoming a serious and widespread issue in all environmental matrices because of accelerated population growth rate, rapid industrialization and urbanization, and other changes which have occurred in most parts of the world in the last few decades [1–3]. The increasingly worrying concern about the presence of PTEs in the environment has attracted considerable attention due to their potential impacts on ecosystem functioning and on public health because of their persistence and biotoxicity [4,5]. PTEs can in fact be transferred into the human body as a consequence of dermal contact, inhalation, and ingestion through the food chain and drinking water [6]. Unfortunately, PTEs are ubiquitous in all environmental compartments, and they have been widely detected worldwide [7,8]. In this context, environmental geochemistry and related subjects are elected matters to investigate, characterize, and reveal the patterns of inorganic elements together with geostatistical computations that are used to identify source patterns of different pollutants related to underlying geological features and/or anthropogenic activities [9,10].

The collection of the research carried out worldwide in this Special Issue "Potentially Toxic Elements Pollution in Urban and Suburban Environments" represents the need of the scientific community to characterize the behavior, transport, fate, and ecotoxicological state of PTEs in environmental matrices in both urban and suburban settings. The published research articles include, in fact, case studies from China to South Africa to Europe and interest various environmental compartments. Tomczyk-Wydrych et al. [11], Famuyiwa et al. [12], and Guagliardi et al. [13] approached, in their researches, the analysis of PTEs pollution in the soil matrix. Particularly, Tomczyk-Wydrych et al. [11] assessed the PTEs contamination of the topsoil in the selected part of the Zduńska Wola Karsznice railroad, in central Poland, and individuated the anthropogenic origin of the high level of copper, cadmium, lead, and others in the surface layer of soil, the risk to living organisms, and the destabilizing effect on ecosystems. The high accumulation index of copper, cadmium, and lead in the surface layer of soil indicated their anthropogenic origin.

Famuyiwa et al. [12] determined the concentrations and potential mobilities of Cr, Cu, Fe, Mn, Ni, Pb, and Zn in soils from public-access areas across the Lagos metropolis (Nigeria) and evaluated the risks associated with human exposure to PTEs, evidencing how the PTE levels are increasing due to rapid urbanization and industrialization. Concentrations of Cr, Cu, Fe, Mn, Ni, Pb, and Zn varied markedly in soils obtained from public-access areas across the megacity of Lagos. The calculations approached in the article also indicated the presence of non-carcinogenic risk for children, as well as carcinogenic risk for both children and adults.

Guagliardi et al. [13] individuated different types of pollution fronts, controlling for a structure of monitoring data sets in an area of Southern Italy, through one of the most powerful techniques for this purpose, the Self-Organizing Maps (SOMs) of Kohonen [14], identifying high-risk areas that can be targeted for environmental risks and public health. A combination of the analysis of major metals, minor metals, and PTEs, with the statistical treatment of SOMs, showed the geolithological formations and anthropogenic pressure on the territory. Anomalies ascribable to anthropogenic input in urban soils, referring to elements such as Pb and Zn, and some geogenic anomalous high values of As, Cr, and V mainly identified in peri-urban areas, were recognized.

For water compartments, both natural spring water and industrial wastewater were studied. More precisely, Infusino et al. [15] classified the spring water types of the Sila Massif (Calabria, Southern Italy) according to their hydrogeochemical features, identifying the factors controlling mineral waters using spatial variables focused on lithological settings and determining the vulnerability to nitrate occurrence in the spring waters. The 59 springs analyzed in this study showed that they were strongly affected by the geological nature of the rocks of the aquifers (granites, magmatites, acid granulites, biotic gneiss, etc.). Indeed, the waters are generally poor in mineral salts (very-low-mineral-content waters) with low calcium and magnesium content (soft).

Parrone et al. [16] analyzed field and laboratory data in order to evaluate the significant connections between aquifer mineralogy and groundwater geochemistry, with particular reference to As, and also to deepen its relationships with other PTEs in groundwater in the Sabatini volcanic apparatus, a volcanic–sedimentary aquifer in Rome, Italy. In this survey, the joint analysis of field data and laboratory observations suggested the existence of a diffused As geochemical background due to the water–rock interactions. This study pointed out that in volcanic areas, although the reductive dissolution of Fe oxyhydroxides could cause an enhanced localized release of As into groundwater, the anion exchange induced by specific exchangers (e.g., phosphates) with As adsorbed on the surface seems to be the most widely diffused process.

Pei and Sun [17] approached the problems of molybdenum pollution and difficult extraction and recovery in industrial wastewater in China, which has the largest reserves of molybdenum in the world. In this study, the influencing factors on the migration characteristics of Mo(VI) from the simulated trade effluent was examined using a newly designed membrane chemical reactor with mixed organic–water solvent (MCR-OW). The results showed that the MCR-OW was able to improve the efficiency of Mo(VI) migration from the simulated trade effluent, using the mixed diesel and NaOH (stripping solution) with the carrier of N-503 as renewal solution.

More attention was placed on an environmental compartment considered a lesser-known air pollutant such as dust, both in a road and a household. In fact, Mugudamani et al. [18] determined the influence of urban informal settlements on trace element accumulation in road dust from Ekurhuleni Metropolitan Municipality, South Africa, and their health implications. The outcomes of this study revealed that informal settlement activities have considerable influences on the accumulation of trace elements in road dust during the winter season. Major elements (SiO_2 and P_2O_5) and trace elements (Cr, Cu, Zn, Zr, Ba, and Pb) were health concerns as they were above their corresponding average shale values.

In addition, Xiao et al. [19] collected road dust samples in Jiayuguan, Hexi Corridor, China and measured the concentration of PTEs to determine the pollution levels. They calculated the enrichment factor and geoaccumulation index of PTEs, identified their sources through principal component analysis, and assessed their health risks using the EPA health risk assessment model. Among the 12 PTEs (V, Cr, Mn, Co, Ni, Cu, Zn, As, Se, Cd, Ba, and Pb) in the road dust samples, the highest concentrations were for Mn, Ba, Zn, and Cr, and the concentration of PTEs in industrial areas was higher than the other three functional areas as a whole. The calculation of the Igeo showed that for most elements, the pollution level of the industrial area was higher than that of the rest of the areas. According to the source apportionment results, the PTEs in Jiayuguan road dust mainly came from geogenic–industrial sources, coal combustion, traffic sources, and oil combustion.

Jeong and Ra [20] evaluated PTE contamination and health risks posed by fine road dust (10–63 μm, <10 μm) collected in Apia City, Samoa. Cr, Co, and Ni concentrations in road dust with a particle size of 10–63 μm were higher than in fine road dust (<10 μm), while Cu, Zn, As, Cd, Sb, Pb, and Hg were present at relatively high concentrations in fine road dust. The PTE concentrations in road dust suggested that anthropogenic sources related to traffic activity rather than industrial emissions affect road dust in the study area. The results of correlation analyses and pollution assessments showed that only Cr, Co, and Ni were greatly affected by natural sources due to the weathering of volcanic parent rocks. In addition, relatively high concentrations were observed in sampling sites with high traffic volume, suggesting that contamination with these PTEs (Cu, Zn, Sb, and Pb) was mainly due to traffic activity. The ingestion exposure route posed a greater health risk than inhalation and dermal contact.

Gad et al. [21] detected the PTE levels in household dust and identified their spatial distribution in Cairo City, individuating the possible sources of PTEs in household dust using multivariate statistical analysis and assessing the potential health risk for children and adults' exposure to PTEs. Their findings revealed that the levels of As, Cd, Cr, Cu, Hg, Mo, Ni, Pb, and Zn surpassed the background values of UCC, indicating anthropogenic influences; New Cairo recorded the slightest degree of contamination, ranging from considerably to very high pollution, while in other Cairo regions, household dust is very highly polluted. Elevated PTE concentrations in Cairo's household dust may be due to industrial activities and heavy traffic emissions; the health risk assessment model showed that the vital route of potential PTE exposure that leads to both noncarcinogenic and carcinogenic risks is ingestion, followed by dermal and inhalation pathways. The noncarcinogenic risk was generally in the safe range for adults' exposure. Children are at risk in some sites.

Li et al. [22] measured the concentrations of heavy metals (including As, Cd, Co, Cr, Cu, Ni, Pb, and Zn) in $PM_{2.5}$ in Yuci College Town, Shanxi, China, for analyzing the source apportionment and assessing the health risks (noncarcinogenic and carcinogenic) of exposure to eight heavy metals, via inhalation exposure, during winter haze periods. Their results indicate that a total of 18% and 100% of the concentrations assessed in the samples for As and Cr, respectively, exceeded the corresponding threshold values of the Chinese Ambient Air Quality Standards and the WHO Global Air Quality Guidelines; the levels of As and Cr (VI) were also higher than those in some areas worldwide. Combustion was the largest contributor to $PM_{2.5}$-bound heavy metals (37.91%), followed by traffic emissions (32.19%) and industrial sources (29.9%).

Finally, Wan et al. [23] studied the discharge patterns of PTEs from coking plants and their relationship with soil PTE contents in the Beijing–Tianjin–Hebei (BTH) region, China. This discharge of PTEs from the coking industry was found to be closely related to the accumulation of PTEs in soil. The discharge of Hg, As, and Cd from coking in the BTH region were relatively higher than those of other metals, which resulted in higher ecological risks of Hg, As, and Cd in this area. As result, scenario analysis also showed that the level of ecological risks would increase rapidly with an extending production time, especially for As, Cr, Ni, and Pb.

The published research articles are all of importance to face PTEs pollution in the environment through in-depth knowledge of the mechanisms that underlie the occurrence, distribution, migration, evolution over time of these pollutants and their effects on health and environmental risks.

Many challenges remain for researchers in environmental sciences, from the assessment of pollutants in all possible compartments, obtaining a harmonization in terms of sampling procedures, matrix characterization, preservation procedures, analytical methods, etc., to the remediation technologies involved in pollutant removal from matrices in contaminated sites, and building awareness for sustainable management of the territories which limit the pollution of environmental matrices.

According to the important results achieved (12 published papers and numerous ones that unfortunately have not passed scientific quality standards after careful peer review processes), my role as Guest Editor of this Special Issue was exciting and stimulating. The success of this Special Issue has really been a motivation and inspiration for me to continue the research in this fundamental topic for the protection of public health and ecosystems trying also to support the decision makers in the field of sustainable development implementation.

I would like to express my appreciation to all the authors for submitting their original contributions to this Special Issue and to reviewers for their essential part in assessing the suitability of the manuscripts and improving their quality. It would not be appropriate if I failed to mention the editors of Toxics for their kind invitation, and in particular Selena Li of the *Toxics* Editorial Office for her precious and tireless support.

Conflicts of Interest: The author declares no conflict of interest.

References

1. Guagliardi, I.; Zuzolo, D.; Albanese, S.; Lima, A.; Cerino, P.; Pizzolante, A.; Thiombane, M.; De Vivo, B.; Cicchella, D. Uranium, thorium and potassium insights on Campania region (Italy) soils: Sources patterns based on compositional data analysis and fractal model. *J. Geochem. Explor.* **2020**, *213*, 106508. [CrossRef]
2. Goh, T.A.; Ramchunder, S.J.; Ziegler, A.D. Low presence of potentially toxic elements in Singapore urban garden soils. *CABI Agric. Biosci.* **2022**, *3*, 60. [CrossRef]
3. Said, I.; Salman, S.A.E.-R.; Samy, Y.; Awad, S.A.; Melegy, A.; Hursthouse, A.S. Environmental factors controlling potentially toxic element behaviour in urban soils, El Tebbin, Egypt. *Environ. Monit. Assess.* **2019**, *191*, 267. [CrossRef] [PubMed]
4. Lei, M.; Li, K.; Guo, G.; Ju, T. Source-specific health risks apportionment of soil potential toxicity elements combining multiple receptor models with Monte Carlo simulation. *Sci. Total Environ.* **2022**, *817*, 152899. [CrossRef]
5. Hanfi, M.Y.; Seleznev, A.A.; Yarmoshenko, I.V.; Malinovsky, G.; Konstantinova, E.Y.; Alsafi, K.G.; Sakr, A.K. Potentially harmful elements in urban surface deposited sediment of Ekaterinburg, Russia: Occurrence, source appointment and risk assessment. *Chemosphere* **2022**, *307*, 135898. [CrossRef]
6. Bai, J.; Zhang, W.; Liu, W.; Xiang, G.; Zheng, Y.; Zhang, X.; Yang, Z.; Sushkova, S.; Minkina, T.; Duan, R. Implications of Soil Potentially Toxic Elements Contamination, Distribution and Health Risk at Hunan's Xikuangshan Mine. *Processes* **2021**, *9*, 1532. [CrossRef]
7. Cicchella, D.; Zuzolo, D.; Albanese, S.; Fedele, L.; Di Tota, I.; Guagliardi, I.; Thiombane, M.; De Vivo, B.; Lima, A. Urban soil contamination in Salerno (Italy): Concentrations and patterns of major, minor, trace and ultra-trace elements in soils. *J. Geochem. Explor.* **2020**, *213*, 106519. [CrossRef]
8. Gaglioti, S.; Infusino, E.; Caloiero, T.; Callegari, G.; Guagliardi, I. Geochemical characterization of spring waters in the Crati River Basin, Calabria (Southern Italy). *Geofluids* **2019**, *2019*, 3850148. [CrossRef]
9. Guagliardi, I.; Rovella, N.; Apollaro, C.; Bloise, A.; De Rosa, R.; Scarciglia, F.; Buttafuoco, G. Modelling seasonal variations of natural radioactivity in soils: A case study in southern Italy. *J. Earth Syst. Sci.* **2016**, *125*, 1569–1578. [CrossRef]
10. Buttafuoco, G.; Guagliardi, I.; Tarvainen, T.; Jarva, J. A multivariate approach to study the geochemistry of urban topsoil in the city of Tampere, Finland. *J. Geochem. Explor.* **2017**, *181*, 191–204. [CrossRef]
11. Tomczyk-Wydrych, I.; Świercz, A.; Przepióra, P. Assessment of the Railroad Transport Impact on Physical and Chemical Soil Properties: The Case Study from Zduńska Wola Karsznice Railway Junction, Central Poland. *Toxics* **2021**, *9*, 296. [CrossRef] [PubMed]
12. Famuyiwa, A.O.; Davidson, C.M.; Ande, S.; Oyeyiola, A.O. Potentially Toxic Elements in Urban Soils from Public-Access Areas in the Rapidly Growing Megacity of Lagos, Nigeria. *Toxics* **2022**, *10*, 154. [CrossRef] [PubMed]
13. Guagliardi, I.; Astel, A.M.; Cicchella, D. Exploring Soil Pollution Patterns Using Self-Organizing Maps. *Toxics* **2022**, *10*, 416. [CrossRef] [PubMed]
14. Kohonen, T. Self-organized formation of topologically correct feature maps. *Biol. Cybern.* **1982**, *43*, 56–69. [CrossRef]
15. Infusino, E.; Guagliardi, I.; Gaglioti, S.; Caloiero, T. Vulnerability to Nitrate Occurrence in the Spring Waters of the Sila Massif (Calabria, Southern Italy). *Toxics* **2022**, *10*, 137. [CrossRef]
16. Parrone, D.; Ghergo, S.; Preziosi, E.; Casentini, B. Water-Rock Interaction Processes: A Local Scale Study on Arsenic Sources and Release Mechanisms from a Volcanic Rock Matrix. *Toxics* **2022**, *10*, 288. [CrossRef]
17. Pei, L.; Sun, L. Impact Factors on Migration of Molybdenum(VI) from the Simulated Trade Effluent Using Membrane Chemical Reactor Combined with Carrier in the Mixed Renewal Solutions. *Toxics* **2022**, *10*, 438. [CrossRef]
18. Mugudamani, I.; Oke, S.A.; Gumede, T.P. Influence of Urban Informal Settlements on Trace Element Accumulation in Road Dust and Their Possible Health Implications in Ekurhuleni Metropolitan Municipality, South Africa. *Toxics* **2022**, *10*, 253. [CrossRef]
19. Xiao, K.; Yao, X.; Zhang, X.; Fu, N.; Shi, Q.; Meng, X.; Ren, X. Pollution Characteristics, Source Apportionment, and Health Risk Assessment of Potentially Toxic Elements (PTEs) in Road Dust Samples in Jiayuguan, Hexi Corridor, China. *Toxics* **2022**, *10*, 580. [CrossRef]
20. Jeong, H.; Ra, K. Pollution and Health Risk Assessments of Potentially Toxic Elements in the Fine-Grained Particles (10–63 μm and <10 μm) in Road Dust from Apia City, Samoa. *Toxics* **2022**, *10*, 683.
21. Gad, A.; Saleh, A.; Farhat, H.I.; Dawood, Y.H.; Abd El Bakey, S.M. Spatial Distribution, Contamination Levels, and Health Risk Assessment of Potentially Toxic Elements in Household Dust in Cairo City, Egypt. *Toxics* **2022**, *10*, 466. [CrossRef] [PubMed]

22. Li, L.; Qi, H.; Li, X. Composition, Source Apportionment, and Health Risk of PM$_{2.5}$-Bound Metals during Winter Haze in Yuci College Town, Shanxi, China. *Toxics* **2022**, *10*, 467. [CrossRef] [PubMed]
23. Wan, X.; Zeng, W.; Gu, G.; Wang, L.; Lei, M. Discharge Patterns of Potentially Harmful Elements (PHEs) from Coking Plants and Its Relationship with Soil PHE Contents in the Beijing–Tianjin–Hebei Region, China. *Toxics* **2022**, *10*, 240. [CrossRef] [PubMed]

Article

Assessment of the Railroad Transport Impact on Physical and Chemical Soil Properties: The Case Study from Zduńska Wola Karsznice Railway Junction, Central Poland

Ilona Tomczyk-Wydrych *, Anna Świercz and Paweł Przepióra

Department of Geomorphology and Geoarchaeology, Institute of Geography and Environmental Sciences, Jan Kochanowski University, 7 Uniwersytecka St., 25-406 Kielce, Poland; swierczag@poczta.onet.pl (A.Ś.); pawelprzepiora1988@gmail.com (P.P.)
* Correspondence: ilonatomczyk@interia.eu

Abstract: Contamination of the soil and water environment with harmful substances can be associated with many activities carried out on the railway. The problem is particularly relevant to liquid fuel loading and refueling facilities as well as to increased traffic at railway junctions. Studies were conducted in the area of railway junction Zduńska Wola Karsznice in central Poland (Łódź Voivodeship). Soil samples were collected from specific research points: from the inter-railway (A), 5 m from the main track (B), from the embankment—10 m from the main track (C), and from the side track (D), at the depth of 0–5 cm (1) and 20 cm (2). The following analyses were made: granulometric composition, pH in H_2O, and percent content of carbonates ($CaCO_3$). PHEs were determined in the fractions: $0.25 \leq 0.5$ mm, $0.1 \leq 0.25$ mm, and $0.05 \leq 0.1$ mm: Pb, Cd, Cr, Co, Cu, Ni, Zn, Sr by inductively coupled plasma mass spectrometry technique (ICP-MS/TOF OPTIMass 9500). The objectives of the study were (1) to assess PHEs (potentially harmful elements) contamination of the topsoil level of railway area, (2) to determine the correlation between the concentration of PHEs and the size of the fraction, and (3) to identify the areas (places) where the highest concentrations of PHEs were recorded. Based on the studied parameters, significant differentiation in soil properties of the areas in Zduńska Wola Karsznice was found. The analyses carried out showed that the accumulation of potentially harmful elements was as follows: Cu > Zn > Sr > Pb > Ni > Cr > Co > Cd. The average concentrations of Cu, Zn, Sr, Pb, Ni, Cr, Co and Cd were 216.0; 152.1; 97.8; 64.6; 15.2; 14.4; 3.1 and 0.2 mg·kg^{-1} d.w., respectively. These contaminations occur in the topsoil layer of the railway embankment, which suggests a railway transport origin. The highest concentrations of PHEs were recorded in samples collected from close to the rails (inter-railway, side track), and in the embankment (10 m from the track) in the very fine sand fraction ($0.05 \leq 0.1$ mm). The high accumulation index of copper, cadmium and lead in the surface layer of soil indicate their anthropogenic origin. The results presented in the paper can be used in local planning and spatial development of this area, taking into account all future decisions about ensuring environmental protection, including groundwater and soils.

Keywords: potentially harmful elements (PHEs); railway area; transport pollutions

Citation: Tomczyk-Wydrych, I.; Świercz, A.; Przepióra, P. Assessment of the Railroad Transport Impact on Physical and Chemical Soil Properties: The Case Study from Zduńska Wola Karsznice Railway Junction, Central Poland. *Toxics* **2021**, *9*, 296. https://doi.org/10.3390/toxics9110296

Academic Editor: Ilaria Guagliardi

Received: 23 September 2021
Accepted: 3 November 2021
Published: 6 November 2021

Publisher's Note: MDPI stays neutral with regard to jurisdictional claims in published maps and institutional affiliations.

1. Introduction

Along with agriculture, industry, and utilities, it is transportation that contributes to a significant increase in pollution levels. Next to road transport, rail is one of the main transport means in the world. In comparison with road transport, the railroads seemed for a long time to be a relatively harmless mode of transport for the environment. This resulted in the fact that most publications were related to studies on soil contamination with potentially harmful elements (PHEs) along roads and highways in many places around the world, i.e., South Korea [1,2], Nepal [3], Tibet [4], Greece [5], Poland [6] or Thailand [7]. In the last two decades, there has been a significant increase in documentation on environmental hazards and destruction associated with rail transportation. Many studies show an increase of pollution in the immediate vicinity of transport routes, especially PHEs detected in soil and plants [8–11]. Scientists are increasingly paying more attention to the source of many contaminants from the railway. The PHEs can be connected with, i.e., abrasion of wheels and rails or even with the materials that conserve railway sleepers [12–16].

The increase in rail network density and traffic volume leads to higher pressure of rail transport on environmental quality, the nuisance of which depends mainly on the transport technologies used, track quality and capacity, and technical solutions used in rail vehicles. In recent years, with the increase in demand for the means of mass transportation of goods, railroads have become one of the main supply lines in the world. As a result, soil contamination along many railroad lines may have further increased. This issue has been addressed in various papers from around the world. Currently, many factors influencing the inflow of pollutants to the environment from railways are indicated, i.e., wheels abrasion [17] or sewage at railway stations [18]. The relationship between the amount of pollution and the intensity and type of transport is also clear [19,20]. However, researchers agree that rail transport is a source of numerous pollutants. Regardless of the amount of pollution, PHEs are accumulated in soil and plants, which has negative effects on the ecosystem [18,21–25], thereby also indicating the necessity of a way to reduce or prevent its further degradation.

The impact of rail transport on the environment is multidirectional and includes the status of soil contamination. An analysis of literature reports indicates that potentially harmful elements are among the major pollutants generated in railroad areas. The specificity of these pollutants is that they are not biodegradable and decompose to simple compounds. They have the ability to bioaccumulate and biomagnify. PHEs emitted into the environment may migrate from soils to plants, thus there is a risk of their transport to higher trophic levels [12,26].

The aim of this study was (1) to assess the PHEs contamination of the topsoil of the selected part of the Zduńska Wola Karsznice railroad, where previously, research was undertaken on a railway embankment for the first time in this place [27]. The objectives of the study were also (2) to determine the correlation between the concentration of PHEs and the size of the fraction, and (3) to identify the areas (places) where the highest concentrations of PHEs were recorded.

Moreover, the aim of this study was to compare the obtained results with tests from other sites in order to determine the probable source of potentially harmful elements origin. The following research hypothesis was evaluated: does the long-term effect of pollutants emitted by rail transport change soil parameters (including the concentration of PHEs), and if the distance from railroad track is important for topsoil contamination.

2. Materials and Methods

2.1. Study Area

The studies on contamination of the topsoil with PHEs were conducted in the area of railroad junction Zduńska Wola Karsznice in central Poland (Łódź Voivodeship) (Figure 1). According to physico-geographical division by Kondracki [28] the area is situated in the south Wielkopolska Lowland macroregion and Łaska Upland mesoregion. The largest part of this area, included in the Łaska Upland, is covered by till, sand and gravel of glacial accumulation of the Central Poland glaciation.

The geological structure in the study area vicinity is not very diverse. The northern part of the area is built mainly by glacial sands and graves on tills, while in the southern part there are only tills. These sediments were accumulated during the Middle-Polish Glaciations [29]. In terms of soils occurring here, this area is situated on agricultural land of class I–IV [30], and usually these soils are of various agricultural suitability complexes. Very good wheat, very good rye and poor grain and fodder complexes are located closest to the examined railway section. The rest of the area is built-up (urbanized), especially to the west of the railway line with dense residential buildings. The site itself is located directly on land of agriculturally unsuitable soil complex [31]. In the Zduńska Wola Karsznice district, a very acidic (<4.5) and acidic pH (4.6–5.5) of the soils is predominant (37 and 43% of the district's area, respectively) [32].

The railroad junction in Zduńska Wola Karsznice is located in the south-eastern part of the town (51°34′35.47″ N, 19°00′36.74″ E), 187 m above sea level (Figure 1). The 2.6 m wide railroad sleeper (inter-railway width 1.6 m) is made of treated wood. The railroad junction in Zduńska Wola Karsznice functions as a junction for freight trains travelling in the direction Wroclaw–Silesia–Central Poland. Railroad junction Zduńska Wola Karsznice is regarded as one of the biggest junctions in the country.

2.2. Historical News of Zduńska Wola Karsznice Railway

The construction of railroads connecting Silesia with ports was subject to economic and political factors. After the First World War, the export of coal and coke was crucial to the Polish economy. It was then decided to build two first-class, normal-track lines for public use. Despite the incomplete execution of works, in 1933 the last section of the Karsznice–Inowrocław main line was put into service. The entire line was launched on 1 August 1935, while the final completion of works took place in the spring of 1938. The railroad line was electrified between 1965 and 1969. In the initial period of the line's operation, 12 pairs of freight trains per day were accompanied by one pair of long-distance passenger trains from Tarnowskie Góry to Gdynia, two on the Katowice–Gdynia route, two local trains from Karsznice to Inowrocław and to Częstochowa, and a connecting train from Zduńska Wola to Zduńska Wola Karsznice. In 1939 two pairs of passenger trains Gdynia–Częstochowa, Gdynia–Katowice and 10 pairs of connecting trains Zduńska Wola–Zduńska Wola Karsznice operated. Thanks to the station's and main line's expansion during the occupation, including the construction of the second track, the line's capacity increased from 21 pairs of trains per day to 43 pairs in 1944. In 1950, seven pairs of passenger trains operated from Karsznice in the southern direction, while only four pairs operated in the northern direction. Sixteen shuttle trains were in use between South Karsznice and Zduńska Wola. Freight traffic reached its peak and the line's capacity limit in 1945–1950. The 1970s saw the greatest development of the Karsznice junction. The capacity increased thanks to the line's electrification. In 1976, 79 regular and 25 additional trains could arrive at Karsznice. Passenger train services were limited to a few pairs of passenger trains and a few, mostly seasonal, long-distance trains. With the changes in the political system, the decline of transport on the Polish State Railways (PKP) network began. In 1990, there were 64 regular and 39 additional trains in one direction, 62 regular and 49 additional in the other direction. Until the mid-1990s the number of trains in passenger traffic did not change much. Long-distance trainsets began to be reduced after 2000. Finally, passenger trains between Katowice and Inowrocław were liquidated at the end of 2008, and replaced by shortened-route connections at the end of 2010. Seasonal trains from Silesia to the seaside and passenger traffic from Zduńska Wola to Częstochowa were then abandoned. Currently, passenger traffic on the coal main line in the region of Zduńska Wola is only occasional [33].

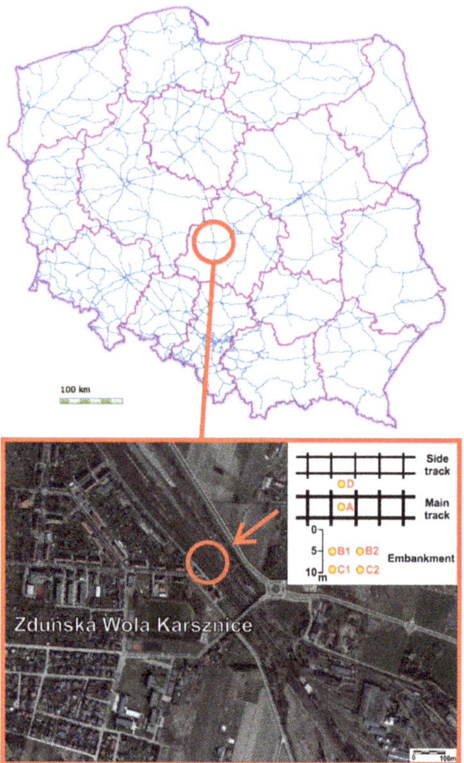

Figure 1. Railway network on the administrative division of Poland [34] with the location of the study area [35] with schematic sampling arrangement.

2.3. Soil Sampling and Chemical Analysis

The study was conducted once in May 2013 during the preparation of the master's dissertation [27]. Soil samples for the study were collected from four locations: inter-railway (A), 5 m from the main track (B), from the embankment—10 m from the main track (C), from the side track (D). Mineral samples were collected from two depths: 0–5 cm (1) and 20 cm (2) (Figure 1). An average of about 1 kg of soil was collected from each site.

The obtained soil material was dried at room temperature and analyzed in the Environmental Research Laboratory of the Department of Environmental Protection and Management, Jan Kochanowski University in Kielce.

The analysis of soil samples was conducted using the following methods: granulometric composition—by sieve method, pH in H_2O—by potentiometric method, percent carbonate content ($CaCO_3$)—by Scheibler method. PHEs determination was carried out in the fraction of $0.25 \leq 0.5$ mm (medium sand), $0.1 \leq 0.25$ mm (fine sand) and $0.05 \leq 0.1$ mm (very fine sand) (according to Polish Society of Soil Sciences, 2008) [36]. Sample masses of 0.1 g of soil from each fraction was weighed for mineralization in Multiwave TM 3000 Anton Paar mineralizer. For this purpose, a sample containing 0.1 g was weighed and mineralized with nitric acid V (Suprapur Merck) and hydrogen peroxide in a ratio of 2.5:1 (microwave power: 1400 W, temperature: 2000 °C, time: 40 min). Certified reference materials were used for evaluating the accuracy attained. The reaction vessel was placed in a rotor which, when closed, was located in a Multiwave 3000 oven. The sample was heated by microwave radiation. Measurement data were continuously transmitted on a display. During the decomposition process, the integrated cooling system generated a flow of cooling air between the reaction vessels and their covers, thus protecting the rotor from excessive thermal load. The blanks and duplicate measurements were performed for quality control. The samples were rinsed with deionized water and the contents of potentially harmful elements (PHEs): Pb, Cd, Cr, Co, Cu, Ni, Zn, Sr were determined in the filtrate using the inductively coupled plasma mass spectrometer time-of-flight (ICP-MS/TOF) OPTIMass 9500 (GBC Scientific Equipment, Melbourne, Australia). In order to control the quality of obtained results, certified reference materials such as ERM-CA713 were used. Limit of detection (LOD) and limit of quantification (LOQ) were found between LOD: 0.05 (Cd, Cr), 0.15 (Cu, Ni, Zn), 0.5 (Pb) mg·kg^{-1} d.w.; LOQ: 0.1 (Cd, Cr), 0.2 (Ni), 0.3 (Zn), 0.5 (Cu), 1.0 (Pb) mg·kg^{-1} d.w., respectively. Precision and accuracy were controlled using certified reference materials with the same matrices for all analysts.

2.4. Statistical Analysis

Statistical analysis of the data was conducted using Microsoft Office Excel software (range, mean, standard deviation) (Table 1).

Table 1. Metals contamination (mg·kg^{-1} d.w.) depending on the sampling location: A (inter-railway, depth 20 cm), B1 (5 m from the main track, depth 0–5 cm), B2 (5 m from the main track, depth 20 cm), C1 (embankment, 10 m from the main track, depth 0–5 cm), C2 (embankment, 10 m from the main track, depth 20 cm), D (side track, depth 20 cm).

Metal	Property	Sampling Location					
		A	B1	B2	C1	C2	D
Pb	Range	22.3–111.1	16.4–60.0	41.9–174.6	47.4–168.1	12.4–63.0	19.9–68.0
	Mean	65.6	35.5	91.3	119.9	35.1	40.5
	SD	44.44	22.29	72.55	63.89	25.68	24.76
Cd	Range	0.2–0.5	0.01–0.2	0.01–0.4	0.1–0.3	0.1–0.2	0.1–0.2
	Mean	0.4	0.1	0.2	0.2	0.1	0.1
	SD	0.17	0.11	0.23	0.12	0.06	0.06
Cr	Range	21.9–71.8	3.0–12.1	6.3–10.1	3.4–9.0	6.8–8.1	8.3–20.3
	Mean	45.0	7.1	7.6	6.6	7.5	12.9
	SD	25.15	4.63	2.16	2.88	0.65	6.47
Co	Range	2.1–7.6	0.01–2.0	0.01–5.2	1.0–2.1	5.4–10.4	1.1–2.1
	Mean	4.6	1.1	2.5	1.6	7.6	1.5
	SD	2.78	1.01	2.60	0.55	2.54	0.55
Cu	Range	379.4–1371.9	65.3–127.2	65.7–131.9	47.0–178.9	31.0–76.1	76.4–270.5
	Mean	795.9	87.7	102.8	106.9	52.6	150.1
	SD	515.12	34.34	51.74	66.77	22.61	105.16
Ni	Range	13.2–38.1	0.01–10.3	6.3–24.8	4.5–11.2	22.2–39.4	6.1–12.5
	Mean	24.7	5.5	14.6	7.7	30.3	8.3
	SD	12.56	5.19	9.39	3.36	8.65	3.64
Zn	Range	121.4–460.6	66.9–180.7	132.0–370.8	41.3–142.7	34.7–98.0	73.2–189.8
	Mean	267.3	116.3	251.4	88.3	67.7	121.5
	SD	174.51	58.35	133.67	51.10	31.74	60.82
Sr	Range	61.5–115.5	18.1–50.5	71.8–120.7	32.0–75.5	195.1–227.7	49.2–92.7
	Mean	114.7	37.6	92.9	59.6	208.9	72.9
	SD	52.75	17.18	25.12	24.02	16.88	22.02

Based on the mean geochemical background values developed by Czarnowska [37], the PHEs accumulation index (AI) in the soil was calculated. Accumulation index is calculated using the equation:

$$AI = \frac{C_i}{B_i}$$

where C_i is the geometric mean content of individual metals (mg·kg^{-1} d.w.); B_i is the geochemical background value of the elements (mg·kg^{-1} d.w.).

Taking into account the values of the geochemical background provided by the author for the examined elements, the average values for the whole of Poland were adopted: Pb 9.8 mg·kg^{-1} d.w., Cd 0.18 mg·kg^{-1} d.w., Cr 27.0 mg·kg^{-1} d.w., Co 4.0 mg·kg^{-1} d.w., Cu 7.1 mg·kg^{-1} d.w., Ni 10.2 mg·kg^{-1} d.w., Zn 30.0 mg·kg^{-1} d.w. [37].

3. Results

3.1. Physicochemical Properties of Soil Samples

The granulometric analysis showed that the grain size composition of investigated samples varied from one test point to another (according to Polish Society of Soil Sciences, PTG, 2008) [36]. From the analysis of the graphs, it was concluded that in Zduńska Wola Karsznice the majority of samples were particles larger than 2 mm (rock fraction, gravel fraction and very coarse sand fraction). Most of the samples showed enrichment in skeletal parts (Figure 2).

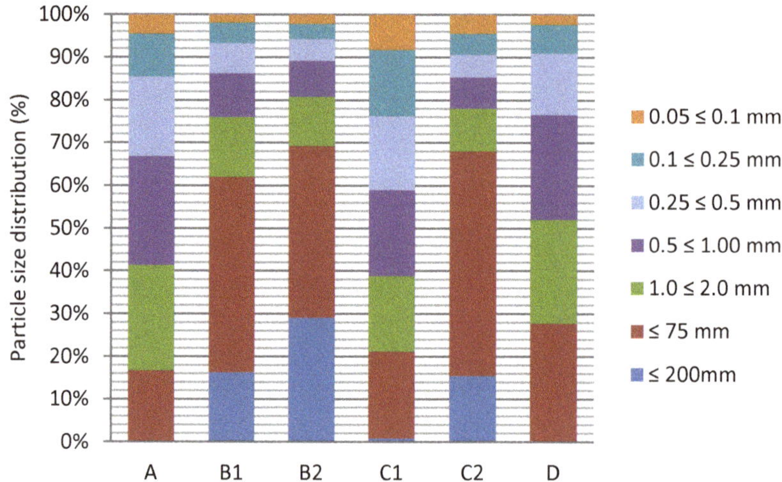

Figure 2. Particle size distribution (%) depending on the sampling location: A (inter-railway, depth 20 cm), B1 (5 m from the main track, depth 0–5 cm), B2 (5 m from the main track, depth 20 cm), C1 (embankment, 10 m from the main track, depth 0–5 cm), C2 (embankment, 10 m from the main track, depth 20 cm), D (side track, depth 20 cm).

The pH values in H$_2$O ranged from 6.46 (slightly acidic) to 7.29 (slightly basic). The lowest values were found in samples collected from the embankment located 10 m from the main track (6.63–6.46). The highest values were found in soils collected from the side track (7.16–7.29) and the inter-railway (6.8–7.01). The coarse sand fraction (0.5 ≤ 1.0 mm) and the very fine sand fraction (0.05 ≤ 0.1 mm) had the highest mean pH values (6.84) while the fine sand fraction (0.1 ≤ 0.25 mm) had the lowest mean pH value (6.78) (Figure 3).

The pH is a major factor affecting the solubility of metal compounds in the environment. For example, the transfer of weighting coefficient between the sorption and desorption processes of metal ions and H$^+$ ions depends on pH. The literature shows that PHEs in neutral and alkaline soils migrate to a small extent to the lower layers of the soil profile due to, among other causes, limited solubility. Generally, solubility of PHEs compounds is low in neutral and alkaline pH, but much higher in acidic pH. Strong acidification of soils may contribute to the release of PHEs bound with minerals, as well as with oxides, e.g., Mn, Fe, and Al. The most mobile is Cd, while the least soluble are Pb, Cr and Hg [38]. Other metals, i.e., Zn or Cu, show increased solubility also at alkaline pH. With increasing pH, the bioavailability of metals decreases. Soils characterized by a high sorption capacity in relation to positive metal ions and containing a large amount of organic matter, show the potential to bind PHEs and retain them in the surface layer [26]. Due to the fact that the studied soils were characterized by a slightly acidic and slightly alkaline pH, the pH was not an element that significantly affected the increased mobility of metals in analyzed soils.

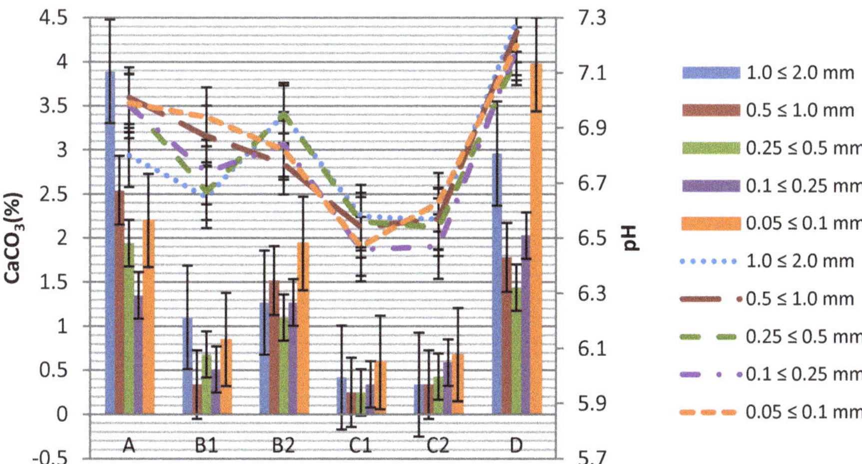

Figure 3. CaCO$_3$ content (%) and pH value in different fractions $1.0 \leq 2.0$; $0.5 \leq 1.0$; $0.25 \leq 0.5$; $0.1 \leq 0.25$; $0.05 \leq 0.1$ (mm) depending on the sampling location with error bars. Designations: A (inter-railway, depth 20 cm), B1 (5 m from the main track, depth 0–5 cm), B2 (5 m from the main track, depth 20 cm), C1 (embankment, 10 m from the main track, depth 0–5 cm), C2 (embankment, 10 m from the main track, depth 20 cm), D (side track, depth 20 cm).

In the soils of the studied railroad junction, the fraction of very coarse sand ($1.0 \leq 2.0$ mm) had the highest CaCO$_3$ content (3.89%), and the fraction of very fine sand ($0.05 \leq 0.1$ mm) 3.97%, while the fraction of medium sand ($0.25 \leq 0.5$ mm) 0.25% and coarse sand ($0.5 \leq 1.0$ mm) (0.25%) had the lowest content (Figure 3).

3.2. Levels of PHEs

The study shows that the highest concentrations of elements occurred in samples from the inter-railway (A) and the side track (D). An increase in pollutant concentrations was observed in the embankment 10 m from the track (C), which may be due to the close proximity to the road. The lowest element concentrations were recorded 5 m from the main track (B).

The average concentration of Pb was 64.6 mg·kg^{-1} d.w. (12.4–174.6 mg·kg^{-1} d.w.). The highest Pb concentrations were found in samples A, B2 and C1, while the lowest concentrations were found in C2. The highest concentration of Cd was found in samples A (0.5 mg·kg^{-1} d.w.), while the lowest contents were recorded in B1 and B2 (0.01 mg·kg^{-1} d.w.). Higher cadmium values were found in the $0.05 \leq 0.1$ mm fraction. The mean Cr concentration was equal to 14.5 mg·kg^{-1} d.w., the maximum, 71.8 mg·kg^{-1} d.w., was recorded in samples A, while the minimum was 3.0 mg·kg^{-1} d.w. in B1. The Co content ranged from 0.01 (B1, B2) to 10.4 mg·kg^{-1} d.w. (C2). Concentrations were comparatively low at the other sites. In the topsoil, the mean copper content was 216.2 mg·kg^{-1} d.w., the maximum 1371.9 mg·kg^{-1} d.w. (A), while the minimum was 31.0 mg·kg^{-1} d.w. (C2). The highest concentrations of Cu were recorded in the inter-railway, and the values decreased with distance from the track. The mean Ni concentration was equal to 15.2 mg·kg^{-1} d.w., with a maximum of 39.4 mg·kg^{-1} d.w., recorded in C2, and was most likely due to proximity to the road. A similar value to the maximum equal to 38.1 mg·kg^{-1} d.w. was also recorded in A. The mean Zn content was 146.3 mg·kg^{-1} d.w. with highly significant variation in results between sites. The minimum and maximum values were 34.7 mg·kg^{-1} d.w.

and 460.6 mg·kg^{-1} d.w., respectively. The highest concentration of zinc in the substrate was found in samples A as well as B1 and B2. The Sr content ranged from 18.1 (B1) to 227.7 mg·kg^{-1} d.w. (C2), with a mean of 97.8 mg·kg^{-1} d.w. High concentrations were also recorded in the inter-railway (167.0 mg·kg^{-1} d.w.) (Figure 4).

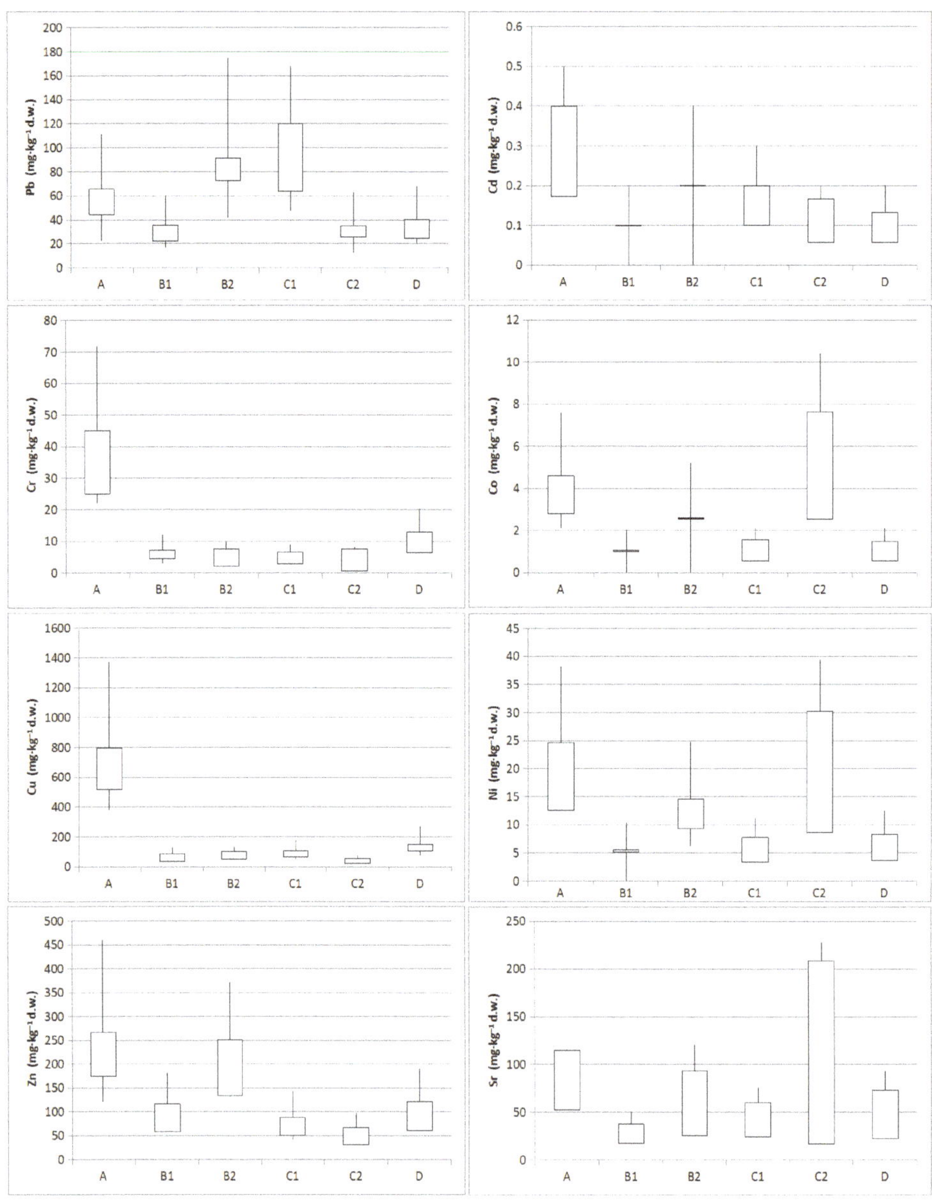

Figure 4. Potentially harmful elements (PHEs) content (mg·kg^{-1} d.w.) (boxplots with min., max., mean, standard deviation) according to sampling location: A (inter-railway, depth 20 cm), B1 (5 m from the main track, depth 0–5 cm), B2 (5 m from the main track, depth 20 cm), C1 (embankment, 10 m from the main track, depth 0–5 cm), C2 (embankment, 10 m from the main track, depth 20 cm), D (side track, depth 20 cm) and fraction diameter (0.05 ≤ 0.1 mm; 0.1 ≤ 0.25 mm; 0.25 ≤ 0.5 mm).

Based on the study, PHEs concentrations were found to be higher in the $0.05 \leq 0.1$ mm fractions and lower in the $0.25 \leq 0.5$ mm fractions. The differences between $0.05 \leq 0.1$ mm and $0.25 \leq 0.5$ mm fractions are statistically significant (Figure 5).

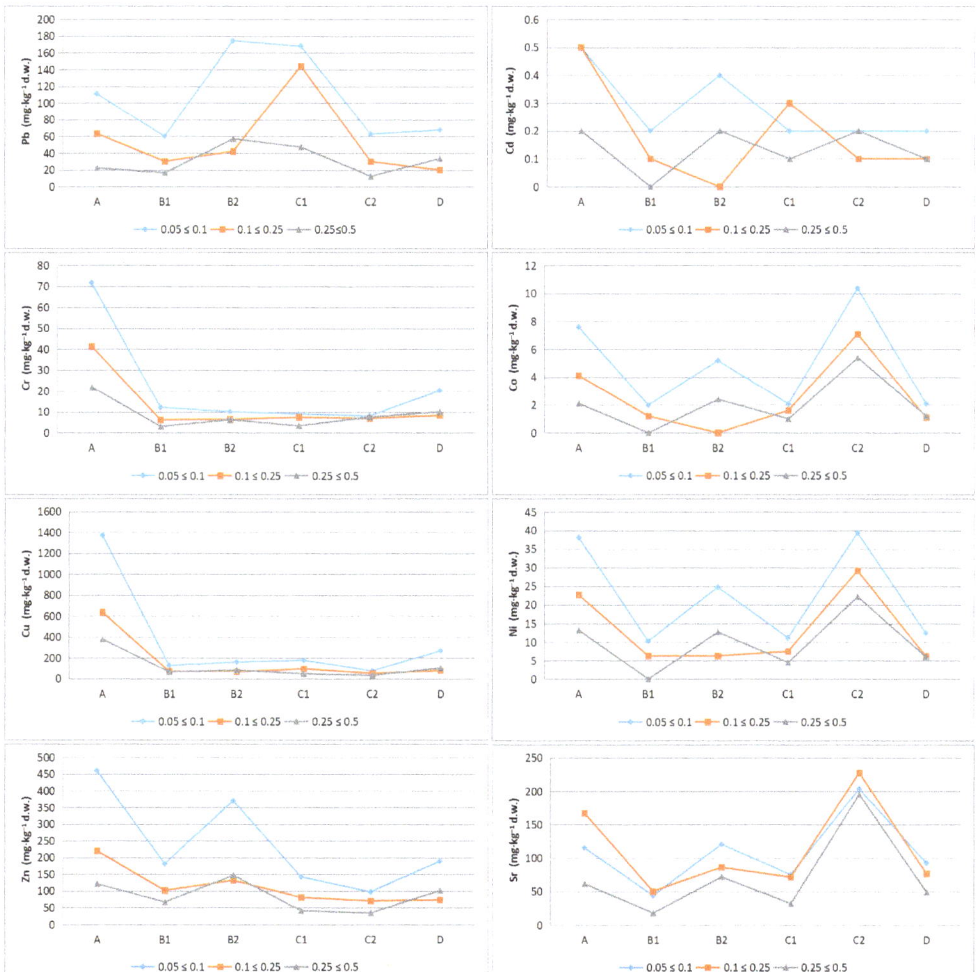

Figure 5. Potentially harmful elements (PHEs) content (mg·kg^{-1} d.w.) in different grain size fractions (mm) according to sampling location: A (inter-railway, depth 20 cm), B1 (5 m from the main track, depth 0–5 cm), B2 (5 m from the main track, depth 20 cm), C1 (embankment, 10 m from the main track, depth 0–5 cm), C2 (embankment, 10 m from the main track, depth 20 cm), D (side track, depth 20 cm).

Based on the mean geochemical background values developed by Czarnowska [37], an index of PHEs accumulation (AI) in the soil was calculated (Table 2). The accumulation index values for Pb, Cd, Cr, Co, Cu, Ni, and Zn ranged from 0.28 (for Cr) to 97.46 (for Cu). It should be noted that the highest accumulation index values of AI = 97.46 for copper; AI = 1.89 for cadmium; AI = 7.69 for zinc; AI = 1.49 for chromium were found in the samples from the inter-railway, the highest accumulation index of AI = 7.64 for lead was found in the samples collected 5 m from the track, and AI = 1.84 for cobalt and AI = 2.89 for nickel were found in the samples collected 10 m from the track. The mean accumulation index values in the investigated soils were arranged in the following series Cu > Zn > Pb > Ni > Cd > Co > Cr.

Table 2. Contamination degree based on accumulation index (AI).

Sampling Location	Index of PHEs Accumulation						
	Pb	Cd	Cr	Co	Cu	Ni	Zn
Inter-railway	5.5	1.89	1.49	1.00	97.46	2.21	7.69
5 m from the main track	7.64	0.53	0.27	0.12	13.39	1.23	6.43
10 m from the main track	2.92	0.88	0.28	1.84	6.95	2.89	2.07
Side track	3.65	0.66	0.44	0.35	18.37	0.77	3.74

4. Discussion

The presence of PHEs in the soil around the railway is a potential threat to the natural environment. The content of metals in soils varies widely. Metals getting into soil may undergo various transformations, ranging from deposition in the form of insoluble compounds with a relatively low impact on plants and microorganisms, to occurrence in a very active ionized form. They may also form chelate connections with humic substances, which provide protection against the toxic effects of a metal ion [38,39]. Liu et al. [40] argue that high organic matter content may contribute to PHEs retention in the immediate vicinity of railroad areas.

According to Kabata-Pendias [26], the limiting content of potentially harmful elements (PHEs) in soils containing only anthropogenic pollutants is for Cd 1, Cu 25, Ni 50, Pb 70, Cr 100, and Zn 150 mg·kg^{-1}. Comparing the obtained results with the limiting values of PHEs accumulated in anthropogenic soils according to Kabata-Pendias [26], it was established that mean concentrations of PHEs in samples from Zduńska Wola Karsznice were exceeded for Cu (at each sampling point), Zn (inter-railway, 5 m from the track) and Pb (10 m from the track).

The analysis of PHEs content in the topsoil of railway areas shows certain regularities in terms of their occurrence and fraction size. Representative studies that are consistent with the conclusions of this study are presented in Table 3. These studies showed that railways have significant influence on PHEs concentrations in surrounding areas, the content of several metals decreased with increasing distance from the railroad, showed increasing trend with increasing operation time of railways, and the concentration of several PHEs increased with the size of the fraction decreases.

In Zduńska Wola Karsznice the highest concentrations of PHEs are found in samples taken in close proximity to the rails, which may be the result of the impact of rail transport. In the operational sphere, trains are susceptible to destructive environmental influences (i.e., atmospheric factors causing corrosion and damage due to lightning, ultraviolet radiation and temperature changes; human activity). All of these factors contribute to the formation of specific wastes that lead to soil contamination. These include: damaged and worn out elements of the vehicle; polluting substances accumulated on the vehicle; waste created in the process of disinfecting compartments and toilets and in the process of heating, ventilating, cleaning; used technical liquids (oils, greases, hydraulic liquids) and in case of freight cars—the remains of transported loads (e.g., wood, cement, coal, liquid chemical products); sludge from domestic and sanitary wastewater (organic and inorganic particles, detergents, soaps, paper, disinfectants, fecal matter); and vapors of solvents, sulfuric and oxalic acid emitted into the atmosphere. Generally, the waste contains steel (Fe, C), non-ferrous metals (Al, Mg, Ti, Cu, Zn, Ni, Pb, Cd), rubber, plastics, chemicals, and glass. Oil and hydraulic fluids leaking from shock absorbers, discs, buffer bushings, draw hook guides and grease (Ca, Cu, Al, Li) found on the current collectors of electric locomotives pulling wagons also contribute to soil pollution [8,40,43,44]. The content of PHEs is higher in the area of the rolling stock cleaning bay and the railroad siding, while the areas of goods handling station and the platform were less contaminated. Malawska et al. [8] and Wiłkomirski et al. [45] found high contents of iron, cobalt, zinc and chromium in samples collected from the rolling stock cleaning bay and high contents of Cu, Mn and Zn at different parts of the railroad junctions. Comparing the scientists' data with the results from Zduńska Wola Karsznice, it should be stated that the content of the mentioned elements was also at the highest level. Antuniassi et al. [46] also showed that herbicides used for conservation purposes on railroad lands contribute to soil degradation.

Table 3. Representative research similar to this study.

Study Area	Main Conclusion	References
Suining Railway Station, Suining, Sichuan Province, China	The concentrations of Cd, Pb showed increasing trend with increasing operation time of railways.	[10]
Railway that connects České Budějovice and Brno cities	The highest copper content was observed in soils taken close to the railroad.	[13]
Railway tracks on the territory of Srem, Serbia	The concentration values of Cu, Ni, Cd, Pb in samples collected from up to 1 km from the railroad line were higher than in samples collected from >1 km.	[14]
Suining, Sichuan, China	The railway transportation caused heavy metal pollution and the degree was Mn > Cd > Cu > Zn > Pb.	[20]
Railroad areas in Tarnowskie Góry	The level of PHEs contamination near railroad sites decreased with increasing distance from the rails.	[41]
Outlet roads in Poznań, Polska	The highest metal concentrations were observed in the smallest fraction (<0.063 mm), which were up to four times higher than those in sand fractions.	[42]

Taking into account the current Regulation of Minister of Environment on the method of assessing the pollution of the Earth's surface of 1 September 2016 (Journal of Laws 2016, item 1395), it was shown that the concentration of Cu was exceeded in samples collected from the inter-railway (A) (>600 mg·kg^{-1} d.w.) [47]. The study conducted on the world's highest railroad line, Qinghai-Tibet, found that the concentrations of Cu, Zn and Cd in the samples from the embankment were seven times higher than the geochemical background of study area [4]. According to Moczarski [44], and Wiłkomirski et al. [43] copper oxides formed from oxidized power collectors, catenary lines and intensive operation of pantographs by rail vehicles.

It can be expected that the increased level of lead in samples taken 5 and 10 m from the tracks may be the result of the impact of road transport, as shown in the studies by Liu et al. [40]. Similar observations were made by Vaiškūnaitė and Jasiūnienė [15], who found that the highest concentrations of Pb and Cd were recorded at a distance of 5.0 m from railroad sleepers in the upper (up to 10 cm) layer of soil.

The increase in lead and chromium concentration in the topsoil adversely affects the microfauna and microflora. As a result of reducing the enzymatic activity of soil microorganisms, the process of decomposition of organic matter slows down, which in turn leads to soil degradation [26].

The study does not support the hypothesis that Pb concentration increases with the depth of soil profile. The study of Łukasiewicz [48] showed that lead was present in the upper part of the profile at 6 of 21 sites, while already deeper at 19 sites. This is not consistent with the results presented in this paper.

Cadmium was characterized by the lowest concentration in the studied soil samples. Concentration of this element does not show any change depending on the depth of sampling. It is accumulated both in shallower and deeper soil profiles. The investigated soil samples contain high concentrations of manganese, which sorbs cadmium, thus decreasing its mobility in the environment and availability to the flora. The content of Zn does not show a strong correlation with the depth of sampling, which confirms that Zn is a common element and its content in soil strongly depends on the content in the source rock [49]. However, the presence of lead is largely the result of anthropogenic activity rather than the result of source rock.

It should be noted that the highest concentrations of Pb, Cd, Cr, Co, Cu, Ni, and Zn occur in $0.05 \leq 0.1$ mm fractions. As the fraction size increases, the concentration of PHEs decreases. In the smallest fractions of the tested soils (≤ 0.1 mm), the content of lead, cadmium, chromium, cobalt, copper, nickel and zinc exceeded 168.1; 0.5; 71.8; 10.4; 1371.9; 39.4; and 460.6 mg·kg^{-1} d.w., respectively. The richest in metals are fine-grained fractions separated from the soils of A (Cd, Cr, Cu, Zn), B2 (Pb), C2 (Co, Ni).

The concentration of PHEs in soil also depends on the duration of exposure to the contamination. Despite the fact that the authors did not attempt to confirm this thesis in these studies, these issues are often discussed in the world literature. According to Chen et al. [10], soils were contaminated with Cd and Pb, and their maximum content was determined in samples from railroad areas with the longest operating time. The concentration levels of Zn, Cu and Fe in soil samples are not affected by anthropogenic activities.

5. Conclusions

This study examined the physical and chemical properties in surface soils from Zduńska Wola Karsznice railway junction. The study found that the pH values of investigated soils ranged from 6.47 (slightly acidic) to 7.29 (slightly alkaline). With increasing distance from the rails, the pH value decreased. Through analysis of accumulation index (AI) the mean values of PHEs in the investigated soils were arranged in the following series Cu > Cd > Pb > Zn > Ni > Co > Cr. The high accumulation index of copper, cadmium and lead in the surface layer of soil indicate their anthropogenic origin. The average concentrations of particular metals in the soil of railroad junction in Zduńska Wola Karsznice were as follows: Cu > Zn > Sr > Pb > Ni > Cr > Co > Cd. Higher PHEs contents were found in the very fine sand fraction ($0.05 \leq 0.1$ mm).

The highest concentrations of potentially harmful elements were recorded in samples collected from close to the rails (inter-railway, side track), and in the embankment (10 m from the track). The high level of PHEs contamination of soils in railroad areas can have a destabilizing effect on ecosystems. The discovery of increased concentrations of PHEs in railway junctions is of great importance, not only cognitive, but also practical, e.g., in the process of remediation of these areas. The authors are well aware of the limitations resulting from a low number of soil samples taken and plan to increase the number of soil samples and to extend the scope of methods. In addition, the authors recommend the determination of PHEs in the smallest fractions during continuous monitoring studies ($0.05 \leq 0.1$ mm). The highest concentration of metals was recorded in these fractions. These findings should contribute to assessing sources for further migration of PHEs into the groundwater, crops, and finally, humans. This will allow the visualization of the potential risk to living organisms. Moreover, zoning plans should consider these results when planning new agricultural lands and single family houses.

Author Contributions: Conceptualization, I.T.-W.; methodology, I.T.-W.; software, I.T.-W., A.Ś., P.P.; validation, I.T.-W., A.Ś., P.P.; formal analysis, I.T.-W., P.P.; investigation, I.T.-W., A.Ś., P.P.; resources, I.T.-W., A.Ś., P.P.; data curation, I.T.-W., A.Ś., P.P.; writing—original draft preparation, I.T.-W.; writing—review and editing, I.T.-W., A.Ś., P.P.; visualization, I.T.-W., A.Ś., P.P.; supervision, I.T.-W., A.Ś., P.P.; project administration, I.T.-W., A.Ś., P.P.; funding acquisition, I.T.-W., A.Ś., P.P. All authors have read and agreed to the published version of the manuscript.

Funding: This research was funded by Jan Kochanowski University, grant number SUPD.RN.21.027 and SUPB.RN.21.260.

Institutional Review Board Statement: Not applicable.

Informed Consent Statement: Not applicable.

Data Availability Statement: Not applicable.

Conflicts of Interest: The authors declare no conflict of interest.

References

1. Duong, T.T.T.; Lee, B.K. Determining contamination level of heavy metals in road dust from busy traffic areas with different characteristics. *J. Environ. Manag.* **2011**, *92*, 554–562. [CrossRef] [PubMed]
2. Jeong, H.; Choi, J.Y.; Ra, K. Potentially toxic elements pollution in road deposited sediments around the active smelting industry of Korea. *Sci. Rep.* **2021**, *31*, 7238. [CrossRef]
3. Yan, X.; Zhang, F.; Zeng, C.; Zhang, M.; Devkota, L.P.; Yao, T. Relationship between heavy metal concentrations in soils and grasses of roadside farmland in Nepal. *Int. J. Environ. Res. Public Health* **2012**, *9*, 3209–3226. [CrossRef]
4. Zhang, H.; Wang, Z.; Zhang, Y.; Ding, M.; Li, L. Identification of traffic-related metals and the effects of different environments on their enrichment in roadside soils along the Qinghai-Tibet highway. *Sci. Total Environ.* **2015**, *521–522*, 160–172. [CrossRef]
5. Golia, E.E.; Papadimou, S.G.; Cavalaris, C.; Tsiropoulos, N.G. Level of Contamination Assessment of Potentially Toxic Elements in the Urban Soils of Volos City (Central Greece). *Sustainability* **2021**, *13*, 2029. [CrossRef]
6. Adamiec, E.; Jarosz-Krzemińska, E.; Wieszała, R. Heavy metals from non-exhaust vehicle emissions in urban and motorway road dusts. *Environ. Monit. Assess.* **2015**, *188*, 369. [CrossRef] [PubMed]
7. Krailertrattanachai, N.; Ketrot, D.; Wisawapipat, W. The Distribution of Trace Metals in Roadside Agricultural Soils, Thailand. *Int. J. Environ. Res. Public Health* **2019**, *16*, 714. [CrossRef]
8. Malawska, M.; Wiłkomirski, B. An analysis of soil and plant (Taraxacum officinale) contamination with heavy metals and polycyclic aromatic hydrocarbons (PAHs) in the area of railway junction Iława Główna, Poland. *Water Air Soil Pollut.* **2001**, *127*, 339–349. [CrossRef]
9. Zhang, H.; Wang, Z.F.; Zhang, Y.L.; Hu, Z.J. The effects of the Qinghai–Tibet railway on heavy metals enrichment in soils. *Sci. Total Environ.* **2012**, *439*, 240–248. [CrossRef]
10. Chen, Z.; Wang, K.; Ai, Y.W.; Li, W.; Gao, H.; Fang, C. The effects of railway transportation on the enrichment of heavy metals in the artificial soil on railway cut slopes. *Environ. Monit. Assess.* **2014**, *186*, 1039–1049. [CrossRef] [PubMed]
11. Wierzbicka, M.; Bemowska-Kałabun, O.; Gworek, B. Multidimensional evaluation of soil pollution from railway tracks. *Ecotoxicology* **2015**, *24*, 805–822. [CrossRef] [PubMed]
12. Lucas, P.S.; de Carvalho, R.G.; Grilo, C. Railway Disturbances on Wildlife: Types, Effects, and Mitigation Measures. *Railw. Ecol.* **2017**, 81–99. [CrossRef]
13. Šeda, M.; Ším, J.; Volavka, T.; Vondruška, J. Contamination of soils with Cu, Na and Hg due to the highway and railway transport. *Eurasian J. Soil Sci.* **2017**, *6*, 59–64. [CrossRef]
14. Stojic, N.; Pucarevic, M.; Stojic, G. Railway transportation as a source of soil pollution. *Transp. Res. Part D Transport. Environ.* **2017**, *57*, 124–129. [CrossRef]
15. Vaiškūnaitė, R.; Jasiūnienė, V. The analysis of heavy metal pollutants emitted by railway transport. *Transport* **2020**, *35*, 213–223. [CrossRef]
16. Samarska, A.; Zelenko, Y.; Kovrov, O. Investigation of Heavy Metal Sources on Railways: Ballast Layer and Herbicides. *J. Ecol. Eng.* **2020**, *21*, 32–46. [CrossRef]
17. Pohrebennyk, V.; Ruda, M.; Paslavskyi, M.; Salamon, I. Consortiums of ecotones of protective type to ensure the environmental safety on railway lines, Acta Facultatis Studiorum Humanitatis Et Naturae Universitatis Prešoviensis. *Nat. Sci.* **2016**, *XLIII*, 173–181.
18. Radziemska, M.; Fronczyk, J.; Mazur, Z.; Vaverková, M. Impact of railway transport on soil and Pleurozium schreberi contamination with heavy metals. *Infrastruct. Ecol. Rural Areas* **2016**, *1*, 45–57. [CrossRef]
19. Alok Gude, L. Heavy metal detection from sewage irrigated soil beside railway tracks in Mumbai. *Kong. Res. J.* **2017**, *4*, 121–128. [CrossRef]
20. Meng, X.; Ai, Y.; Li, R.; Zhang, W. Effects of heavy metal pollution on enzyme activities in railway cut slope soils. *Environ. Monit. Assess.* **2018**, *190*, 197. [CrossRef] [PubMed]
21. Mohsen, M.; Ahmed, M.B.; Zhou, J.L. Particulate matter concentrations and heavy metal contamination levels in the railway transport system of Sydney, Australia. *Transp. Res. D Transp. Environ.* **2018**, *62*, 112–124. [CrossRef]
22. Samarska, A.V.; Zelenko, Y.V. Assessment of the railway influence on the heavy metal accumulation in soil. *Sci. Transp. Prog. Bull. Dnipropetr. Natl. Univ. Railw. Transp.* **2018**, *4*, 76. [CrossRef]
23. Chen, Z.; Liu, X.; Ai, Y.; Chen, J.; Luo, X.; Chen, J.; Zhong, S. Effects and mechanisms of revegetation modes on cadmium and lead pollution in artificial soil on railway rock-cut slopes. *Sci. Total Environ.* **2018**, *644*, 1602–1611. [CrossRef]
24. Goth, A.; Michelsen, A.; Rousk, K. Railroad derived nitrogen and heavy metal pollution does not affectnitrogen fixation associated with mosses and lichens at a tundra site in Northern Sweden. *Environ. Pollut.* **2019**, *247*, 857–865. [CrossRef] [PubMed]
25. Radziemska, M.; Gusiatin, Z.M.; Kowal, P.; Bęś, A.; Majewski, G.; Jeznach-Steinhagen, A.; Mazur, Z.; Liniauskienė, E.; Brtnický, M. Environmental impact assessment of risk elements from railway transport with the use of pollution indices, a biotest and bioindicators. *Hum. Ecol. Risk Assess. Int. J.* **2021**, *27*, 517–540. [CrossRef]
26. Kabata-Pendias, A.; Pendias, H. *Biogeochemistry of Trace Elements*; PWN: Warszawa, Poland, 1999. (In Polish)

27. Tomczyk, I. Ocena Zanieczyszczenia Metalami Ciężkimi Wierzchniego Poziomu Gruntu Terenów Kolejowych na Przykładzie Węzła Kolejowego Zduńska Wola Karsznice (Centralna Polska). Master's Thesis, Jan Kochanowski University in Kielce, Kielce, Poland, 2014.
28. Kondracki, J. *Geografia Regionalna Polski*; Wydawnictwo Naukowe PWN: Warszawa, Poland, 2002. (In Polish)
29. Klatkowa, H. *Szczegółówa Mapa Geologiczna Polski, 1:50,000, Arkusz Łask*; Wydawnictwo Geologiczne: Warszawa, Poland, 1987.
30. Maćków, A.; Gruszecki, J. *Mapa Geośrodowiska Polski, 1:50,000, Arkusz Łask*; PIG, MŚ: Warszawa, Poland, 2003.
31. Geoportal Województwa Łódzkiego, the Agricultural Soil Suitability Complexes. Available online: http://www.geoportal.lodzkie.pl (accessed on 22 September 2021).
32. Wyżkowski, T.; Michalak, J. *Raport Wojewódzkiego Inspektoratu Ochrony Środowiska w Łodzi*; IV Ochrona Powierzchni Ziemi Okręgowa Stacja Chemiczno-Rolnicza w Łodzi: Łódź, Poland, 2008.
33. Pecyna, I. Magistrala węglowa w regionie Szadkowskim. *Biul. Szadkowski* **2016**, *16*, 257–273. [CrossRef]
34. Railway Network on the Administrative Division of Poland. Available online: http://www.mapa.plk-sa.pl (accessed on 22 September 2021).
35. Orthophoto Map. Available online: http://www.geoportal.gov.pl (accessed on 22 September 2021).
36. Polskie Towarzystwo Gleboznawcze. Klasyfikacja uziarnienia gleb i utworów mineralnych—PTG 2008, Particle size distribution and textural classes of soils and mineral materials-classification of Polish Society of Soil Sciences 2008. *Roczniki Gleboznawcze* **2009**, *60*, 5–16.
37. Czarnowska, K. Ogólna zawartość metali ciężkich w skałach macierzystych jako tło geochemiczne gleb. *Gleboznawcze* **1996**, *XLVII*, 43–50.
38. Zajęcka, E.; Świercz, A. Biomonitoring of the Urban Environment of Kielce and Olsztyn (Poland) Based on Studies of Total and Bioavailable Lead Content in Soils and Common Dandelion (Taraxacum officinale agg.). *Minerals* **2021**, *11*, 52. [CrossRef]
39. Kozłowski, R.; Szwed, M.; Żelezik, M. Environmental Aspect of the Cement Manufacturing in the Świętokrzyskie Mountains (Southeastern Poland). *Minerals* **2021**, *11*, 277. [CrossRef]
40. Liu, H.; Chen, L.P.; Ai, Y.W.; Yang, X.; Yu, Y.H.; Zuo, Y.B. Heavy metal contamination in soil alongside mountain railway in Sichuan, China. *Environ. Monit. Assess.* **2009**, *152*, 25–33. [CrossRef]
41. Malawska, M.; Wiłkomirski, B. Soil and plant contamination with heavy metals in the area of old railway junction Tarnowskie Góry and near two main railway routes. *Rocz. Panstw. Zakl. Hig.* **2000**, *51*, 259–267. [PubMed]
42. Ciazela, J.; Siepak, M. Environmental factors affecting soil metals near outlet roads in Poznań, Poland: Impact of grain size, soil depth, and wind dispersal. *Environ. Monit. Assess.* **2016**, *188*, 6. [CrossRef] [PubMed]
43. Wiłkomirski, B.; Sudnik-Wójcikowska, B.; Galera, H.; Wierzbicka, M.; Malawska, M. Railway transportation as a serious source of organic and inorganic pollution. *Water Air Soil Pollut.* **2011**, *218*, 333–345. [CrossRef] [PubMed]
44. Moczarski, M. Obsługiwanie pojazdów szynowych a zagrożenie środowiska. Rail vehicle maintenance—Environmental risk. *Probl. Kolejnictwa* **2006**, *143*, 59–86.
45. Wiłkomirski, B.; Galera, H.; Sudnik-Wójcikowska, B.; Staszewski, T.; Malawska, M. Railway tracks—Habitat conditions, contamination, floristic settlement—A review. *Environ. Nat. Resour. Res.* **2012**, *2*, 86–95. [CrossRef]
46. Antuniassi, U.R.; Velini, E.D.; Nogueira, H.C. Soil and weed survey for spatially variable herbicide application on railways. *Precis. Agric.* **2004**, *5*, 27–39. [CrossRef]
47. Minister of the Environment. Regulation of the Minister of the Environment of 1 September 2016 on the Method of the Contamination Assessment of the Earth Surface. In *Journal of Laws of 2016*; Item 1395; Wydawnictwo Sejmowe: Warsaw, Poland, 2016. (In Polish)
48. Łukasiewicz, S. Struktura fizyczna gruntu, zawartość substancji organicznej oraz skład chemiczny gleb w podłożach 21 stanowisk zieleni miejskiej na terenie Poznania. *Bad. Fizjogr. R. III Ser. Seria A Geogr. Geogr.* **2011**, *A63*, 49–75. [CrossRef]
49. Rapalska, M. Analiza zawartości metali ciężkich w glebach Tatrzańskiego Parku Narodowego w części słowackiej (TANAP). In Proceedings of the V Krakowska Konferencja Młodych Uczonych, Kraków, Poland, 23–25 September 2010; pp. 327–335.

Article

Vulnerability to Nitrate Occurrence in the Spring Waters of the Sila Massif (Calabria, Southern Italy)

Ernesto Infusino [1], Ilaria Guagliardi [2,*], Simona Gaglioti [1] and Tommaso Caloiero [2]

[1] Department of Environmental Engineering (DIAm), University of Calabria, Via P. Bucci 41C, 87036 Rende, Italy; ernesto.infusino@unical.it (E.I.); simonagaglioti@gmail.com (S.G.)
[2] National Research Council—Institute for Agricultural and Forest Systems in Mediterranean (CNR-ISAFOM), Via Cavour 4/6, 87036 Rende, Italy; tommaso.caloiero@isafom.cnr.it
* Correspondence: ilaria.guagliardi@isafom.cnr.it; Tel.: +39-0984-841427

Abstract: Knowledge of spring waters' chemical composition is paramount for both their use and their conservation. Vast surveys at the basin scale are required to define the nature and the location of the springs and to identify the hydrochemical facies of their aquifers. The present study aims to evaluate the hydrochemical facies and the vulnerability to nitrates of 59 springs falling in the Sila Massif in Calabria (southern Italy) and to identify their vulnerability through the analysis of physicochemical parameters and the use of the Langelier–Ludwig diagram. A spatial analysis was performed by the spline method. The results identified a mean value of 4.39 mg NO_3^-/L and a maximum value of 24 mg NO_3^-/L for nitrate pollution in the study area. Statistical analysis results showed that the increase in electrical conductivity follows the increase in alkalinity values, a correlation especially evident in the bicarbonate Ca-Mg waters and linked to the possibility of higher nitrate concentrations in springs. These analyses also showed that nitrate vulnerability is dependent on the geological setting of springs. Indeed, the Sila igneous–metamorphic batholith, often strongly affected by weathering processes, contributes to not buffering the nitrate impacts on aquifers. Conversely, anthropogenic activities, particularly fertilization practices, are key factors in groundwater vulnerability.

Keywords: spring waters; hydrogeochemical characterization; statistical analysis; correlation; nitrate vulnerability

Citation: Infusino, E.; Guagliardi, I.; Gaglioti, S.; Caloiero, T. Vulnerability to Nitrate Occurrence in the Spring Waters of the Sila Massif (Calabria, Southern Italy). *Toxics* **2022**, *10*, 137. https://doi.org/10.3390/toxics10030137

Academic Editor: Danny D. Reible

Received: 8 February 2022
Accepted: 10 March 2022
Published: 12 March 2022

Publisher's Note: MDPI stays neutral with regard to jurisdictional claims in published maps and institutional affiliations.

1. Introduction

In spring waters, nitrogen is one of the principal biogenous elements. It naturally occurs in the environment or derives from anthropogenic input, commonly found in fertilizers and animal and human wastes. Usually, human activities are altering all biogeochemical cycles [1,2] and tend to particularly affect the natural nitrogen cycle either directly (e.g., industrial, residential, agricultural, farming discharge) or indirectly, by altering soil degradation processes [3–8]. Furthermore, the dynamics of the hydrogeological systems and their water resource quality can be affected by climate change [9–12].

Once nitrogen is in the environment and converted to nitrate form, it will completely dissolve in water and move easily with water to aquatic ecosystems, where it can cause undesirable effects. Indeed, epidemiological studies evidenced that high levels of nitrate in water are a human health concern [13], considering that nitrate is classified as a probable human carcinogen [14]. Several studies have been conducted on the potential onset of stomach or brain cancer in people caused by exposure to nitrates through drinking water [7,15–17]. More precisely, as a result of a reaction into an acid environment, when nitrates encounter the amines contained within food products, they produce nitrosamines, classified as carcinogenic substances [18]. Furthermore, nitrates have been demonstrated to be responsible for "blue baby syndrome" (methemoglobinemia) [19], especially in infants, whereby reduced intestinal acidity facilitates the proliferation of bacterial flora transforming nitrates into nitrites. The latter enter the blood and cause the modification of hemoglobin into methemoglobin, which is thus unable to transport oxygen to tissues, resulting in cyanosis. The "blue baby syndrome" occurs when methemoglobin levels exceed a 10% concentration in the blood [20].

According to the WHO, the maximum permissible concentration level of nitrates in drinking waters is 50 mg/L [21], while the standard level is 25 mg/L. The European Directive 98/83 CE sets the maximum concentration level for nitrates at 50 mg/L. At the national scale, the Legislative Decree No. 30/2009 [22] defines the criteria and procedure for assessing the chemical status of groundwater. It reports the environmental quality standards established at the community level for nitrates and identifies for a specific set of parameters, the threshold values adopted at the national level for the purpose of assessing the chemical status of groundwater. This decree sets the nitrate value for groundwaters at 50 mg/L for the determination of nitrate-vulnerable areas. Conversely, as to what concerns mineral waters [23], the limit value for nitrates is 45 mg/L, which becomes a precautionary 10 mg/L value for those mineral waters consumed during pregnancy and infancy.

For this health risk, studying spring water vulnerability to nitrates is paramount to solve possible aquifer contamination and/or have a clear historical picture for further evaluations. Deterministic or geostatistical methods are the right approach for studying the spatial structure of nitrate and for mapping spring water quality. Indeed, the application of these techniques to the interpretation of the relationship between the aquifer lithology and the chemical composition of spring water and of the contamination of groundwater has provided good results so far. Furthermore, other studies have carried out spring water analysis at a large scale, collecting considerable data and contributing to the water management decision making

Therefore, taking into account all previous considerations, from the intrinsic public health risk in water supply nitrate sources to geochemical and environmental implications of nitrate contents in water, the present work is aimed at classifying the spring water types of the Sila Massif (Calabria, southern Italy) according to their hydrogeochemical features, identifying the factors controlling mineral waters using spatial variables focused on lithological settings and determining the vulnerability to nitrate occurrence in the spring waters. The study area is characterized by many spring waters, and its groundwater is largely used for drinking purposes, thus representing a natural and socioeconomically important resource. However, few analyses on few samples have been carried out in the past years to assess the quality and the degradation processes affecting the aquifers of the Calabria region [24,25].

It is expected that the outcomes of this work could help to develop effective strategies for integrated water resource management, particularly by providing scientific evidence for decisions on and management of groundwater sources used for drinking water supply.

2. Study Area and Data

The Sila Massif, located in the central-northern area of the Calabria region, is the study area (Figure 1). It extends for 150,000 hectares, which include mountains, plateaus, thick wooded areas, rivers and high-altitude lakes, through three different provinces of the region. From north to south it is usually divided into three subranges: Sila Greca, Sila Grande and Sila Piccola. The main rivers of Calabria originate from this massif, and some of these rivers feed artificial lakes [26].

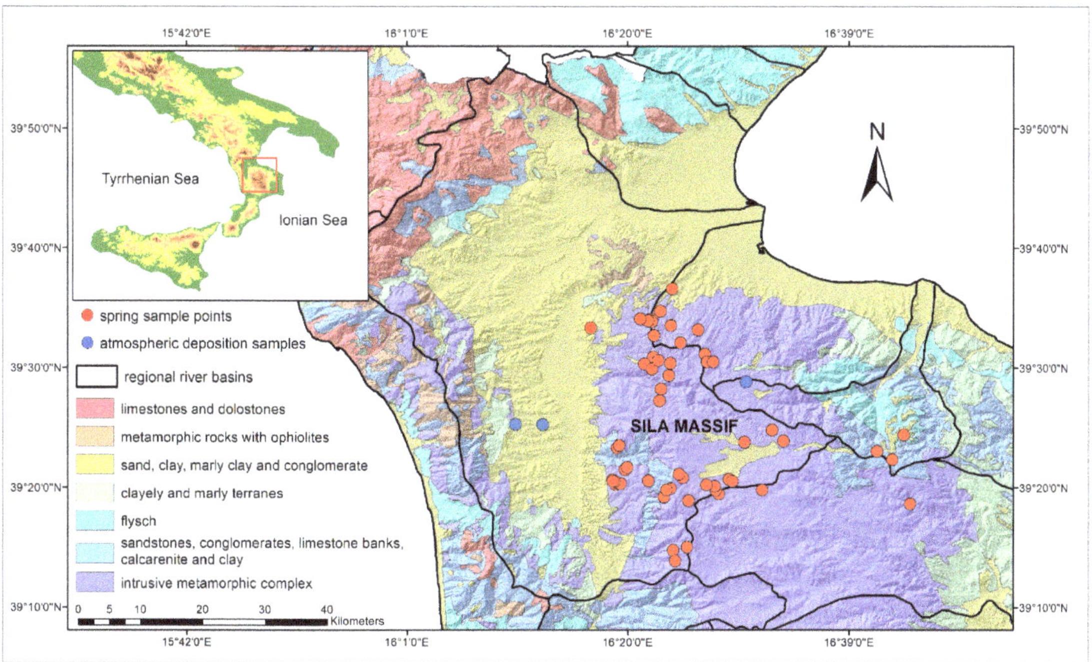

Figure 1. Map of the geographical and geological setting of the study area, with the locations of the samples.

The Sila Massif is part of a Paleogene orogenic belt (Calabrian Arc), thrusted over the Apennine Chain during Miocene times [27–29]. From a geological point of view, the area is a batholith formed by late Hercynian intrusive rocks and Paleozoic medium- to high-grade metamorphites, which are often strongly affected by weathering processes [30]: it underlies Mesozoic and Miocene to Quaternary sedimentary terranes [31,32]. The Paleozoic complex is composed mainly of paragneiss, biotite schists and grey phyllitic schists (with dominant quartz, chlorite and muscovite). The intense tectonics that molded the site, the subsequent and rapid uplifting affecting the area and an unfavorable set of climatic factors often reduced the crystalline rocks to sand [33,34].

The Quaternary is represented by Pliocene sediments, made up of light brown and red sands and gravels, blue-grey silty clays with silt interlayers, Pleistocene to Holocene alluvial sands and gravels and very small outcrops of Miocene carbonate rocks [35].

The dominant altitude of the central part of the Sila Massif is about 1300 m a.s.l., with the most representative peaks at almost 2000 m a.s.l.

Intense potato and cereal crops occupy the mountainous areas, while on the slopes, at the lower altitudes, there are pastures and forests of *Pinus laricio* (*Pinus nigra* subsp. *laricio calabrica*), many of which result from reforestation in the 1950s [36], and at the higher altitudes, beech (*Fagus sylvatica* L.) and silver fir (Abies alba) prevail.

Following the Köppen–Geiger classification [37], the climate of the study area can be identified as hot-summer Mediterranean, but due to its mountainous nature, colder winters with snow and cooler summer with some precipitation can be observed [38,39].

The study area is characterized by numerous springs with a density of 1.16/km^2 but of reduced flow rate (many of which have a flow rate lower than 1 L/s), typical of schistose–granitic settings which are almost impermeable.

In this survey, 59 spring waters were identified and sampled in the Sila Massif (Figure 1) together with 3 atmospheric deposition locations. For each site, an identification form was compiled accompanied by photographic material and contained all the initial information collected. The sampling was carried out between the months of February and May 2016, a period in which the springs' maximum flow rates are generally recorded due to aquifer recharge following snow melting at high altitude and/or more simply following rainfall.

3. Methodology

3.1. Laboratory Analysis

The samples collected were analyzed at the Genesis of Pollutants in the Water Cycle (GICA) laboratory at the University of Calabria. To evaluate the geochemical and hydrodynamic characteristics of the springs and in order to classify the investigated areas [40–42], a consolidated methodological procedure in geochemical approach was followed, analyzing the so-called labile parameters (temperature, pH and specific EC compensated at 20 °C) at the time of sampling by using portable apparatus including a pH meter and a conductivity meter and analyzing the main cations and anions once the samples reached the laboratory in clean high-density polyethylene (PE-HD) bottles with screw caps.

The major cations (Ca^{2+}, Mg^{2+}, Na^+, K^+) were determined by using the atomic absorption spectroscopy (AAS) method, major anions (NO_{3-}, SO_4^{2-}, Cl^-, HCO_3^-) were determined by using the UV spectrophotometry method, alkalinity with determined by titration with 0.1 N HCl on unfiltered samples (expressed as mg HCO_3^-/L) and hardness was determined by using complexometric EDTA titration.

Total dissolved solutes (TDS) expressed in mg/L is calculated as the sum of all major anions (Cl^-, NO_3^-, SO_4^{2-} and alkalinity as HCO_3^-) and cations (Na^+, K^+, Mg^{2+} and Ca^{2+}).

A quality control analysis was performed evaluating sampling and analytical duplicates. Analytical precision was calculated by the 20% analysis (in duplicate) of randomly chosen samples and was found to be within accepted international standards (<5%).

3.2. Facies Classification Analysis

In literature, there are several methods to identify the dominant facies via a comparison of the most salient chemical characteristics, including by means of graphic analysis such as qualitative diagrams, which offer a good visual interpretation but no indication of the actual water mineralization, and quantitative diagrams, which, instead, provide an indication of mineralization [43].

The analyses carried out have allowed classifying the spring waters according to Italian Legislative Decree No. 31/2001, which categorizes waters as follows:

(a)　TDS (mg/L):

- $0 < TDS < 50$—very low mineral content water (or light mineral water);
- $50 < TDS < 500$—low mineral content water;
- $500 < TDS < 1500$—medium mineral content water;
- $TDS > 1500$—waters rich in mineral salts.

(b)　Hardness (H) due to the presence of calcium and magnesium ions expressed in French degrees (1° f = 10 mg/L = 10 mg/kg) or the equivalents of $CaCO_3$:

- $H < 15$—light or soft water;
- $15 < H < 30$—average hard water;
- $H > 30$—hard water.

To determine the tendency of water to attack and solubilize certain minerals contained in rocks and soils, the aggressiveness index I_A proposed by the American Water Works Association [44,45] was determined. This index, a simplified form of the Langelier Index, valid within the temperature range 4.5–26.5 °C, is expressed by the formula:

$$I_A = pH + log_{10}(A \cdot H) \tag{1}$$

where A is the total alkalinity (mg/L HCO_3) and H is the calcium hardness (mg/L $CaCO_3$).

According to this index, the more aggressive the water is, the lower the value I_A:

- $I_A < 10$—aggressive water;
- $10 < I_A < 12$—moderately aggressive water;
- $I_A > 12$—nonaggressive water.

In order to graphically visualize the water chemistry, various diagrams have been developed over the years [46,47]. Among others, in this paper, the hydrochemical facies were determined according to the Langelier–Ludwig (LL) square diagram [48], which has been largely applied in literature [49–52]. In particular, Langelier and Ludwig proposed a diagram in which rectangular coordinates are used to represent patterns and correlations between major cations and anions for multiple samples, thus allowing separation between Ca from Mg facies and Cl from SO_4 facies. More precisely, with the LL diagram, the water classification is ensured by positioning the water sample in one of the four sectors of which the diagram is composed:

I.   Chloride–sulfate Ca-Mg waters;
II. Chloride–sulfate alkaline waters;
III. Bicarbonate alkaline waters;
IV. Bicarbonate Ca-Mg waters.

The identified ion concentrations are recalculated from mg/L to meq/L. Therefore, knowing all meq/L ion concentrations, the dimensionless value R can be calculated for each i-th cation C or anion A, according to the following relationships, and plotted on the LL diagram:

$$RC_i = \frac{50 \cdot C_i}{\sum C_{tot}} \tag{2}$$

$$RA_i = \frac{50 \cdot A_i}{\sum A_{tot}} \tag{3}$$

The alignment of the sample points into the different sectors allows the identification of possible phenomena such as mixing or evolutionary processes of water [25].

3.3. Atmospheric Depositions Input Analysis

The presence of nitrogen compounds in the air is due both to natural causes (soil erosion, volcanic eruptions, etc.) and anthropogenic activities (fires, industrial plants, motor vehicles, heating, nitrogenous fertilizers). Nitric acid (HNO_3) in the atmosphere mainly results from the transformation of NO_2 reacting with free radicals. NO_2 in turn is formed by the reaction of NO with ozone. Nitric acid has a high deposition rate. In an aqueous medium, this compound turns into nitrite and nitrate ions. Nitric acid reacts with ammonia to form ammonium nitrate. Ammonium nitrate is in an equilibrium state with nitric acid and ammonia for relative humidity lower than 60%. Ammonium nitrate, which has low sedimentation, can be transported away from its originating source, and if humidity exceeds 60%, it turns back into nitrate and ammonia, contributing to the formation of acidity even in distant places.

To quantify the contribution of background nitrogen concentration from the atmosphere, total depositions at three different sites (Figure 1) were measured. Specifically, the three sites are denominated according to their location: San Antonello in Montalto Uffugo (175 m a.s.l.), Settimo in Montalto Uffugo (250 m a.s.l.) and Bonis (1090 m a.s.l.).

3.4. Data Analysis

A preliminary analysis of the dataset was performed through the box plot test in order to identify probable outliers.

Spatial interpolation is the procedure to predict the value of attributes at unobserved points within a study region using existing observations [53].

Spatial interpolation covers a variety of methods including trend surface models, Thiessen polygons, kernel density estimation, inverse distance weighting, splines and kriging. Among these several methods, mostly used in environmental analysis, in water resources management and in hydro-geochemical forecasting, the spline interpolation method was adopted for the analysis of the spatial variation of nitrate concentration in the spring waters of the Sila Massif. The spline interpolation model was chosen since the distribution of the sampled points did not allow a homogeneous coverage of the study area, and therefore the model that was least affected by the effect of inhomogeneity in the distribution and spacing between the points was applied. In fact, the spline method can generate sufficiently accurate surfaces from only a few sampled points, and the generated surfaces retain small features.

Spline interpolation is a deterministic interpolation method that fits a mathematical function through input data to create a smooth surface [54,55]. Splines are piecewise polynomial functions that are fitted together to provide a complete, yet complex, representation of the surface between the measurement points. Functions are fitted exactly to a small number of points while, at the same time, ensuring that the joins between different parts of the curve are continuous and have no disjunctions.

Spline interpolation creates a surface that passes through the control points and has the least possible changes in slope at all the points. This method uses a mathematical function with input points:

$$Q(x,y) = \sum_{i=1}^{N} A_i d_i^2 logd_i + a + bx + cy \tag{4}$$

$$d_i^2 = (x - x_i)^2 + (y - y_i)^2 \tag{5}$$

where x and y are the coordinates of the point to be interpolated, and x_i and y_i are the coordinates of the control point.

This function minimizes the overall surface curvature, resulting in a smooth curvature passing exactly through the input points.

A regularized spline yields a smooth surface and smooth first derivatives. With the regularized option, higher values used for the weight parameter produce smoother surfaces. The values entered for this parameter must be equal to or less than zero. Typical values that may be used are 0, 0.001, 0.01, 0.1 and 0.5. A tension spline is flatter than a regularized spline of the same sample points, forcing the estimate to stay closer to the sample data. It produces a surface more rigid according to the character of the modeled phenomenon. With the tension option, higher values entered for the weight parameter result in somewhat coarser surfaces, but surfaces that closely conform to the control points. The values entered must be equal to or greater than zero. The typical values are 0, 1, 5 and 10. The tension spline was evaluated as the best method for generating surfaces that vary gently, such as modeling of aquifer levels or groundwater pollution concentrations.

To determine the interpolation errors, the values of RMSE, %RMSE and MRE were assessed by Equation (6), Equation (7) and Equation (8), respectively. Values of RMSE and MRE close to zero indicate a good estimate of the model used to assess the unknown parameters:

$$RMSE = \sqrt{\frac{\sum_{j=1}^{n} \left(x(P)_j - x(m)_j \right)^2}{n}} \tag{6}$$

$$\%RMSE = \frac{RMSE}{\overline{X}} * 100 \tag{7}$$

$$MRE = \frac{1}{n} \sum_{i=1}^{n} \frac{|z^*(x_i) - z(x_i)|}{z(x_i)} \tag{8}$$

where $x(P)$ is the estimated value of each component, $x(m)$ is the measurement of water parameter, n is sample number, X is the mean of a measured parameter, $z(x_i)$ is the observed value at location i, $z^*(x_i)$ is the interpolated value at location i and n is the sample size.

The spatial analysis tool used for this analysis was ArcGIS 9.3.1 software.

4. Results and Discussion

A summary of physicochemical parameters of spring water samples and their statistics is provided in Table 1.

Table 1. Basic statistics for analyzed water parameters.

Parameter	Min	Max	Mean	Median	Lower Quartile	Upper Quartile	Standard Deviation	Skewness	Kurtosis	CV (%)
pH	5.46	8.6	6.86	6.82	6.5	7.24	0.58	0.36	0.51	8.48
EC (μS/cm)	51.3	710.61	179.37	152	102.9	236	110.7	2.2	8.13	61.72
H (°f)	2.01	33.6	7.7	6	3.7	9.8	5.5	2.23	7.7	71.35
TDS (mg/L)	0.12	533	132.73	113	78.11	175.9	83.97	2.1	7.92	63.27
Ca^{2+} (mg/L)	3.61	110	16.42	14.4	8.61	20	14.53	4.77	29.92	88.48
Mg^{2+} (mg/L)	0.97	20.47	7.92	6.8	3.1	12.07	5.5	0.6	−0.81	69.7
K^+ (mg/L)	0.009	2.92	1.46	1.56	1.1	1.85	0.55	−0.3	0.25	37.3
Na^+ (mg/L)	4.51	59	11.27	9.05	6.96	12.5	8.3	3.87	19.25	73.7
HCO_3^- (mg/L)	8	209.82	63.08	35.12	17	98.94	56.99	0.97	−0.11	90.34
Cl^- (mg/L)	7.80	85.12	18.07	14.9	10.72	21.14	11.97	3.62	17.41	66.26
SO_4^{2-} (mg/L)	0.07	25	2.08	0.6	0.31	0.9	4.52	3.67	14.66	216.62
NO_3^- (mg/L)	0.06	24	4.38	2.8	0.7	5.65	5.38	2.14	5	124.11

CV (%) = coefficient of variation; EC = electrical conductivity; H = hardness; TDS = total dissolved solutes.

The range of pH of spring waters is between 5.46 and 8.6 (mean = 6.86).

The mean water temperature of the analyzed springs is 10.51 °C with variability between 6–16 °C and a decreasing trend with altitude Z (m a.s.l.) according to the following correlation:

$$T(°C) = -0.0052Z + 16.257 \qquad (9)$$

The thermal gradient of −0.0052 °C/m indicates the presence of relatively shallow aquifers, influenced by the external air temperature. The spring temperature seems to be mainly influenced by vertical zonation because there are no statistically significant differences in temperature between springs located in different geological formations at the same altitude. In addition, the behavior of temperature vs. altitude is probably due to recharge from snow melting during springtime, which can promote water temperature decrease, especially in springs located at higher altitudes.

Electrical conductivity ranged between 51.3 and 710.61 μS/cm with a mean of 179.37 μS/cm.

The coefficient of variation (CV, %) representing the degree of scattering, showed a wide range among ions, with values for Na^+ (73.7%), Cl^- (66.26%), SO_4^{2-} (216.62%), Ca^{2+} (88.48%), Mg^+ (69.7%), K^+ (37.3%), HCO_3^- (90.34%) and NO_3^- (124.11%) (Table 1).

The major cation concentrations were detected for Ca^{2+} and Na^+, while the major anions ones were for HCO_3^- and Cl^-. The nitrate (NO_3^-) concentration had a mean value of 4.38 mg/L, and no sample exceeded the threshold value of 50 mg/L of WHO [21] and Italian Legislative Decree [22] recommendations. These results are in accordance with previous surveys, where the nitrate concentration did not generally exceed 2 mg/L in uncontaminated spring water aquifers [56,57]; anyway, in case of contamination, nitrates can reach extremely high levels. Nitrate pollution is increasing in many countries [58]. One example is in the Campania region (Italy), where nitrate values exceeding 200 mg/L in variable concentrations both in space and time have been found in spring waters and wells [59].

The spring water samples can be classified as very low mineral content (TDS < 50 mg/L) and low mineral content (50 < TDS < 500 mg/L), with a single sample that reaches the TDS value of 533 mg/L. These results indicate that the geochemistry of spring waters in the Sila Massif is strongly affected by the mineralogical composition of the local rocks and controlled by hydrolysis of sodium and/or calcium silicate minerals. The samples generally present an increase in low TDS with a decrease in altitude (Figure 2a). As regards the hardness, a trend similar to that of TDS can be observed (Figure 2b). With few exceptions, the springs analyzed can be classified as light or soft waters, with a hardness value lower than 15 °f. There is a good correlation between hardness and TDS.

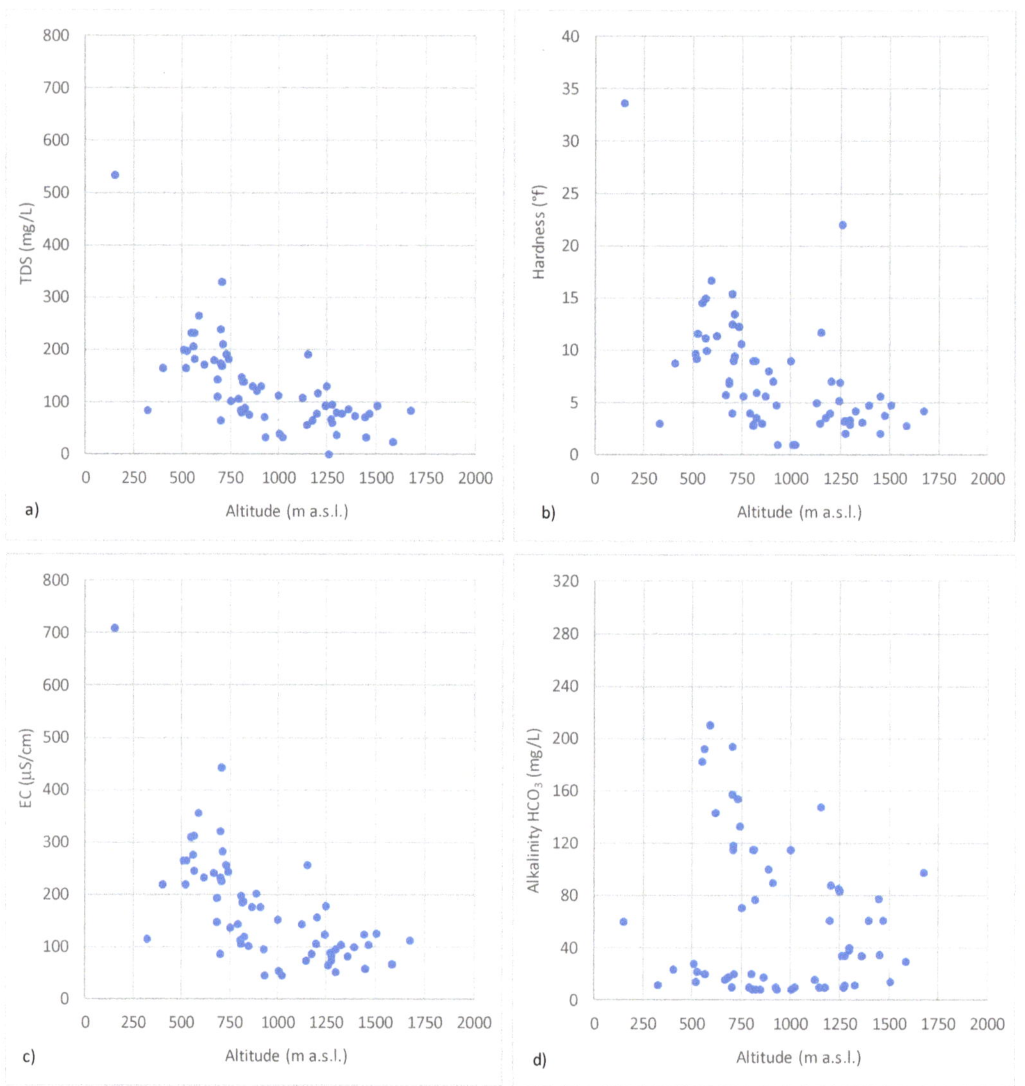

Figure 2. Bivariate plots of TDS (**a**), H (**b**), EC (**c**) and A (**d**) vs. altitude.

According to the results of the aggressiveness index I_A, it was possible to classify 78% of the analyzed springs as aggressive ($I_A < 10$) and the remaining 22% as moderately aggressive ($10 < I_A < 12$), the latter having a higher conductivity than the former. Figure 3 shows the I_A classification of the spring waters.

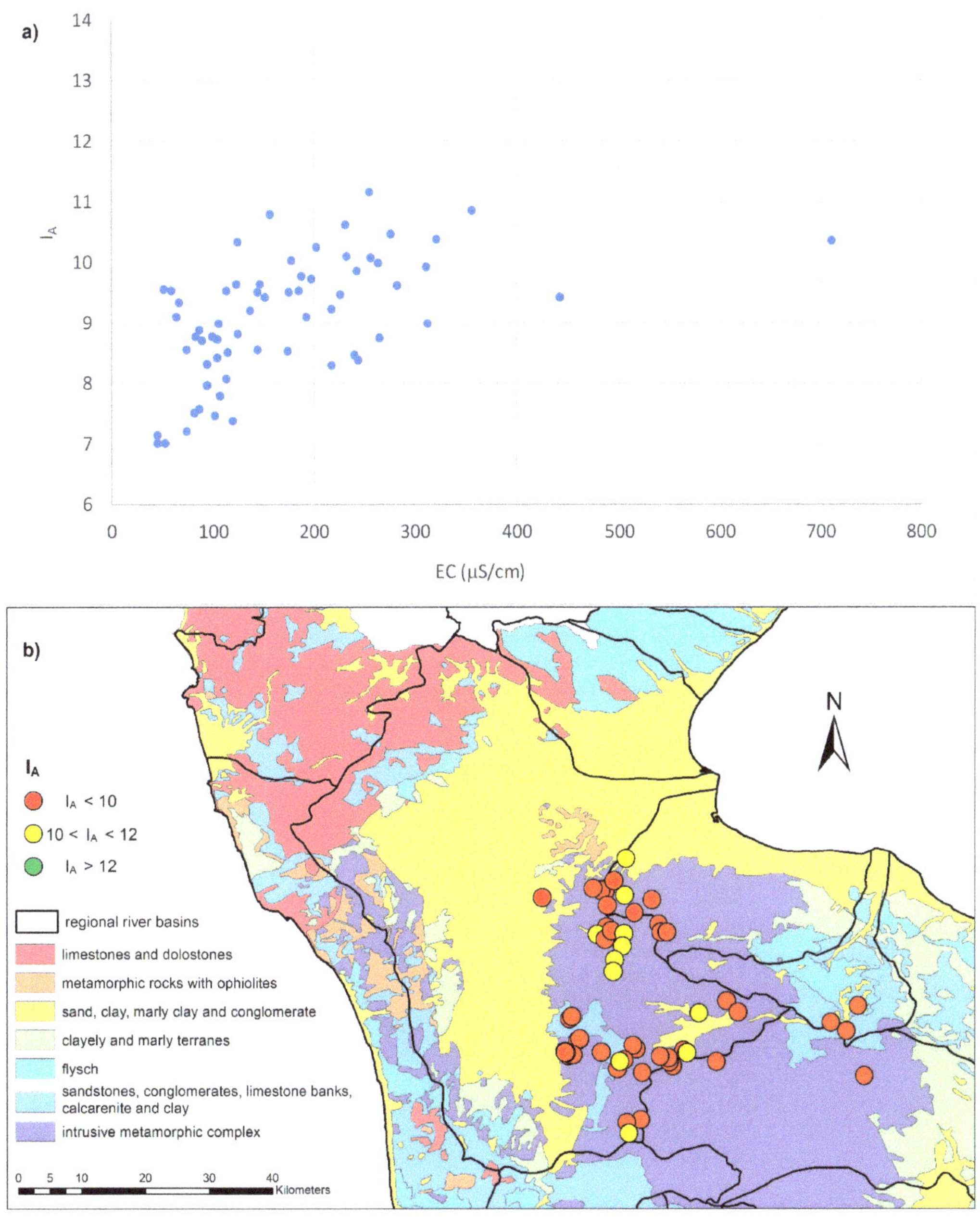

Figure 3. Bivariate plots of I_A vs. EC (**a**) Bivariate plots of I_A vs. EC and (**b**) I_A classification of the spring waters according to geology.

Aggressiveness is mainly due to the presence of carbon dioxide (CO_2), in the form of carbonic acid, as well as the presence of sulfates and nitrates which reduce the water pH. The acids, not being buffered by the acid rocks of the aquifers, act as a solubilizing agent on some minerals, which are washed out.

A correlation matrix of hydrochemical parameters was constructed and is shown in Table 2.

Table 2. Pearson correlation coefficients for analyzed parameters.

	EC	TDS	pH	H	Ca_2^+	Mg_2^+	Na^+	K^+	Cl^-	SO_4^{2-}	NO_3^-	A	Altitude
EC	**1**												
TDS	**0.99**	**1**											
pH	0.11	0.08	**1**										
H	**0.82**	**0.78**	0.12	**1**									
Ca_2^+	**0.85**	**0.85**	0.17	**0.83**	**1**								
Mg_2^+	**0.78**	**0.77**	0.14	**0.71**	**0.53**	**1**							
Na^+	**0.71**	**0.70**	0.01	0.47	0.45	**0.52**	**1**						
K^+	**0.51**	**0.51**	0.05	0.37	0.39	0.36	0.23	**1**					
Cl^-	**0.76**	**0.76**	−0.04	**0.53**	**0.68**	0.33	**0.56**	**0.63**	**1**				
SO_4^{2-}	0.32	0.32	−0.08	0.18	0.19	0.18	**0.60**	−0.03	0.22	**1**			
NO_3^-	**0.57**	**0.57**	−0.10	0.31	0.38	0.29	**0.52**	0.33	**0.65**	0.26	**1**		
A	0.44	0.43	0.15	0.48	0.41	**0.59**	**0.52**	−0.11	−0.07	0.37	0.02	**1**	
Altitude	**−0.67**	**−0.68**	0.05	**−0.53**	**−0.48**	**−0.66**	**−0.43**	**−0.56**	**−0.50**	−0.15	**−0.43**	−0.16	**1**

Coefficients are significant at the 0.05 level (2-tailed), and those higher than 0.5 are shown in bold.

Because geological properties are closely related to characteristics of natural springs, including discharge, chemical composition, and temperature, hydrochemical parameter analysis between the occurrence of spring water and geological factors can be useful. Indeed, extensive monitoring campaigns are necessary, which can determine the natural cycle of spring waters resulting from the interaction between rocks, atmosphere, mixing with older aquifers and the effects of human activities. Groundwaters, because of their interaction with various matrices, become rich in gases such as carbon dioxide, oxygen and nitrogen; carbonic acid salts and strong bases, e.g., calcium, magnesium, sodium and potassium carbonates; and strong acid salts and strong bases, such as sulfates, chlorides, calcium, magnesium, sodium and potassium nitrates.

Accordingly, in this survey, major cations including Na^+ and Mg^{2+} showed a linear relationship with alkalinity. This can be explained by the acidic hydrolysis of minerals. The hydrolysis reaction consumes water and acid that might have originated from CO_2 and increases the pH, alkalinity and cation concentration of water.

Further evaluations can be detected by observation of electrical conductivity values according to altitude for each zone (Figure 2c): although the altitude decreasing is combined with the conductivity increasing (with the exception of the three spring waters in which higher NO_{3-} concentration was detected), this last one remains low, thus evidencing a typical behavior of the igneous lithology and the presence of relatively shallow aquifers affected by atmospheric depositions [60].

The Langelier–Ludwig (LL) square plot, obtained by positioning the representative points of the individual springs within the four standard quadrants, is represented in Figure 4a.

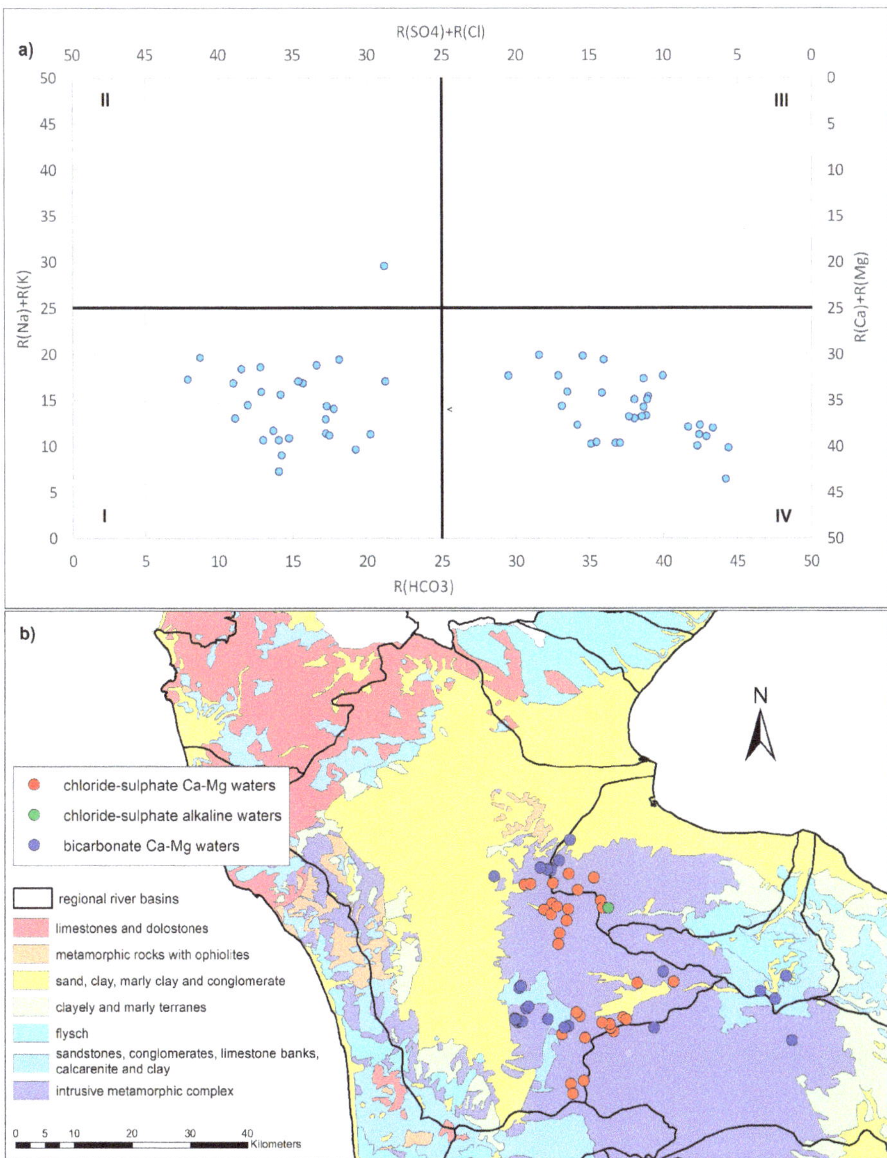

Figure 4. Langelier–Ludwig (LL) diagram (**a**) and classification of the spring waters (LL quadrants) according to geology (**b**).

According to their chemical compositions, the samples fall into three of the four quadrants of the LL diagram:

- Chloride–sulfate Ca-Mg waters (I quadrant). The waters belonging to these facies mainly spring from the slopes forming the crown of the Sila Massif (Figure 4b). These springs fall into geological units made up of acid rocks (acid granulites, biotic gneisses). They have a low alkalinity which is almost independent of altitude. For them, good correlations between conductivity EC (μS/cm) and alkalinity A in terms of HCO_3 (mg/L) and between chloride and sodium ions were observed.

- Chloride–sulfate alkaline waters (II quadrant). A limited number of springs belong to these facies, falling in metamorphic geological units located at the extreme offshoots of the investigated area, arising from metamorphic rocks. An exception is a spring in another setting (the municipality of Acri) classifiable within these facies due to its high content of sodium, chlorides, etc., although the aquifer rocks are granite like those of the IV quadrant. The waters are rich in sodium and potassium ions and show a low calcium and magnesium content. Alkalinity does not differ from that described for the I quadrant.
- Bicarbonate Ca-Mg waters (IV quadrant). They are the springs arising in the central part of the Sila Massif, from acid rocks (acid granulites, biotic gneisses, granites, granodiorites, magmatites). Alkalinity A increases with decrease in altitude Z. For these waters, an excellent correlation between conductivity EC (μS/cm) and alkalinity A was found (Figure 5a).

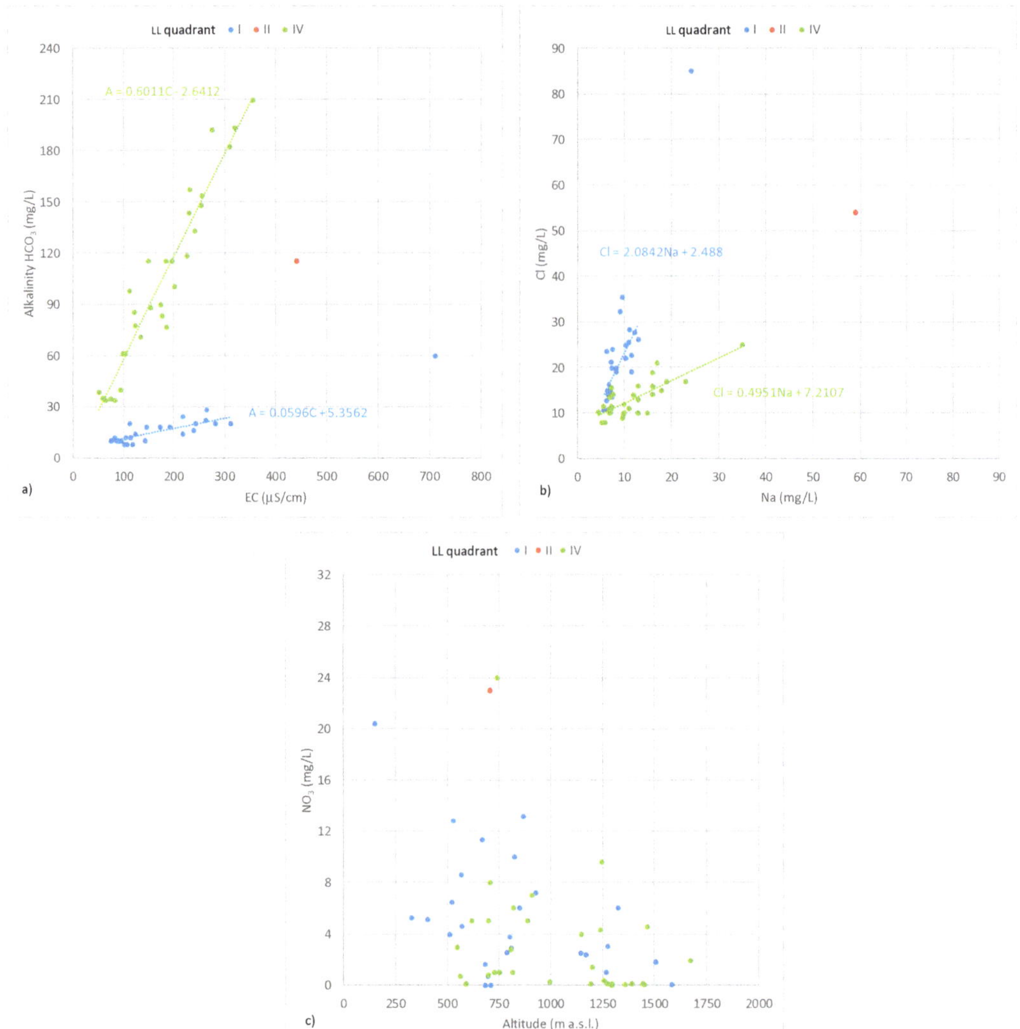

Figure 5. (a) Correlations between EC and A, (b) Na$^+$ and Cl$^-$ and (c) NO$_3^-$ and altitude according to classification of water springs in LL diagram.

A good correlation between chloride and sodium ions was observed (Figure 5b).

Waters with alkalinity close to 40 mg/L show a change in their gradient, becoming very similar to that of the I quadrant.

The sampled waters located in the north and east of the Sila Massif, which fall in the facies of chloride–sulfate alkaline waters, show a NO_3^- concentration above 10 mg/L. This value is considerably higher than the other samples of about 2 mg/L found in springs unaffected by anthropogenic activities.

The increase in EC follows the increase in alkalinity values, a correlation especially evident in the bicarbonate Ca-Mg waters and linked to the possibility of higher nitrate concentrations in springs (Figure 5a).

Generally, the analyzed waters have a nitrate content lower than 5 mg/L, especially the bicarbonate Ca-Mg waters (IV quadrant), which show a mean NO_3^- content of 2.45 mg/L, as well as the chloride–sulfate alkaline waters (II quadrant) with a mean NO_3^- content of 1.8 mg/L. Instead, the chloride–sulfate Ca-Mg waters (I quadrant) are generally richer with a mean NO_3^- content of 4.8 mg/L with a maximum value of 13 mg/L.

The results of statistical data analysis have detected the absence of outliers. As shown in Figure 5c, three water springs arising at hilly altitudes, falling in the municipalities of Acri and San Demetrio Corone, exhibit an anomalous behavior compared to the other analyzed water springs in terms of conductivity and presence of ions (Cl^-, Na^+, etc.). These three anomalous spring waters have nitrate concentrations ranging between 20 and 24 mg/L. These samples also present an anomalous content of the other analyzed parameters, such as a high conductivity, indicating probable aquifer contamination mainly due to widespread wild farming and pastures in the area, most practiced in the municipalities in which they fall. This factor could induce a greater exposure to anthropogenic contaminants and result in higher nitrate concentration.

Table 3 shows the mean concentrations of NH_4^+, NO_2^- and NO_3^- recorded in the three sampling sites for the atmospheric deposition analysis. They allow us to define the extent of the atmospheric contribution into soil, considering that the concentration trend is seasonal with the maximum in spring. From previous studies, the nitrate concentration in the Bonis area (Sila Massif) is about 2 mg/L [61]; it shows lower concentrations than those found in the Crati valley (Sant'Antonello and Settimo) in Montalto Uffugo.

Table 3. Mean concentrations of NH_4^+, NO_2^- and NO_3^- detected in the three sampling sites.

Measurement Site	Coordinates	NH_4^+ (mg/L)	NO_2^- (mg/L)	NO_3^- (mg/L)
Sant'Antonello in Montalto Uffugo	39°28′47″ N 16°14′28″ E	0.8296	0.0462	3.3287
Settimo in Montalto Uffugo	39°25′15″ N 16°12′38″ E	0.6304	0.0150	1.1121
Bonis	39°28′50″ N 16°30′12″ E	2.1325	0.0249	1.8664

Factors that may account for the higher deposition rates in these sites include higher rainfall amounts, local atmospheric inputs of N from volatilization of ammonia from fertilizers and animal wastes and N from the combustion of fossil fuels.

Cross-validation results for predicted versus measured nitrate values evidenced a good performance of the spline model in the study area. In fact, both MRE and RMSE values showed that the spline model can predict the data accurately.

Figure 6 shows the spatial distribution of nitrates in the study area mapped through the spline method in five different ranges of values. The results were represented using the digital terrain model in the background.

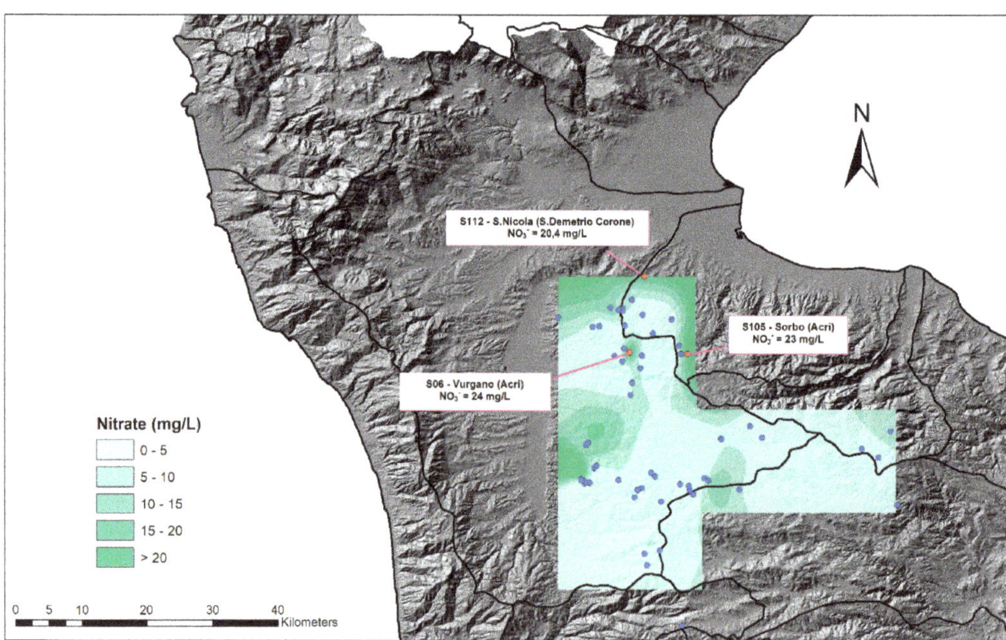

Figure 6. Spatial distribution of nitrates in the study area.

The obtained results were consistent with other studies that emphasized the spline method was one of the most suitable techniques for mapping nitrate concentrations in spring water. Indeed, the use of the spline method for water quality spatial analysis has been verified by numerous scientists. Guo et al. [62] investigated the five typical spatial interpolation methods, namely kriging, natural neighborhood, IDW, spline and trend surface, and found that the IDW and spline methods are appropriate for platform and small undulating areas. The national neighbor method and spline method provided the most accurate estimates of nitrates in the aquifer in the Qazvin plain [63]. The results of Zabihi et al. [64] indicate that the spline method is a good estimator of groundwater spring potential in the Iran area. The performance of the spline method was excellent in a case study in the Khalkhal region (Iran) [65].

The spatial distribution of nitrates provides some insight into the geochemical processes during the evolution of groundwater in the study area. The maximum level of nitrate concentrations was located in springs at lower altitudes in the north and western parts of the study area, in which lower precipitation and more intense agricultural activities occur compared to the eastern area. This factor could induce a greater exposure to anthropogenic contaminants and result in higher nitrate concentration.

The minimum level of nitrate, instead, can be attributed to the high-altitude area, which has no human activity and in which the lithological distribution may give an explanation for these values, as mentioned above.

In addition, in the Sila Massif, it is necessary to underline the subsequent reforestations since the postwar period and the related cleaning and maintenance activities of the woods that have deprived the subsoil of the organic substrate necessary for denitrifying bacterial colonies.

The study area was selected for its characteristics, along with the type of land use and human pressure, which are common to other wide Mediterranean areas, and therefore, despite its relatively small extension, it can well represent wider territories.

Moreover, the investigated area represents an extraordinary case of geological homogeneity, with the occurrence of purely acid metamorphic rocks. In it, there are few and no significant anthropogenic presences, and the only inputs that contribute to the nitrate vulnerability in spring waters are attributable to the agricultural uses of soil and especially to forestry as well as to atmospheric depositions.

In addition, in the study area, investigations for vulnerability to nitrate occurrence in spring waters are scarce. The present survey confirmed the importance of assessing nitrate pollution as an environmental issue with multiple implications in terms of water quality deterioration, biodiversity loss, human health problems, global carbon-cycle alterations and climate change.

5. Conclusions

Springs perform numerous tasks, including ensuring sources of potable water and providing recreational, ecological and cultural value; moreover, they provide a way to assess groundwater quality since they are the connection between groundwater and surface water and give direct information about the state of groundwater in the aquifers that feed them.

The main focuses of this survey were to classify the spring water types of the Sila Massif (Calabria, southern Italy) according to their hydrogeochemical features and thus identify the factors controlling mineral waters using spatial variables focused on lithological settings and finally determine the vulnerability to nitrate occurrence in the spring waters.

The characterization of the waters of 59 springs in the study area has shown that they are strongly affected by the geological nature of the rocks of the aquifers (granites, magmatites, acid granulites, biotic gneiss, etc.). Indeed, the waters are generally poor in mineral salts (very low mineral content waters) with low calcium and magnesium content (soft).

The hydrochemical facies of the aquifers from which the water springs draw, according to the classification proposed by Langelier–Ludwig, can be classified as chloride–sulfate alkaline waters, those arising mostly from the mountainous slopes that crown the Sila Massif, and bicarbonate Ca-Mg waters that flow mostly in the central part of the Sila.

According to these considerations, most of the sampled points, having lower nitrate values, demonstrate the overall good quality of the spring waters in the Sila Massif. Therefore, the monitoring of springs in the Sila Massif revealed the scarce vulnerability to a potential alteration of their groundwaters by nitrates.

Author Contributions: Conceptualization, E.I.; methodology, E.I. and I.G.; software, S.G. and T.C.; validation, I.G.; formal analysis, E.I., S.G. and I.G.; investigation, E.I., S.G. and I.G.; data curation, S.G. and I.G.; writing—original draft preparation, E.I., I.G. and T.C.; writing—review and editing, I.G. and T.C.; visualization, I.G. and T.C.; supervision, E.I. All authors have read and agreed to the published version of the manuscript.

Funding: This research received no external funding.

Institutional Review Board Statement: Not applicable.

Informed Consent Statement: Not applicable.

Data Availability Statement: The data presented in this study are available on request.

Conflicts of Interest: The authors declare no conflict of interest.

References

1. Kalvāns, A.; Popovs, K.; Priede, A.; Koit, O.; Retiķe, I.; Bikše, J.; Dēliņa, A.; Babre, A. Nitrate vulnerability of karst aquifers and associated groundwater-dependent ecosystems in the Baltic region. *Environ. Earth. Sci.* **2021**, *80*, 628. [CrossRef]
2. Zuzolo, D.; Cicchella, D.; Lima, A.; Guagliardi, I.; Cerino, P.; Pizzolante, A.; Thiombane, M.; de Vivo, B.; Albanese, S. Potentially toxic elements in soils of Campania region (Southern Italy): Combining raw and compositional data. *J. Geochem. Explor.* **2020**, *213*, 106524. [CrossRef]
3. Buttafuoco, G.; Guagliardi, I.; Tarvainen, T.; Jarva, J. A multivariate approach to study the geochemistry of urban topsoil in the city of Tampere, Finland. *J. Geochem. Explor.* **2017**, *181*, 191–204. [CrossRef]

4. Cicchella, D.; Zuzolo, D.; Albanese, S.; Fedele, L.; di Tota, I.; Guagliardi, I.; Thiombane, M.; de Vivo, B.; Lima, A. Urban soil contamination in Salerno (Italy): Concentrations and patterns of major, minor, trace and ultra-trace elements in soils. *J. Geochem. Explor.* **2020**, *213*, 106519. [CrossRef]

5. Fernández-Martínez, M.; Corbera, J.; Domene, X.; Sayol, F.; Sabater, F.; Preece, C. Nitrate pollution reduces bryophyte diversity in Mediterranean springs. *Sci. Total Environ.* **2020**, *705*, 135823. [CrossRef] [PubMed]

6. Galloway, J.N.; Townsend, A.R.; Erisman, J.W.; Bekunda, M.; Cai, Z.; Freney, J.R.; Martinelli, L.A.; Seitzinger, S.P.; Sutton, M.A. Transformation of the nitrogen cycle: Recent trends, questions, and potential solutions. *Science* **2008**, *320*, 889–892. [CrossRef] [PubMed]

7. Kazakis, N.; Matiatos, I.; Ntona, M.M.; Bannenberg, M.; Kalaitzidou, K.; Kaprara, E.; Mitrakas, M.; Ioannidou, A.; Vargemezis, G.; Voudouris, K. Origin, implications and management strategies for nitrate pollution in surface and ground waters of Anthemountas basin based on a δ^{15}N-NO$_3$− and δ^{18}O-NO$_3$−isotope approach. *Sci. Total Environ.* **2020**, *724*, 138211. [CrossRef] [PubMed]

8. Xiao, Q.; Wu, K.; Shen, L. Nitrate fate and origin in epikarst springs in Jinfo Mountain area, Southwest China. *Arab. J. Geosci.* **2016**, *9*, 483. [CrossRef]

9. Buttafuoco, G.; Caloiero, T.; Guagliardi, I.; Ricca, N. Drought assessment using the reconnaissance drought index (RDI) in a southern Italy region. In Proceedings of the 6th IMEKO TC19 Symposium on Environmental Instrumentation and Measurements, Reggio Calabria, Italy, 24–25 June 2016; pp. 52–55.

10. Caloiero, T.; Guagliardi, I. Climate change assessment: Seasonal and annual temperature analysis trends in the Sardinia region (Italy). *Arab. J. Geosci.* **2021**, *14*, 2149. [CrossRef]

11. Mas-Pla, J.; Menció, A. Groundwater Nitrate Pollution and Climate Change: Learnings from a Water Balance-Based Analysis of Several Aquifers in a Western Mediterranean Region (Catalonia). *Environ. Sci. Pollut. Res.* **2019**, *26*, 2184–2202. [CrossRef]

12. Pellicone, G.; Caloiero, T.; Guagliardi, I. The De Martonne aridity index in Calabria (Southern Italy). *J. Maps* **2019**, *15*, 788–796. [CrossRef]

13. European Environment Agency (EEA). *Europe's Environment: State of the Environment Report No. 3/2003*; EEA: Copenhagen, Denmark, 2003.

14. Darvishmotevalli, M.; Moradnia, M.; Noorisepehr, M.; Fatehizadeh, A.; Fadaei, S.; Mohammadi, H.; Salari, M.; Jamali, H.A.; Daniali, S.S. Evaluation of carcinogenic risks related to nitrate exposure in drinking water in Iran. *MethodsX* **2019**, *6*, 1716–1727. [CrossRef] [PubMed]

15. Mohammadi, A.A.; Zarei, A.; Majidi, S.; Ghaderpoury, A.; Hashempour, Y.; Saghi, M.H.; Alinejad, A.; Yousefi, M.; Hosseingholizadeh, N.; Ghaderpoori, M. Carcinogenic and non-carcinogenic health risk assessment of heavy metals in drinking water of Khorramabad, Iran. *MethodsX* **2019**, *6*, 1642–1651. [CrossRef] [PubMed]

16. Rezaei, H.; Jafari, A.; Kamarehie, B.; Fakhri, Y.; Ghaderpoury, A.; Karami, M.A.; Ghaderpoori, M.; Shams, M.; Bidarpoor, F.; Salimi, M. Health-risk assessment related to the fluoride, nitrate, and nitrite in the drinking water in the Sanandaj, Kurdistan County, Iran. *Hum. Ecol. Risk Assess. Int. J.* **2018**, *25*, 1242–1250. [CrossRef]

17. Rehman, J.U.; Ahmad, N.; Ullah, N.; Alam, I.; Ullah, H. Health Risks in Different Age Group of Nitrate in Spring Water Used for Drinking in Harnai, Balochistan, Pakistan. *Ecol. Food. Nutr.* **2020**, *59*, 462–471. [CrossRef] [PubMed]

18. International Agency for Research on Cancer. *IARC Monographs on the Evaluation of Carcinogenic Risks to Humans. Volume 94—Ingested Nitrate and Nitrite, and Cyanobacterial Peptide Toxins*; World Health Organization: Geneva, Switzerland, 2010.

19. Massoudinejad, M.; Ghaderpoori, M.; Jafari, A.; Nasehifar, J.; Malekzadeh, A.; Ghaderpoury, A. Data on nitrate and nitrate of Taham dam in Zanjan (Iran). *Data Brief.* **2018**, *17*, 431–437. [CrossRef] [PubMed]

20. Knobeloch, L.; Salna, B.; Hogan, A.; Postle, J.; Anderson, H. Blue babies and nitrate-contaminated well water. *Environ. Health Perspect.* **2000**, *108*, 675–678. [CrossRef] [PubMed]

21. World Health Organization (WHO). *Guidelines for Drinking-Water Quality*, 2nd ed.; WHO: Geneva, Switzerland, 1993; Volume 1.

22. Decreto Legislativo (D.Lgs.) 16 Marzo 2009, n. 30 (Pubblicato nella Gazz. Uff. 4 Aprile 2009, n. 79). Attuazione della direttiva 2006/118/CE, Relativa Alla Protezione Delle Acque Sotterranee Dall'inquinamento e dal Deterioramento. Available online: https://www.gazzettaufficiale.it/atto/serie_generale/caricaDettaglioAtto/originario?atto.dataPubblicazioneGazzetta=2013-04-04&atto.codiceRedazionale=13G00075 (accessed on 7 February 2022).

23. Decreto Legislativo (D.Lgs.) 2 Febbraio 2001, n. 31. (Pubblicato nella Gazz. Uff. 3 Marzo 2001, n. 52). Attuazione della Direttiva 98/83/CE, Relativa alla Qualità delle Acque Destinate al Consumo Umano. Available online: https://www.camera.it/parlam/leggi/deleghe/01031dl.htm (accessed on 7 February 2022).

24. Gaglioti, S.; Infusino, E.; Caloiero, T.; Callegari, G.; Guagliardi, I. Geochemical Characterization of Spring Waters in the Crati River Basin, Calabria (Southern Italy). *Geofluids* **2019**, *2019*, 3850148. [CrossRef]

25. Guagliardi, I.; Caloiero, T.; Infusino, E.; Callegari, G.; Ricca, N. Environmental Estimation of Radiation Equivalent Dose Rates in Soils and Waters of Northern Calabria (Italy). *Geofluids* **2021**. [CrossRef]

26. Balestrieri, A.; Remonti, L.; Smiroldo, G.; Prigioni, C.; Reggiani, G. Surveying otter Lutra Lutra distribution at the southern limit of its Italian range. *Hystrix It. J. Mamm.* **2008**, *19*, 165–173.

27. Iovine, G.; Guagliardi, I.; Bruno, C.; Greco, R.; Tallarico, A.; Falcone, G.; Lucà, F.; Buttafuoco, G. Soil-gas radon anomalies in three study areas of Central-Northern Calabria (Southern Italy). *Nat. Hazards* **2017**, *91*, 193–219. [CrossRef]

28. Bonardi, G.; Cavazza, W.; Perrone, V.; Rossi, S. Calabria-Peloritani terrane and northern Ionian Sea. In *Anatomy of an Orogen: The Apennines and Adjacent Mediterranean Basins*; Vai, G.B., Martini, I.P., Eds.; Springer: Dordrecht, The Netherlands, 2001.

29. Vignaroli, G.; Minelli, L.; Rossetti, F.; Balestrieri, M.L.; Faccenna, C. Miocene thrusting in the eastern Sila Massif: Implication for the evolution of the Calabria-Peloritani orogenic wedge (southern Italy). *Tectonophysics* **2012**, *538–540*, 105–119. [CrossRef]
30. Le Pera, E.; Arribas, J.; Critelli, S.; Tortosa, A. The effects of source rocks and chemical weathering on the petrogenesis of siliciclastic sand from the Neto River (Calabria, Italy): Implications for provenance studies. *Sedimentology* **2001**, *48*, 357–378. [CrossRef]
31. Liotta, D.; Caggianelli, A.; Kruhl, J.H.; Festa, V.; Prosser, G.; Langone, A. Multiple injections of magmas along a Hercynian mid-crustal shear zone (Sila Massif, Calabria, Italy). *J. Struct. Geol.* **2008**, *30*, 1202–1217. [CrossRef]
32. Critelli, S.; Muto, F.; Tripodi, V.; Perri, F. Relationships between lithospheric flexure, thrust tectonics and stratigraphic sequences in foreland setting: The Southern Apennines foreland basin system, Italy. In *New Frontiers in Tectonic Research at the Midst of Plate Convergence*; Schattner, U., Ed.; Intech Open: London, UK, 2011; pp. 121–170.
33. Scarciglia, F.; Critelli, S.; Borrelli, L.; Coniglio, S.; Muto, F.; Perri, F. Weathering profiles in granitoid rocks of the Sila Massif uplands, Calabria, southern Italy: New insights into their formation processes and rates. *Sediment. Geol.* **2016**, *336*, 46–67. [CrossRef]
34. Guagliardi, I.; Cicchella, D.; Rosa, R. A Geostatistical approach to assess concentration and spatial distribution of heavy metals in urban soils. *Water Air Soil Pollut.* **2012**, *223*, 5983–5998. [CrossRef]
35. Fabbricatore, D.; Robustelli, G.; Muto, F. Facies analysis and depositional architecture of shelf-type deltas in the Crati Basin (Calabrian Arc, south Italy). *Ital. J. Geosci.* **2014**, *133*, 131–148. [CrossRef]
36. Gallo, M.; Iovino, F. A brief history of the forest changes in the Sila Greca mountains. In *Forest History International Studies on Socioeconomic and Forest Ecosystem Change. Report No. 2 of the IUFRO Task Force on Environmental Change*; Agnoletti, M., Anderson, S., Eds.; CABI Publishing: Wallingford, UK, 2000; pp. 289–305.
37. Beck, H.E.; Zimmermann, N.E.; McVicar, T.R.; Vergopolan, N.; Berg, A.; Wood, E.F. Present and future Köppen-Geiger climate classification maps at 1-km resolution. *Sci. Data* **2018**, *5*, 180214. [CrossRef]
38. Coscarelli, R.; Gaudio, R.; Caloiero, T. Climatic trends: An investigation for a Calabrian basin (southern Italy). In Proceedings of the International Symposium The Basis of Civilization, Rome, Italy, 3–6 December 2003; Rodda, J.C., Ubertini, L., Eds.; IAHS: Wallingford, CT, USA; pp. 255–266.
39. Buttafuoco, G.; Caloiero, T.; Ricca, N.; Guagliardi, I. Assessment of drought and its uncertainty in a southern Italy area (Calabria region). *Measurement* **2018**, *113*, 205–210. [CrossRef]
40. Bianucci, G.; Ribaldone, E.; Bianucci, E. *Composizione Chimica Delle Rocce, La Chimica Delle Acque Sotterranee*; Ulrico, H., Ed.; Hoepli: Milan, Italy, 1985; pp. 76–83.
41. Appelo, C.A.J.; Postma, D. *Geochemistry. Groundwater and Pollution*; AA Balkema: Rotterdam, The Netherlands, 2005.
42. Ako, A.A.; Shimada, J.; Hosono, T.; Kagabu, M.; Ayuk, A.R.; Nkeng, G.E.; Takem, G.E.E.; Takounjou, A.L.F. Spring water quality and usability in the Mount Cameroon area revealed by hydrogeochemistry. *Environ. Geochem. Health* **2012**, *34*, 615–639. [CrossRef] [PubMed]
43. Callegari, G.; Cantasano, N.; Infusino, E.; Callegari, G.; Cipriani, M.G. Hydrogeological and geochemical assessment of some spring waters in the municipal area of Chiaravalle Centrale (Calabria, southern Italy). *Rend. Online Soc. Geol. Ital.* **2012**, *21*, 869–871.
44. AWWA. AWWA standard for asbestos-cement transmission pipe, 18 in. through 42 in., for water and other liquids. *J. Am. Water Works Assoc.* **1975**, *67*, 462–467. [CrossRef]
45. AWWA. AWWA Standard for Asbestos-Cement Distribution Pipe. 4 in. Through 16 in. (100 mm Through 16 in.) (100 mm Through 400 mm). 1980. Available online: https://engage.awwa.org/PersonifyEbusiness/Store/Product-Details/productId/18631 (accessed on 7 February 2022).
46. Collins, W.D. Graphic representation of water analyses. *Ind. Eng. Chem.* **1923**, *15*, 394. [CrossRef]
47. Piper, A.M. A graphic procedure in the geochemical interpretation of water-analyses. *Trans. AGU* **1944**, *25*, 914. [CrossRef]
48. Langelier, W.F.; Ludwig, F. Graphical methods for indicating the mineral character of natural waters. *J. Am. Water Work. Assoc.* **1942**, *34*, 335–352. [CrossRef]
49. Cao, W.G.; Yang, H.F.; Liu, C.L.; Li, Y.J.; Bai, H. Hydrogeochemical characteristics and evolution of the aquifer systems of Gonghe Basin, Northern China. *Geosci. Front.* **2018**, *9*, 907–916. [CrossRef]
50. Ravikumar, P.; Prakash, K.L.; Somashekar, R.K. Evaluation of water quality using geochemical modeling in the Bellary Nala Command area, Belgaum district, Karnataka State, India. *Carbonates Evaporites* **2013**, *28*, 365–381. [CrossRef]
51. Rao, N.S.; Subrahmanyam, A.S.; Kumar, S.R.; Srinivasulu, N.; Rao, G.B.; Rao, P.S.; Reddy, G.V. Geochemistry and quality of groundwater of Gummanampadu sub-basin, Guntur District, Andhra Pradesh, India. *Environ. Earth Sci.* **2012**, *67*, 1451–1471.
52. Rao, N.S.; Vidyasagar, G.; Rao, P.S.; Bhanumurthy, P. Chemistry and quality of groundwater in a coastal region of Andhra Pradesh, India. *Appl. Water Sci.* **2017**, *7*, 285–294. [CrossRef]
53. Waters, N.M. Expert Systems and Systems of Experts, Chapter 12. In *Geographical Systems and Systems of Geography*; Coffey, W.J., Ed.; University of Western Ontario: London, UK, 1988.
54. Hutchinson, M.F.; Gessler, P.E. Splines—More than just a smooth interpolator. *Geoderma* **1994**, *62*, 45–67. [CrossRef]
55. Apaydin, H.; Sonmez, F.K.; Yildirim, Y.E. Spatial interpolation techniques for climate data in the GAP region in Turkey. *Clim. Res.* **2004**, *28*, 31–40. [CrossRef]

56. Mueller, D.K. *Nutrients in Ground Water and Surface Water of the United States: An Analysis of Data through 1992*; US Department of the Interior, US Geological Survey: Washington, DC, USA, 1995; Volume 95.

57. Infusino, E.; Callegari, G.; Cantasano, N. Release of nutrients into a forested catchment of southern Italy. *Rend. Fis. Acc. Lincei.* **2016**, *27*, 127–134. [CrossRef]

58. Padilla, F.M.; Gallardo, M.; Manzano-Agugliaro, F. Global trends in nitrate leaching research in the 1960–2017 period. *Sci. Total Environ.* **2018**, *643*, 400–413. [CrossRef] [PubMed]

59. Accardo, V.; Angeletti, V.; Cocozziello, B.; D'Arienzo, V.; Aquino De Gennaro, V.; De Angelis, C.; De Rosa, E.; Di Meo, T.; Imperatrice, M.I.; Mainolfi, P.; et al. Il monitoraggio dell'inquinamento idrico da nitrati degli acquiferi della Campania. In Proceedings of the Conference Accettabilità Delle Acque per usi Civili e Agricoli, Rome, Italy, 5 June 2002.

60. Wilcox, J.C.; Holland, W.D.; McDougald, J.M. Relation of elevation of a mountain stream to reaction and salt content of water and soil. *Can. J. Soil Sci.* **1956**, *37*, 11–20. [CrossRef]

61. Callegari, G.; Frega, G.; Infusino, E. Deposizioni atmosferiche nell'area urbana di Cosenza in Calabria. In Proceedings of the Conference Qualità Dell'aria Nelle Città Italiane, Rome, Italy, 6 June 2005.

62. Guo, B.; Yang, F.; Wu, H.; Zhang, R.; Zang, W.; Wei, C.; Zhang, H. How the variations of terrain factors affect the optimal interpolation methods for multiple types of climatic elements? *Earth Sci. Inform.* **2021**, *14*, 1021–1032. [CrossRef]

63. Kazemi, E.; Karyab, H.; Mohammad-Mehdi Emamjome, M. Optimization of interpolation method for nitrate pollution in groundwater and assessing vulnerability with IPNOA and IPNOC method in Qazvin plain. *J. Environ. Health Sci. Engineer.* **2017**, *21*, 15–23. [CrossRef] [PubMed]

64. Zabihi, M.; Pourghasemi, H.R.; Pourtaghi, Z.S.; Behzadfar, M. GIS-based multivariate adaptive regression spline and random forest models for groundwater potential mapping in Iran. *Environ. Earth Sci.* **2016**, *75*, 665. [CrossRef]

65. Naghibi, S.A.; Moradi, D.M. Evaluation of four supervised learning methods for groundwater spring potential mapping in Khalkhal region (Iran) using GIS-based features. *Hydrogeol. J.* **2017**, *25*, 169–189. [CrossRef]

Article

Potentially Toxic Elements in Urban Soils from Public-Access Areas in the Rapidly Growing Megacity of Lagos, Nigeria

Abimbola O. Famuyiwa [1,2], Christine M. Davidson [1,*], Sesugh Ande [1,3] and Aderonke O. Oyeyiola [4]

1   Department of Pure and Applied Chemistry, University of Strathclyde, 295 Cathedral Street, Glasgow G1 1XL, UK; abimbola.famuyiwa@gmail.com (A.O.F.); sesughande@gmail.com (S.A.)
2   Department of Science Laboratory Technology, Moshood Abiola Polytechnic, Abeokuta, Ogun State, Nigeria
3   Department of Chemistry, University of Agriculture, Makurdi, Benue State, Nigeria
4   Department of Chemistry, University of Lagos, Akoka-Yaba, Lagos State, Nigeria; aderonkeoyeyiola@gmail.com
*   Correspondence: c.m.davidson@strath.ac.uk; Tel.: +44-(0)141-548-2134

Abstract: Rapid urbanization can lead to significant environmental contamination with potentially toxic elements (PTEs). This is of concern because PTEs are accumulative, persistent, and can have detrimental effects on human health. Urban soil samples were obtained from parks, ornamental gardens, roadsides, railway terminals and locations close to industrial estates and dumpsites within the Lagos metropolis. Chromium, Cu, Fe, Mn, Ni, Pb and Zn concentrations were determined using inductively coupled plasma mass spectrometry following sample digestion with aqua regia and application of the BCR sequential extraction procedure. A wide range of analyte concentrations was found—Cr, 19–1830 mg/kg; Cu, 8–11,700 mg/kg; Fe, 7460–166,000 mg/kg; Mn, 135–6100 mg/kg; Ni, 4–1050 mg/kg; Pb, 10–4340 mg/kg; and Zn, 61–5620 mg/kg—with high levels in areas close to industrial plants and dumpsites. The proportions of analytes released in the first three steps of the sequential extraction were Fe (16%) < Cr (30%) < Ni (46%) < Mn (63%) < Cu (78%) < Zn (80%) < Pb (84%), indicating that there is considerable scope for PTE (re)mobilization. Human health risk assessment indicated non-carcinogenic risk for children and carcinogenic risk for both children and adults. Further monitoring of PTE in the Lagos urban environment is therefore recommended.

Keywords: soil contamination; heavy metals; sequential extraction; health risk assessment

Citation: Famuyiwa, A.O.; Davidson, C.M.; Ande, S.; Oyeyiola, A.O. Potentially Toxic Elements in Urban Soils from Public-Access Areas in the Rapidly Growing Megacity of Lagos, Nigeria. *Toxics* **2022**, *10*, 154. https://doi.org/10.3390/toxics10040154

Academic Editor: Ilaria Guagliardi

Received: 31 January 2022
Accepted: 21 March 2022
Published: 23 March 2022

Publisher's Note: MDPI stays neutral with regard to jurisdictional claims in published maps and institutional affiliations.

1. Introduction

Urban soil pollution has become a major environmental concern in recent decades. Increased migration from rural to urban areas, in particular in the developing world, has resulted in high population density and rapid increase in anthropogenic activities [1]. Over half of the world's population now lives in urban areas [2], and this puts substantial pressure on environmental resources, such as soil and water.

With a population of at least 20 million people and an estimated population growth rate of about 600,000 persons per annum, Lagos is one of the most densely populated cities in the world [3]. Within the metropolis, there is rapid industrialization, continuous infrastructural development, and a high prevalence of vehicular traffic congestion. Incessant demand for land means that recreational open spaces can be found in proximity to dumpsites, and schools and housing are often co-located with industrial estates. Therefore, Lagos residents are subjected to an array of potential pollution sources that may have adverse effects on their health.

Numerous studies have reported evidence of anthropogenic inputs of potentially toxic elements (PTE) to urban soils. Extensive work has been carried out in developed countries [4–7]. Less attention has been paid to urban soils of developing nations, although their importance is increasingly being recognized [8–12]. Potentially toxic elements are one of the most studied soil contaminants because they are ubiquitous and persistent.

Metals are non-biodegradable and accumulative in nature; emission and deposition over a long period of time can lead to enrichment in surface environments. The prolonged presence of PTEs in urban soils, together with their proximity to human populations, can lead to exposure via inhalation, ingestion, and dermal contact [13,14]. Because these contaminants become hazardous when present in soil above certain concentrations, this can have significant health implications.

Measurement of total or pseudo-total (aqua-regia soluble) PTE concentrations in soil can provide important information about distribution and enrichment, but it is likely to overestimate human health risk because only a fraction of the total content is usually bioavailable [15]. The mobility and availability of PTEs in soil depends on complex interactions between multiple factors, including solubility, the availability of binding sites, complexation, pH and redox processes [16]. Sequential extraction involves the treatment of solid environmental samples with a series of reagents to partition PTE content into various fractions, nominally corresponding to major soil mineral phases but more accurately representing operationally defined reservoirs with the potential to become mobile under changes in environmental conditions, such as pH and redox potential. Several sequential extraction schemes have been developed and applied [17–20]. Amongst the most popular is the harmonized Community Bureau of Reference of the European Commission (BCR) sequential extraction procedure [21], summarised in Table 1. Advantages of this protocol are that it incorporates an internal quality check—comparison of the sum of the steps, Σ(step 1–4), with results of a separate pseudo-total digestion—and that dedicated certified reference material (CRM)—BCR 701—is available.

Table 1. BCR sequential extraction [21].

Step	Label	Reagents	Nominal Target Phase(s)
1	Exchangeable	0.11 mol L^{-1} CH$_3$COOH	Exchangeable, water- and acid-soluble PTE
2	Reducible	0.5 mol L^{-1} NH$_2$OH.HCl at pH 1.5	PTE bound to iron and manganese oxyhydroxides
3	Oxidisable	H$_2$O$_2$ (85 °C) then 1.0 mol L^{-1} CH$_3$COONH$_4$ at pH 2	PTE bound to organic matter and sulphides
4	Residual	aqua regia	Remaining PTE not bound within refractory/primary silicates

Previous studies have investigated PTE in urban soils of Lagos State. However, their scope has either been limited to a particular land use—for example, roadside soils [22,23] or soils from school playgrounds [8,24]—or featured assessment of contamination in proximity to specific industrial plants [25], dumpsites [26] or electronic waste (e-waste) processing sites [27–29]. Much of the work is at least a decade old [22–25] and therefore may not accurately reflect the current status of the rapidly growing metropolis. Few works have considered PTE mobility. Adeyi et al. [30] applied Tessier sequential extraction [17] to residential soils from Lagos and Ibadan as part of their study of the potential health impacts of Cd and Pb associated with the use of metal-rich paints. Oyeyiola et al. [31] employed the BCR procedure [21] in their investigation of partitioning, mobility and ecotoxicology of Cd, Cr, Cu, Pb and Zn in sediment from the Lagos Lagoon and three trans-urban rivers. However, there is a need for more comprehensive evaluation of both levels and potential mobility of PTE in Lagos urban soils, in particular soils that local citizens are most likely to interact with.

The aims of the current study were therefore to determine the concentrations and potential mobilities of Cr, Cu, Fe, Mn, Ni, Pb and Zn in soils from public-access areas across the Lagos metropolis and to evaluate the risks associated with human exposure to PTE in Lagos urban soils.

2. Materials and Methods

2.1. Study Area

Lagos lies in the Nigerian sector of the Dahomey (or Benin) basin. The area is characterized by sediments of Cretaceous to recent origin underlain by Precambrian basement rocks of granitic composition [32]. The Cretaceous and Tertiary sediments include sands, marine shales and limestone. Quaternary sediments consist of coastal plain sands (>100 m in thickness) with alluvial deposits in the river valleys. Lagos is Nigeria's most populous city and the seventh fastest-growing city in the world. Lagos State Government estimated the population of Lagos as 17.5 million during a parallel census conducted in 2006, with more than 12 million people living in the urban areas. A more recent report [3] estimated its population as 21 million, making Lagos the largest city in Africa. Lagos experiences rainy and dry seasons, with the latter accompanied by hot, dry and dusty winds. It represents the most industrialized area in Nigeria, with over 60% of total industrial activities.

2.2. Sampling and Sample Preparation

We selected 20 urban from a larger set of 92 samples collected as part of a previous study [33]. Sampling locations are shown in Figure 1. These included examples of different types of public-access areas (parks and open spaces, ornamental gardens, roadsides, industrial estates, railway terminals and locations in the vicinity of dumpsites). At each sampling location, a composite soil sample was collected to a depth of 10 cm. This consisted of 4–8 sub-samples taken 2 m apart in a square grid (the number of sub-samples depending on the shape of the area). Grass, leaves, paper and plastic debris present in the samples were discarded. Wet soil samples were air-dried for 3 days in the laboratory of the Lagos Ministry of Environment. Then, approximately 500 g of each soil was placed in a sealed, labelled polythene bag and transported from Lagos, Nigeria to the University of Strathclyde, Scotland, UK under a Scottish Government soil import license (IMP/SOIL/24/2014) for further processing and analysis.

Figure 1. Map of the Lagos area showing sampling points (prepared using Google Maps).

On arrival, the soil samples were air-dried for 14 days, and then sieved through a 2 mm nylon mesh sieve before grinding and homogenization with mortar and pestle. Test portions for digestion or extraction were obtained by coning and quartering. All glass and plasticware was soaked in 5% (v/v) nitric acid overnight (general-purpose-grade reagent, Sigma Aldrich, Gillingham, UK) and then washed thoroughly with distilled water before use.

2.3. Microwave-Assisted Pseudo-total Digestion

Samples were digested with aqua regia prepared by mixing extra-pure hydrochloric (HCl) and nitric (HNO_3) acids (Sigma-Aldrich, Gillingham, UK) in the ratio 3:1 (v/v). Each soil sample (1 g) was weighed into a high-pressure vessel, and 20 mL of freshly prepared aqua regia was added. This was left to stand overnight in a fume cupboard to allow any vigorous reaction to subside. Then, the vessels were placed in a MarsXpress™ (CEM Microwave Technology, Ltd., Middle Slade, Buckingham, UK) microwave digestion system and heated to 160 °C using 1600 W power for 40 min (comprising 20 min ramp to temperature and 20 min hold at temperature). Digests were then allowed to cool and were filtered. Filtrates were made up to 100 mL with deionised water and stored at 4 °C in a refrigerator prior to analysis. Replicate samples (n = 3) were digested, along with procedural blanks.

2.4. Sequential Extraction

The BCR sequential extraction procedure was carried out as described by Rauret et al. [21]. The experimental protocol is summarised below. Samples were analysed in triplicate, along with procedural blanks.

Step 1: Exchangeable phase

Approximately 1 g of soil was weighed into a 100 mL centrifuge tube, and 40 mL of 0.11 M acetic acid added. The mixture was shaken for 16 h (overnight) using a GFL 3040 mechanical end-over-end shaker (GFL, Burgwedel, Germany). The extract was separated from the residue by centrifuging at 3000× g for 20 min in an Allegra 21 centrifuge (Beckman Coulter Ltd., High Wycombe, UK). The supernatant was decanted and stored in a polyethylene bottle at 4 °C in a refrigerator prior to analysis. The residue was washed by adding 20 mL of distilled water and shaking for 15 min. Following centrifugation, the supernatant was decanted and discarded.

Step 2: Reducible phase

A volume of 40 mL of freshly prepared 0.5 M hydroxylamine hydrochloride solution was added to the washed residue from step 1 in the same centrifuge tube. The mixture was shaken and centrifuged, the supernatant was recovered, and the residue was washed, as described in step 1.

Step 3: Oxidisable phase

A volume of 10 mL of 8.8 M hydrogen peroxide solution was added slowly, in small aliquots to avoid losses due to possible violent reaction, to the washed residue from step 2. The centrifuge tube was loosely covered with its cap, and the contents were digested at room temperature for 1 h with occasional manual shaking. The digestion was continued for another 1 h at 85 ± 2 °C in a water bath, with occasional manual shaking for the first 30 min. Then, the sample mixture was reduced in volume to about 3 mL by further heating the uncovered tube. Another 10 mL of hydrogen peroxide solution (8.8 M) was added, and the covered sample was heated for a further 1 h at 85 ± 2 °C. Subsequently, the cap of the centrifuge tube was removed, and the volume was reduced to about 1 mL, with care not to take to complete dryness. A volume of 50 mL of 1.0 M ammonium acetate solution was added to the cool, moist residue, and the mixture was shaken for 16 h (overnight). The sample was centrifuged, and the supernatant was recovered as described in step 1.

Step 4: Residual phase

A volume of 20 mL of aqua regia was used to wash the residue from step 3 into a pressure vessel, where it was digested as described in Section 2.3.

2.5. Analysis of Sample Digests and Extracts

Analyte concentrations were measured in soil digests and extracts using a Model 7700x inductively coupled plasma mass spectrometry system (Agilent Technologies, Cheadle, UK). Commercially available stock solutions (from Qmx Laboratories, Thaxted, UK) were used to prepare reagent-matched multielement standard solutions for instrument calibration in the range of 0–1600 µg/L for Cr, Cu, Mn, Pb, Ni and Zn and 0–100,000 µg/L for Fe. The internal standard was ^{115}In. Before each batch of analyses, the instrument was tuned to verify mass resolution and maximise sensitivity. Collision cell technology mode was used for the determination of ^{52}Cr, ^{63}Cu, ^{56}Fe, ^{55}Mn, ^{60}Ni, ^{208}Pb and ^{64}Zn. During analysis, a mid-range calibration standard (800 µg/L) was checked after every tenth sample measured. The calibration curves for the determined PTEs gave a linear fit with regression coefficient of at least 0.999.

2.6. Determination of Soil pH and Organic Matter

pH was determined in a suspension of 5 g soil in 25 mL of deionised water using a pH meter (SG2-ELK-SevenGO^{TM} pH, Mettler Toledo, Leicester, UK) according to the British standard method [34]. Soil organic matter was estimated by loss on ignition of dry matter [35]. A muffle furnace (Elite Thermal Systems Box Furnace, model number BSF12/6-2416CG, Market Harborough, UK) ramped at 10 °C per min and held at 550 °C for 8 h was used for this purpose.

2.7. Quality Control

Analytical quality was assessed using CRMs BCR 143R (sewage sludge amended soil) for pseudo-total digestion and BCR 701 (lake sediment) for sequential extraction (Table 2).

Table 2. Analysis of certified reference materials BCR 143R and BCR 701 (mg/kg dry weight).

Step	Parameter	Cr	Cu	Fe	Mn	Ni	Pb	Zn
				BCR 143R				
	Found	422 ± 27	137 ± 3	29,700 ± 12	863 ± 7	297 ± 4	178 ± 3	1110 ± 10
	Certified [†]	426 ± 12	131 ± 2		858 ± 11	296 ± 4	174 ± 5	1063 ± 16
	Recovery (%)	99	105		101	100	102	104
				BCR 701				
Step 1	Found	2.0 ± 1	42 ± 9	66 ± 8	186 ± 25	14 ± 9	3 ± 1	187 ± 25
Exchangeable	Certified	2.26 ± 0.16	49.3 ± 1.7			15.4 ± 0.9	3.18 ± 0.21	205 ± 6
	Recovery (%)	88	85			91	94	91
Step 2	Found	45 ± 4	124 ± 15	370 ± 3	5 ± 1	25 ± 1	125 ± 8	104 ± 4
Reducible	Certified	45.7 ± 2.0	124 ± 3			26.6 ± 1.3	126 ± 3	114 ± 5
	Recovery (%)	98	100			94	99	91
Step 3	Found	109 ± 9	39 ± 7	246 ± 43	21 ± 0.1	14 ± 0.2	3 ± 0.3	33 ± 1
Oxidisable	Certified	143 ± 7	55 ± 4			15.3 ± 0.9	9.3 ± 20	46 ± 4
	Recovery (%)	76	71			89	32	71
Step 4	Found	80 ± 0.3	44 ± 3	21,400 ± 81	261 ± 20	40 ± 1	13 ± 2	118 ± 9
Residual	Indicative	63 ± 8	39 ± 12			41 ± 4	11 ± 6	95 ± 13
	Recovery (%)	128	114			96	122	124
Σ(steps 1–4)	Found	236	249	22,000	474	93	144	442
	Indicative	253	267			98.7	149	461
	Recovery (%)	93	93			94	97	96

[†] CRM 143R certified vales are for aqua regia-soluble PTE content, except for Cu, for which only the total content is available.

Agreement between found and certified values for BCR 143R was excellent (100 ± 5%). For BCR 701, the recoveries of PTEs in exchangeable and reducible phases were 100 ± 15%, whereas recoveries in the oxidisable and residual phases were generally 100 ± 30%, except for Pb in step 3 (32%). In his review of results reported for BCR 701 over a 10-year period, Sutherland [36] highlighted other instances of low Pb recovery (<50% of the certified value) in step 3, together with the high degree of imprecision associated with the measurement of Pb in this step during the BCR 701 certification process. Overall, recoveries tended to be low in step 3, relative to certified or indicative values, but high in step 4. However, summations of the amounts of analyte released in steps 1 to 4 of the sequential extraction agreed (within ±20%) with results of pseudo-total measurement in BCR 701. This suggests that the quality of the extraction was adequate.

2.8. Potential Health Risk of PTE

Potential non-carcinogenic and carcinogenic health risks were determined using United States Environmental Protection Agency [37–40] methods and exposure parameters recommended for management of contaminated land in South Africa [41]. Exposure assessment was carried out by calculating the average daily intake (*ADI*) of each PTE through ingestion, inhalation and dermal contact for adults and children (Equations (1)–(3)). Adults and children are considered separately because of their behavioural and physiological differences [42].

$$ADI_{ing} = \frac{C \times IR \times EF \times ED}{BW \times AT} \times 10^{-6} \tag{1}$$

$$ADI_{inh} = \frac{C \times Inh \times EF \times ED}{PEF \times BW \times AT} \tag{2}$$

$$ADI_{dermal} = \frac{C \times SA \times AF \times ABS \times EF \times ED}{BW \times AT} \times 10^{-6} \tag{3}$$

where ADI_{ing} is the average daily intake of a PTE from soil via ingestion in mg per kg per day (mg/kg/day), C is the concentration of PTE in the soil in mg/kg, IR is the ingestion rate, EF is the exposure frequency, ED is the exposure duration, BW is the body weight of the exposed individual and AT is the time period over which the dose is averaged. ADI_{inh} is the average daily intake of PTE from soil via inhalation in mg/kg/day, Inh is the inhalation rate and PEF is the particulate emission factor. ADI_{dermal} is the average daily intake of PTE from soil via dermal contact in mg/kg/day, SA is the skin surface area, AF is the soil-to-skin adherence factor and ABS is the fraction of the applied dose absorbed across the skin.

Hazard quotient (HQ) and hazard index (HI) were used to estimate the non-carcinogenic risk of PTEs in soil [40]. HQ characterizes the health risk of non-carcinogenic adverse effects due to exposure to toxicants and is defined as the quotient of ADI or dose divided by the toxicity threshold value, which is referred to as the chronic reference dose (RfD) in mg/kg/day for a specific PTE, as shown in Equation (4).

$$HQ = ADI/RfD \tag{4}$$

To assess the overall potential of non-carcinogenic effects posed by a PTE, the calculated values of HQ are summed to give HI (Equation (5)).

$$HI = HQ_{ing} + HQ_{inh} + HQ_{dermal} \tag{5}$$

For carcinogens, the risks are estimated as the incremental probability of an individual developing cancer over a lifetime (assumed to be 70 years) because of exposure to the potential carcinogen, CR, calculated using Equation (6).

$$CR = ADI \times CSF \tag{6}$$

where CSF (mg/kg/day) is the cancer slope factor, which converts the estimated daily intake of a PTE to an incremental risk of an individual developing cancer [39]. The total excess lifetime cancer risk for an individual is ultimately calculated from the average contribution of the individual PTE across all exposure pathways using Equation (7).

$$CR_{total} = CR_{ing} + CR_{inh} + CR_{dermal} \tag{7}$$

where CR_{ing}, CR_{inh}, and CR_{dermal} are risk contributions through ingestion, inhalation, and dermal pathways, respectively.

Values for all of the parameters used in the risk calculations are presented in Table 3.

Table 3. Parameters used in risk calculations.

Parameter	Unit	Child	Adult
Ingestion rate (*IR*)	mg/day	200	100
Exposure frequency (*EF*)	days/year	350	350
Exposure duration (*ED*)	years	6	24
Body weight (*BW*)	kg	15	70
Average time (*AT*)	days		
Inhalation rate (*Inh*)	m^3/day	10	20
Particulate emission factor (*PEF*)	m^3/kg	1.3×10^9	1.3×10^9
Skin surface area (*SA*)	cm^2	2100	5800
Soil–skin adherence factor (*AF*)	mg/cm^2	0.2	0.07
Dermal absorption factor (*ABS*)	none	1	1
For carcinogens		365×70	365×70
For non-carcinogens		$365 \times ED$	$365 \times ED$

3. Results and Discussion

3.1. Soil Characterstics and Pseudo-total PTE Concentrations

The pH of the soil samples ranged from 5.8 to 10, with an average of 7.5 (Table 4). Loss on ignition (LOI) values were low (0.07–5.0, with an average of 1.5). This is consistent with the hot and humid climatic conditions in Lagos, which typically deplete soil organic matter [43]. Findings were in agreement with previous work on Lagos residential soils [30] and playgrounds [24], which generally reported organic matter content <2% and slightly alkaline pH values.

Pseudo-total PTE concentrations (Table 4) reflected varying degrees of soil contamination, as expected, given the various types of land use represented. Samples A1 to A7, obtained from gardens and open spaces, were less contaminated than samples A8 to A20, (soil from industrial estates; A8 to A10), railway terminals (A11 to A13) and dumpsites (A14 to A20). Soil guideline values have not yet been defined specifically for use in Nigeria. However, compared with the frequently cited Dutch soil quality standards [44], more than half of the samples contained Cu, Pb and Zn (the 'urban metals') [45,46] at concentrations greater than target values [44], whereas three soils—A10, A16 and A19—contained all three elements at concentrations sometimes considerably in excess of the intervention values [44]. The first of these, soil A10, was from an industrial estate where high PTE levels may be attributed to emissions from zinc smelting, steel production and metal foundry plants. Soils A16 and A19 were from the vicinity of dumpsites. Typical dumpsites in Lagos receive large volumes of domestic, industrial and e-waste daily, and there are also a number of auto repair workshops in proximity to these specific locations, all of which are likely to have contributed to the enhanced soil PTE contents observed.

As mentioned above, literature data on Lagos soils is limited to values for a few elements measured at a few sites. However, average concentrations found in the current study were higher than those reported in previous studies [22–26,30], which likely represents PTE accumulation over time as urbanization and industrialization have progressed. Exceptions were some high Cu, Pb and Zn concentrations reported recently at e-waste recycling locations in Owutu, Ikorodu [28] and Alaba International Market [29], which were of a similar magnitude to results for sites A16 and A19. Levels of PTE in the current study were generally substantially higher than PTE measured in urban soils of cities in other developing countries, such as Kampala, Uganda [47]; Karachi, Pakistan [48]; and Sunyani, Ghana [49].

Table 4. Pseudo-total PTE concentrations in Lagos urban soils (mg/kg dry weight).

Sampling Site	Land Use	pH	% LOI	Cr	Cu	Fe	Mn	Ni	Pb	Zn
A1	PO	7.2	3.6	22 ± 1	25 ± 4	12,600 ± 485	233 ± 22	8 ± 1	87 ± 15	239 ± 36
A2	PO	10	5.0	25 ± 2	27 ± 8	12,000 ± 1310	179 ± 26	6 ± 0.1	26 ± 5	175 ± 4
A3	PO	8.1	1.4	30 ± 2	10 ± 2	10,000 ± 178	212 ± 59	8 ± 4	24 ± 4	99 ± 45
A4	PO	7.0	0.36	*19 ± 1*	*8.0 ± 1*	*7460 ± 284*	*135 ± 38*	*4 ± 0.2*	*10 ± 1*	*61 ± 3*
A5	PO	7.0	0.92	34 ± 12	29 ± 12	14,100 ± 2110	199 ± 49	8 ± 18	41 ± 12	165 ± 25
A6	PO	6.5	0.11	51 ± 1	18 ± 1	22,300 ± 480	359 ± 19	16 ± 0.2	57 ± 3	165 ± 3
A7	PO	8.3	2.2	38 ± 6	20 ± 1	11,100 ± 1680	159 ± 17	8 ± 1	29 ± 5	132 ± 17
A8	IE	7.0	1.4	49 ± 4	71 ± 6	28,600 ± 4280	375 ± 32	17 ± 5	400 ± 154	1080 ± 123
A9	IE	8.1	0.17	111 ± 18	71 ± 8	60,200 ± 1060	437 ± 73	39 ± 5	144 ± 14	433 ± 37
A10	IE	5.8	4.3	175 ± 12	759 ± 406	146,000 ± 15,600	2570 ± 311	109 ± 9	536 ± 62	3240 ± 475
A11	RT	5.9	1.6	71 ± 16	218 ± 7	46,300 ± 1780	413 ± 14	29 ± 3	182 ± 10	241 ± 37
A12	RT	7.4	2.0	94 ± 4	168 ± 2	69,100 ± 787	1220 ± 39	44 ± 3	321 ± 14	1200 ± 35
A13	RT	6.9	0.67	46 ± 3	243 ± 17	36,100 ± 8370	401 ± 70	21 ± 2	212 ± 53	1220 ± 283
A14	DS	7.0	2.6	290 ± 42	182 ± 64	47,600 ± 3810	899 ± 34	139 ± 24	76 ± 4	880 ± 202
A15	DS	7.1	0.85	79 ± 32	133 ± 19	41,100 ± 2730	566 ± 59	29 ± 8	153 ± 29	546 ± 6
A16	DS	8.0	1.2	**1830 ± 76**	**11,700 ± 1780**	**166,000 ± 2530**	1540 ± 62	**1050 ± 240**	**4340 ± 974**	2810 ± 70
A17	DS	8.3	0.07	202 ± 30	82 ± 20	99,900 ± 4250	1600 ± 107	38 ± 3	102 ± 26	353 ± 12
A18	DS	8.6	0.77	47 ± 19	108 ± 14	14,700 ± 1020	261 ± 75	12 ± 1	315 ± 45	511 ± 222
A19	DS	8.0	0.16	108 ± 13	611 ± 153	57,600 ± 5230	753 ± 82	50 ± 68	802 ± 305	**5620 ± 362**
A20	DS	8.2	0.08	602 ± 53	108 ± 21	52,600 ± 5340	**6100 ± 1750**	306 ± 31	22 ± 7	520 ± 20
Mean		7.5	1.5	196	755	49,600	930	97	394	985
Dutch target [44]				100	36			35	85	140
Dutch intervention [44]				380	190			210	530	720

PO = parks, gardens and open spaces; IE = industrial estates; RT = railway terminals; DS = dumpsites. The lowest concentration for each metal is indicated in italics, and the highest concentration is indicated in bold.

3.2. Sequential Extraction and PTE Mobility

Sequential extraction was performed on the soil samples to assess the potential for PTE (re)mobilization. Results are presented in Supplementary Tables S1–S7. To check the quality of the data obtained, the amounts of analyte recovered—Σ(step 1–4)—were compared with those released by aqua regia digestion. A total of 57 (of 140) of the sequential extraction results fell within 10% of corresponding pseudo-total values. A further 63 were either 70–90% or 110–130% of pseudo-total concentrations, i.e., overall, in 86% of cases, the sum of the steps of the sequential extractions was $100 \pm 30\%$ of the aqua regia-soluble content. Only three recoveries were either <50% or >150% of the aqua regia-soluble content: Ni at site A20 (39%) and Pb at sites A1 (48%) and A11 (169%). Site A20 is close to a foundry, and it is possible that there are metal-rich particles heterogeneously distributed within it. This may also be the case for Pb at A11. As well as a railway, this site is close to a major bus depot where mechanical work is undertaken (including removal and servicing of vehicle batteries). Site A1 is an open space in a wealthier part of Lagos metropolis and not highly contaminated with PTEs. There is a relatively large uncertainty in the Pb pseudo-total concentration (17%, $n = 3$) and hence in the recovery calculated.

The fractionation patterns obtained using the BCR procedure are shown in Figures 2 and 3. Chromium was predominantly associated with step 4, the residual phase, in most of the samples (Figure 2a). Three-quarters of the soils studied contained more than 70% of their Cr content in the residual phase. This is in agreement with other urban soil studies [50–52]. Wu et al. [53] found 92% of Cr in the residual phase of urban soils of Guiyang City, China. The presence of Cr in residual forms suggests that the element is strongly bound to soil minerals; therefore, mobilization is unlikely to occur under typical environmental conditions. However, where Cr concentrations were highest, a larger proportion was associated with the reducible phase (step 2). This is worrisome because it indicates that were the soil to become waterlogged and anoxic, there is potential for Cr remobilization due to reduction and dissolution of iron and manganese oxyhydroxides.

Copper was released at various steps in the sequential extraction (Figure 2b), with the highest proportion associated with the reducible phase in most samples. The association of Cu with iron and manganese oxides and hydroxides—the target phase of step 2 of the BCR protocol—has been well documented in polluted urban soils, dusts and sediments [50,54–59]. For some samples, an appreciable amount of Cu was also released in step 2 which, again, has been reported in previous studies; for example, Szolnoki et al. [55] found 24% of Cu in the oxidisable fraction of urban vegetable garden soils from Szeged, Hungary. Because the majority of the overall Cu content was in non-residual forms, there is considerable potential for Cu mobilization under changing environmental conditions. Of particular concern is the most contaminated dumpsite, A16, where >3000 mg/kg of Cu was found in the easily mobilized exchangeable phase.

Almost all the Fe in Lagos urban soils was associated with the reducible and residual phases (Figure 2c), which was expected because step 2 of the BCR protocol targets iron and manganese oxyhydroxides, and ferrous minerals constitute a major structural component of soil. Similar findings have been reported for urban soils from public-access areas of five European cities [52], as well as urban vegetable garden soil [60]. The predominance of Fe in the residual phase indicates low mobility and bioavailability.

Manganese was found in all four phases (Figure 2d), with residual and reducible forms generally dominating for similar reasons to Fe. Manganese is one of the most abundant elements in the earth crust [57], and the hydroxylamine–hydrochloride reagent employed in step 2 principally targets Fe-Mn oxyhydroxides. Previous work on urban soils from five European cities reported [52] a similar manganese distribution between the fractions defined by sequential extraction. Significant amounts of Mn were also located in the most labile, exchangeable phase. This suggests that Mn is relatively mobile in Lagos soils.

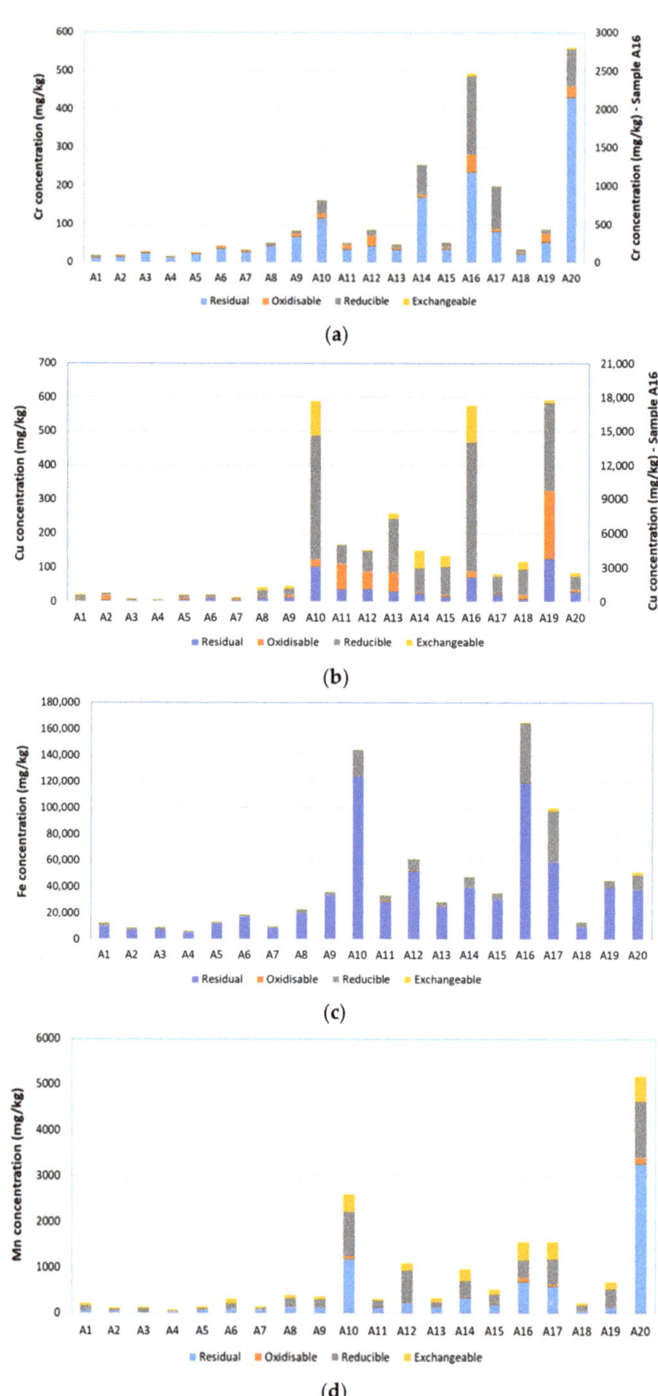

Figure 2. Fractionation of (**a**) Cr, (**b**) Cu, (**c**) Fe and (**d**) Mn according to BCR sequential extraction. Note that Cr and Cu are plotted against a secondary axis for site A16.

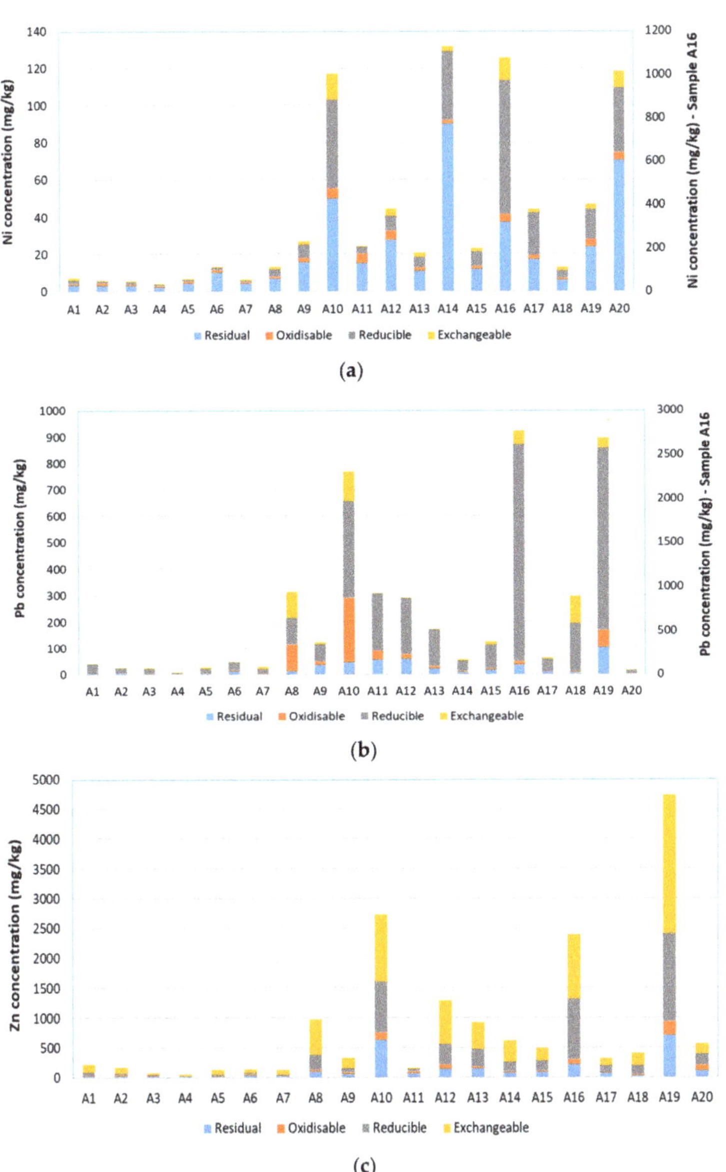

Figure 3. Fractionation of (**a**) Ni, (**b**) Pb and (**c**) Zn according to BCR sequential extraction. Note that Ni and Pb are plotted against a secondary axis for site A16.

Nickel was mainly associated with the reducible and residual phases (Figure 3a). This is in agreement with sequential extraction results of other studies, which found the largest proportions of Ni in the residual phase of urban vegetable garden soils [55], urban soils [52] and urban street dusts [51]. Similarly to Cr, the amounts of Ni found in the reducible phase were generally larger where pseudo-total concentrations were highest, and similarly to Cu, some of the dumpsite soils contained more exchangeable Ni than soils from other locations.

The reducible fraction was most important for Pb at all sites (Figure 3b), likely reflecting the element's ability to form stable complexes with Fe-Mn oxides [50,61]. Umoren et al. [54]

also found the largest percentage of extractable Pb (75%) associated with the reducible fraction in their study of refuse dump soils in Uyo, southern Nigeria. Similar observations have been reported in other urban areas [52,53]. However, Adeyi et al. [30] found lead fractionation varied between residential soils from different parts of Lagos. In high-income areas, Pb concentration was low (6–17 mg/kg), and the Fe-Mn oxide-bound fraction was dominant; in low-income areas, Pb concentration was higher (90 mg/kg), with higher proportions of Pb found in both more refractory and more labile phases. Particularly large amounts of Pb in the current study were associated with the reducible phase in dumpsite soils A16 (2460 mg/kg) and A19 (688 mg/kg), which is of concern because of the health risk this might pose if the element were mobilized under reducing conditions. Lead associated with the reducible phase can also be liberated by erosion processes of top soils and transported to a new environment, such as road surfaces [62].

Zinc was mainly associated with the exchangeable phase, followed by the reducible phase (Figure 3c). A previous study on sediment from three urban rivers and Lagos Lagoon [31] also found that a high proportion (40 to 87%) of Zn content was released in step 1 of the BCR sequential extraction. This is relevant to the current work because trans-urban water bodies are likely to contain considerable amounts of soil-derived material; therefore, trace elements may have similar speciation. A fractionation pattern similar to that reported in the current study were reported for soils and dusts collected in other urban areas [51,52,54,60]. Like Cu, the majority of Zn was present in non-residual forms, which is of concern from the point of view of potential mobilization and transport.

The relative availability of the analytes based on the average proportion found in the exchangeable phase was Fe (1.1%) < Cr (1.3%) < Ni (9.2%) < Pb (12%) < Cu (15%) < Mn (23%) < Zn (48%). Based on the proportion found in the three most labile phases (step 1 + step 2 + step 3), the relative availability was Fe (16%) < Cr (30%) < Ni (46%) < Mn (63%) < Cu (78%) < Zn (80%) < Pb (84%). Elements found to be mainly of lithogenic origin in previous urban soil studies (Fe [50,52] and Cr [4,51]) were similarly less available for mobilization in the Lagos samples; therefore, it might be expected that they would have lesser environmental or human health impact than elements likely of mainly anthropogenic origin (Cu, Pb, Zn). Where concentrations of the urban metals were highest—site A16 for Cu and Pb and site A19 for Zn—they were also more labile, which is clearly of concern.

3.3. Human Health Impact

Results obtained for human health risk assessment are presented in Table 5 and Figure 4. For non-carcinogenic risk, comparison of HQ values indicates that ingestion was the most significant exposure route for both children and adults, followed by dermal contact and inhalation. The HI (summation of HQ across the three exposure pathways) values were less than 1 for Cr, Ni and Zn, indicating that there is no significant non-carcinogenic risk associated with exposure to the average concentrations of the above PTEs in Lagos urban soils. In contrast, the HI for Cu, Mn and Pb were greater than 1 in children, which suggests that there is a risk of non-carcinogenic health effects. Among the PTEs studied, Pb was the largest contributor to non-carcinogenic risks. This is of particular concern, given the impact that this element can have on children's development, even at low concentrations. Risk from the other metals followed the order Cu > Mn > Ni > Zn > Cr for children and Mn > Cu > Ni > Zn > Cr for adults (Figure 3). Although similar trends were observed in both groups, non-carcinogenic risk was greater in children than in adults. This is expected, given their lower body mass and immature physiology, as has been reported in many previous studies [63,64].

Carcinogenic risk was calculated for Cr, Ni and Pb based on their respective cancer slope factors (the other analytes are not considered a cancer risk, so slope factor data are not available). For Cr, a slope factor of 0.5 (the lower of the values commonly cited in literature [41]) was used because results of the sequential extraction suggested that the element was not readily available. Risks greater than 1×10^{-4} are considered unacceptable, those between 10^{-4} and 10^{-6} acceptable and those less than 1×10^{-6} unlikely to lead to any detrimental health outcomes. In the current study, total carcinogenic risk levels for children were greater than those for adults, and values for both Cr and Ni exceeded the threshold for unacceptable risk in both age groups. In contrast, values were in the range of 10^{-4} to 10^{-6} for Pb; therefore, the additional probability of developing cancer over a 70-year lifespan due to exposure to this element at the average concentration found in Lagos soils is considered acceptable.

Table 5. Human health risk assessment for children and adults.

Element	HQ_{ing}	HQ_{inh}	HQ_{dermal}	HI Children	HQ_{ing}	HQ_{inh}	HQ_{dermal}	HI Adult	CR_{total} Children	CR_{total} Adult
Cr	1.67×10^{-2}	3.23×10^{-5}	3.51×10^{-3}	5.21×10^{-3}	1.79×10^{-4}	1.23×10^{-5}	6.49×10^{-4}	8.40×10^{-4}	$\mathbf{3.33 \times 10^{-4}}$	$\mathbf{2.33 \times 10^{-4}}$
Cu	2.60×10^{0}	2.96×10^{-5}	5.47×10^{-1}	$\mathbf{3.15 \times 10^{0}}$	2.79×10^{-2}	1.32×10^{-5}	1.18×10^{-1}	1.46×10^{-1}		
Mn	4.95×10^{-1}	3.05×10^{-6}	1.04×10^{0}	$\mathbf{1.53 \times 10^{0}}$	5.30×10^{-2}	9.26×10^{-7}	1.52×10^{-1}	2.05×10^{-1}		
Ni	6.19×10^{-2}	7.70×10^{-6}	1.30×10^{-1}	1.92×10^{-1}	6.63×10^{-3}	2.92×10^{-6}	2.38×10^{-2}	3.05×10^{-2}	$\mathbf{2.76 \times 10^{-4}}$	$\mathbf{1.93 \times 10^{-4}}$
Pb	1.39×10^{0}	3.56×10^{-6}	2.93×10^{0}	$\mathbf{4.33 \times 10^{0}}$	1.49×10^{-1}	1.60×10^{-6}	6.38×10^{-1}	7.88×10^{-1}	1.13×10^{-5}	7.96×10^{-6}
Zn	4.19×10^{-2}	1.28×10^{-5}	8.81×10^{-2}	1.30×10^{-1}	4.49×10^{-3}	5.64×10^{-6}	1.87×10^{-2}	2.32×10^{-2}		

Values in bold indicate an HI value > 1 or a CR value > 1×10^{-4}.

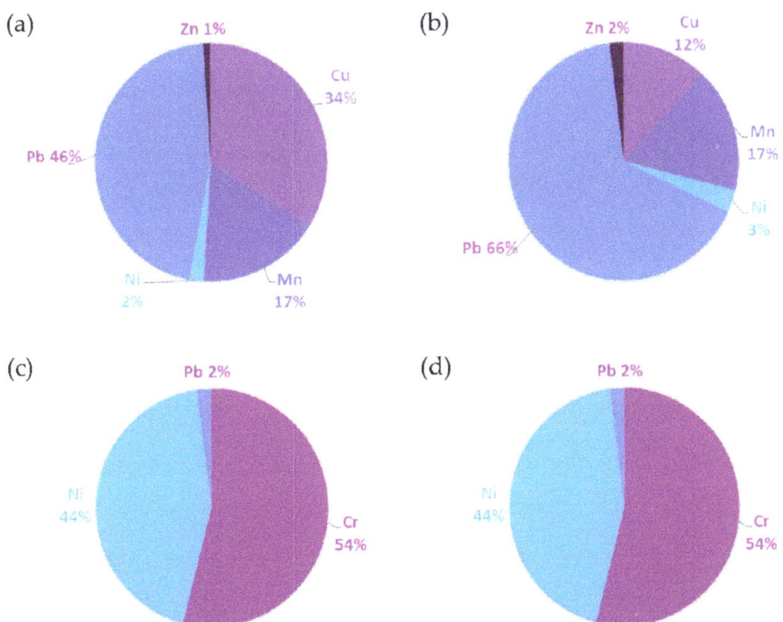

Figure 4. Contribution of PTE to risk: (**a**) non-carcinogenic risks in children, (**b**) non-carcinogenic risks in adults, (**c**) carcinogenic risks in children and (**d**) carcinogenic risks in adults.

However, it is important to emphasize that these risk calculations are based on average PTE concentrations. Given the remarkably wide range of analyte concentrations found and the fact that a few values markedly exceeded the mean for each element, it is probable that results overestimate risk for the majority of Lagos residents.

4. Conclusions

Concentrations of Cr, Cu, Fe, Mn, Ni, Pb and Zn varied markedly in soils obtained from public-access areas across the megacity of Lagos. The highest values were found in proximity to known pollution sources, such as dumpsites, but there is evidence that general ambient PTE levels are increasing as rapid urbanization and industrialization occurs. The urban metals—Cu, Pb and Zn—were generally found in more labile forms than elements such as Cr, Fe and Ni and are therefore more susceptible to (re)mobilization and transport under changing environmental conditions. Calculations indicated the presence of non-carcinogenic risk for children, as well as carcinogenic risk for both children and adults, although this is likely associated mainly with sites where PTE concentrations were highest. Nevertheless, further monitoring and assessment of the status of Lagos urban soils is recommended, together with the development and implementation of an appropriate regulatory framework to protect soil quality and public health.

Supplementary Materials: The following supporting information can be downloaded at: https://www.mdpi.com/article/10.3390/toxics10040154/s1, Table S1: Results for sequential extraction of chromium (mg/kg); Table S2: Results for sequential extraction of copper (mg/kg); Table S3: Results for sequential extraction of iron (mg/kg); Table S4: Results for sequential extraction of manganese (mg/kg); Table S5: Results for sequential extraction of nickel (mg/kg); Table S6: Results for sequential extraction of lead (mg/kg); Table S7: Results for sequential extraction of zinc (mg/kg).

Author Contributions: Conceptualization, A.O.F., C.M.D. and A.O.O.; methodology, A.O.F., S.A. and C.M.D.; formal analysis, A.O.F.; investigation, A.O.F.; resources, C.M.D.; writing—original draft preparation, A.O.F.; writing—review and editing, C.M.D., S.A. and A.O.O.; supervision, C.M.D. and A.O.O.; funding acquisition, A.O.F. All authors have read and agreed to the published version of the manuscript.

Funding: A.O.F. and S.A. wish to thank the Tertiary Education Trust Fund of Nigeria for awards of Ph.D. scholarships to study at the University of Strathclyde, Glasgow, UK.

Institutional Review Board Statement: Not applicable.

Informed Consent Statement: Not applicable.

Data Availability Statement: Data are available on request from abimbola.famuyiwa@gmail.com.

Acknowledgments: The authors wish to thank Alexander Clunie for assistance with ICP-MS and Michael Davidson for preparation of figures in Excel. We are grateful to the Lagos Ministry of Environment for logistical assistance during the soil sampling campaign.

Conflicts of Interest: The authors declare no conflict of interest.

References

1. Department of Economics and Social Affairs, Population Division, United Nations, 2015, Revision. Available online: http://esa.un.org/unpd/wpp/publications/files/key_findings_wpp_2015.pdf (accessed on 12 August 2021).
2. Wong, C.S.; Li, X.; Thornton, I. Urban environmental geochemistry of trace metals. *Environ. Pollut.* **2006**, *142*, 1–16. [CrossRef]
3. Available online: https://lagosstate.gov.ng/ (accessed on 11 January 2022).
4. Ajmone-Marsan, F.; Biasioli, M.; Kralj, T.; Grčman, H.; Davidson, C.; Hursthouse, A.; Madrid, L.; Rodrigues, S. Metals in particle-size fractions of the soils of five European cities. *Environ. Pollut.* **2008**, *152*, 73–81. [CrossRef] [PubMed]
5. Elom, N.I.; Entwistle, J.; Dean, J.R. Human health risk from Pb in urban street dust in northern UK cities. *Environ. Chem. Lett.* **2014**, *12*, 209–218. [CrossRef]
6. Filippelli, G.M.; Adamic, J.; Nichols, D.; Shukle, J.; Frix, E. Mapping the urban lead exposome: A detailed analysis of soil metal concentrations at the household scale using citizen science. *Int. J. Environ. Res. Public Health* **2018**, *15*, 1531. [CrossRef]
7. Imperato, M.; Adamo, P.; Naimo, D.; Arienzo, M.; Stanzione, D.; Violante, P. Spatial distribution of heavy metals in urban soils of Naples city (Italy). *Environ. Pollut.* **2003**, *124*, 247–256. [CrossRef]
8. Famuyiwa, A.O.; Davidson, C.M.; Oyeyiola, A.O.; Ande, S.; Lanre-Iyanda, Y.; Babajide, S.O. Pollution characteristics and health risk assessment of potentially toxic elements in school playground soils: A case study of Lagos, Nigeria. *Hum. Ecol. Risk Assess. Int. J.* **2018**, *25*, 1729–1744. [CrossRef]
9. Panghal, V.; Singh, A.; Kumar, R.; Kumari, G.; Kumar, P.; Kumar, S. Soil heavy metals contamination and ecological risk assessment in Rohtak urban area, Haryana (India). *Environ. Earth Sci.* **2021**, *80*, 731. [CrossRef]
10. Defo, C.; Yerima, B.P.K.; Noumsi, I.M.K.; Bemmo, N. Assessment of heavy metals in soils and groundwater in an urban watershed of Yaoundé (Cameroon-West Africa). *Environ. Monit. Assess.* **2015**, *187*, 77. [CrossRef] [PubMed]
11. Ogunbanjo, O.; Onawumi, O.; Gbadamosi, M.; Ogunlana, A.; Anselm, O. Chemical speciation of some heavy metals and human health risk assessment in soil around two municipal dumpsites in Sagamu, Ogun state, Nigeria. *Chem. Speciat. Bioavailab.* **2016**, *28*, 142–151. [CrossRef]

12. Banat, K.M.; Howari, F.M.; Al-Hamad, A.A. Heavy metals in urban soils of central Jordan: Should we worry about their environmental risks? *Environ. Res.* **2005**, *97*, 258–273. [CrossRef]
13. Mielke, H.W.; Reagan, P.L. Soil is an important pathway of human lead exposure. *Environ. Health Perspect.* **1998**, *106*, 217–229. [CrossRef] [PubMed]
14. Boyd, H.B.; Pedersen, F.; Cohr, K.-H.; Damborg, A.; Jakobsen, B.M.; Kristensen, P.; Samsøe-Petersen, L. Exposure Scenarios and Guidance Values for Urban Soil Pollutants. *Regul. Toxicol. Pharmacol.* **1999**, *30*, 197–208. [CrossRef] [PubMed]
15. Poggio, L.; Vrščaj, B.; Schulin, R.; Hepperle, E.; Marsan, F.A. Metals pollution and human bioaccessibility of topsoils in Grugliasco (Italy). *Environ. Pollut.* **2009**, *157*, 680–689. [CrossRef] [PubMed]
16. Kabata-Pendias, H.A.; Mukherjee, A.B. *Trace Elements from Soil to Human*; Springer: Berlin/Heidelberg, Germany, 2007; ISBN 978-3-540-32714-1. [CrossRef]
17. Tessier, A.; Campbell, P.G.C.; Bisson, M. Sequential extraction procedure for the speciation of particulate trace metals. *Anal. Chem.* **1979**, *51*, 844–851. [CrossRef]
18. Ure, A.M. Trace elements in soil: Their determination and speciation. *Anal. Bioanal. Chem.* **1990**, *337*, 577–581. [CrossRef]
19. Filgueiras, A.V.; Lavilla, I.; Bendicho, C. Chemical sequential extraction for metal partitioning in environmental solid samples. *J. Environ. Monit.* **2002**, *4*, 823–857. [CrossRef]
20. Bacon, J.R.; Davidson, C.M. Is there a future for sequential chemical extraction? *Analyst* **2008**, *133*, 25–46. [CrossRef] [PubMed]
21. Rauret, G.; López-Sánchez, J.F.; Sahuquillo, A.; Rubio, R.; Davidson, C.; Ure, A.; Quevauviller, P. Improvement of the BCR three step sequential extraction procedure prior to the certification of new sediment and soil reference materials. *J. Environ. Monit.* **1999**, *1*, 57–61. [CrossRef] [PubMed]
22. Olukanni, D.O.; Adeoye, D.O. Heavy metal concentrations in road side soils from selected locations in the Lagos Metropolis, Nigeria. *Int. J. Eng. Technol.* **2012**, *2*, 1743–1752.
23. Awofolu, O.R. A survey of trace metals in vegetation, soil and lower animal along some selected major roads in metropolitan city of Lagos. *Environ. Monit. Assess.* **2005**, *105*, 431–447. [CrossRef]
24. Popoola, O.E.; Bamgbose, O.; Okonkwo, O.J.; Arowolo, T.A.; Odukoya, A.; Popoola, A.O. Heavy metals content in play-ground topsoil of some public primary schools in metropolitan Lagos, Nigeria. *Res. J. Env. Earth Sci.* **2012**, *4*, 434–439.
25. Njoku, K.L.; Akinola, M.O.; Zighadina, T.M. A study on the spatial distribution of heavy metal in industrial area of Ikorodu, Lagos State, Nigeria. *J. Res. Environ. Sci. Toxicol.* **2012**, *2*, 64–70.
26. Olorunfemi, A.O.; Alao-Daniel, A.B.; Adesiyan, T.A.; Onah, C.E. Geochemical assessment of heavy metal impact on soil around Ewu-Elepe Dumpsite, Lagos State, Nigeria. *Ife J. Sci.* **2021**, *22*, 119–138. [CrossRef]
27. Adeyi, A.A.; Oyeleke, P. Heavy metals and polycyclic aromatic hydrocarbons in soil from e-waste dumpsites in Lagos and Ibadan, Nigeria. *J. Health Pollut.* **2017**, *7*, 71–84. [CrossRef]
28. Sawyerr, H.O.; Oladeji, F.O. Evaluation of the distribution of heavy metals in soil around electronic dumpsite in Owutu, Ikorodu, Lagos State, Nigeria. *Asian J. Chem. Sci.* **2020**, 24–31. [CrossRef]
29. Anselm, O.H.; Cavoura, O.; Davidson, C.M.; Oluseyi, T.O.; Oyeyiola, A.O.; Togias, K. Mobility, spatial variation and human health risk assessment of mercury in soil from an informal e-waste recycling site, Lagos, Nigeria. *Environ. Monit. Assess.* **2021**, *193*, 1–13. [CrossRef] [PubMed]
30. Adeyi, A.A.; Babalola, B.A. Lead and cadmium levels in residential soils of Lagos and Ibadan, Nigeria. *J. Health Pollut.* **2017**, *7*, 42–55. [CrossRef] [PubMed]
31. Oyeyiola, A.O.; Davidson, C.M.; Olayinka, K.O.; Alo, B.I. Fractionation and ecotoxicological implication of potentially toxic metals in sediments of three urban rivers and the Lagos Lagoon, Nigeria, West Africa. *Environ. Monit. Assess.* **2014**, *186*, 7321–7333. [CrossRef]
32. Falufosi, M.O.; Osinowo, O.O. Geology and hydrocarbon potential of the Nigerian sector of Dahomey Basin. *J. Sediment. Environ.* **2021**, *6*, 335–358. [CrossRef]
33. Famuyiwa, A.O.; Lanre-Iyanda, Y.A.; Osifeso, O. Impact of Land Use on Concentrations of Potentially Toxic Elements in Urban Soils of Lagos, Nigeria. *J. Health Pollut.* **2018**, *8*, 180904. [CrossRef]
34. *BS ISO 10390:2005*; Soil Quality. Determination of pH. British Standard International: London, UK, 2005; 1–7.
35. Soil, Waste, Treated Biowaste and Sludge. Determination of Loss on Ignition. Available online: https://www.en-standard.eu/bs-en-15935-2021-soil-waste-treated-biowaste-and-sludge-determination-of-loss-on-ignition/ (accessed on 25 January 2022).
36. Sutherland, R.A. BCR®-701: A review of 10-years of sequential extraction analyses. *Anal. Chim. Acta* **2010**, *680*, 10–20. [CrossRef] [PubMed]
37. United States Environmental Protection Agency. *Soil Screening Guidance Technical Background Document*; EPA/540 95; Office of Solid Waste and Emergency Response: Washington, DC, USA, 1996.
38. United States Environmental Protection Agency. *Exposure Factors Handbook*; EPA/600/P-95/002Fa-c U.S.; EPA National Center for Environmental Assessment, Office of Research and Development: Washington, DC, USA, 1997.
39. United States Environmental Protection Agency. *Supplemental Guidance for Developing Soil Screening Levels for Superfund Sites*; OSWER 9355.4e24; Office of Solid Waste and Emergency Response: Washington, DC, USA, 2001.
40. United States Environmental Protection Agency. Risk assessment guidance for superfund. In *Part A: Human Health Evaluation Manual; Part E, Supplemental Guidance for Dermal Risk Assessment; Part F, Supplemental Guidance for Inhalation Risk Assessment*; United States Environmental Protection Agency: Washington, DC, USA, 2011; Volume 1.

41. DEA. Framework for the Management of Contaminated Land (Part 1 of 3). Department of Environmental Affairs, Republic of South Africa. 2010. Available online: https://sawic.environment.gov.za/documents/562.pdf (accessed on 27 February 2022).
42. Kamunda, C.; Mathuthu, M.; Madhuku, M. Health risk assessment of heavy metals in soils from witwatersrand gold mining Basin, South Africa. *Int. J. Environ. Res. Public Health* **2016**, *13*, 663. [CrossRef]
43. Karmakar, R.; Das, I.; Dutta, D.; Rakshit, A. Potential effects of climate change on soil properties: A review. *Sci. Int.* **2016**, *4*, 51–73. [CrossRef]
44. Dutch Ministry of Housing, Spatial Planning and Environment. *Circular on Target and Intervention Values for Soil Remediation*; Netherlands Government Gazette: Hague, The Netherlands, 2000; pp. 1–51.
45. Marsan, F.A.; Biasioli, M. Trace elements in soils of urban areas. *Water Air Soil Pollut.* **2010**, *213*, 121–143. [CrossRef]
46. McIlwaine, R.; Doherty, R.; Cox, S.F.; Cave, M. The relationship between historical development and potentially toxic element concentrations in urban soils. *Environ. Pollut.* **2017**, *220*, 1036–1049. [CrossRef] [PubMed]
47. Nabulo, G.; Oryem-Origa, H.; Diamond, M. Assessment of lead, cadmium, and zinc contamination of roadside soils, surface films, and vegetables in Kampala City, Uganda. *Environ. Res.* **2006**, *101*, 42–52. [CrossRef] [PubMed]
48. Karim, Z.; Qureshi, B.A. Health risk assessment of heavy metals in urban soil of Karachi, Pakistan. *Hum. Ecol. Risk Assess. Int. J.* **2014**, *20*, 658–667. [CrossRef]
49. Asamoah, B.D.; Asare, A.; Okpati, S.W.; Aidoo, P. Heavy metal levels and their ecological risks in surface soils at Sunyani magazine in the bono region of Ghana. *Sci. Afr.* **2021**, *13*, e00937. [CrossRef]
50. Lu, Y.; Zhu, F.; Chen, J.; Gan, H.; Guo, Y. Chemical fractionation of heavy metals in urban soils of Guangzhou, China. *Environ. Monit. Assess.* **2007**, *134*, 429–439. [CrossRef]
51. Li, X.; Feng, L. Geostatistical analyses and fractionation of heavy metals in urban soil from industrial district in Weinan, NW China. *Environ. Earth Sci.* **2012**, *67*, 2129–2140. [CrossRef]
52. Davidson, C.M.; Urquhart, G.J.; Ajmone-Marsan, F.; Biasioli, M.; Duarte, A.D.C.; Díaz-Barrientos, E.; Grčman, H.; Hossack, I.; Hursthouse, A.S.; Madrid, L.; et al. Fractionation of potentially toxic elements in urban soils from five European cities by means of a harmonised sequential extraction procedure. *Anal. Chim. Acta* **2006**, *565*, 63–72. [CrossRef]
53. Wu, Y.; Liu, C.; Tu, C. Distribution and sequential extraction of some heavy metals in urban soils of Guiyang City, China. *Chin. J. Geochem.* **2008**, *27*, 401–406. [CrossRef]
54. Umoren, I.U.; Udoh, A.P.; Udousoro, I.I. Concentration and chemical speciation for the determination of Cu, Zn, Ni, Pb and Cd from refuse dump soils using the optimized BCR sequential extraction procedure. *Environmentalist* **2007**, *27*, 241–252. [CrossRef]
55. Szolnoki, Z.; Farsang, A. Evaluation of metal mobility and bioaccessibility in soils of urban vegetable gardens using sequential extraction. *Water Air Soil Pollut.* **2013**, *224*, 1–16. [CrossRef]
56. Sayadi, M.H.; Rezaei, M.R.; Rezaei, A. Fraction distribution and bioavailability of sediment heavy metals in the environment surrounding MSW landfill: A case study. *Environ. Monit. Assess.* **2014**, *187*, 4110. [CrossRef]
57. Yu, R.L.; Hu, G.R.; Wang, L.J. Speciation and ecological risk of heavy metals in intertidal sediments of Quanzhou Bay, China. *Environ. Monit. Assess.* **2009**, *163*, 241–252. [CrossRef]
58. Nemati, K.; Abu Bakar, N.K.; Abas, M.R.; Sobhanzadeh, E. Speciation of heavy metals by modified BCR sequential extraction procedure in different depths of sediments from Sungai Buloh, Selangor, Malaysia. *J. Hazard. Mater.* **2011**, *192*, 402–410. [CrossRef]
59. Wang, X.-S.; Qin, Y.; Chen, Y.-K. Leaching characteristics of arsenic and heavy metals in urban roadside soils using a simple bioavailability extraction test. *Environ. Monit. Assess.* **2007**, *129*, 221–226. [CrossRef]
60. Tokalıoğlu, Ş.; Kartal, Ş.; Gültekın, A. Investigation of heavy-metal uptake by vegetables growing in contaminated soils using the modified BCR sequential extraction method. *Int. J. Environ. Anal. Chem.* **2006**, *86*, 417–430. [CrossRef]
61. Gleyzes, C.; Tellier, S.; Astruc, M. Fractionation studies of trace elements in contaminated soils and sediments: A review of sequential extraction procedures. *TrAC Trends Anal. Chem.* **2002**, *21*, 451–467. [CrossRef]
62. Sutherland, R.A.; Tolosa, C.A. Multi-element analysis of road-deposited sediment in an urban drainage basin, Honolulu, Hawaii. *Environ. Pollut.* **2000**, *110*, 483–495. [CrossRef]
63. Doyi, I.N.Y.; Isley, C.F.; Soltani, S.N.; Taylor, P.M. Human exposure and risk associated with trace element concentrations in indoor dust from Australian homes. *Environ. Int.* **2019**, *133*, 105125. [CrossRef] [PubMed]
64. Famuyiwa, A.O.; Entwistle, J.A. Characterising and communicating the potential hazard posed by potentially toxic elements in indoor dusts from schools across Lagos, Nigeria. *Environ. Sci. Process. Impacts* **2021**, *23*, 867. [CrossRef] [PubMed]

Article

Discharge Patterns of Potentially Harmful Elements (PHEs) from Coking Plants and Its Relationship with Soil PHE Contents in the Beijing–Tianjin–Hebei Region, China

Xiaoming Wan [1,2,*], Weibin Zeng [1,2], Gaoquan Gu [1,2], Lingqing Wang [1,2] and Mei Lei [1,2]

[1] Institute of Geographic Sciences and Natural Resources Research, Chinese Academy of Sciences, Beijing 100101, China
[2] University of Chinese Academy of Sciences, Beijing 100049, China
* Correspondence: wanxm.06s@igsnrr.ac.cn; Tel.: +86-1064888087

Abstract: The Beijing–Tianjin–Hebei (BTH) region in China is a rapid development area with a dense population and high-pollution, high-energy-consumption industries. Despite the general idea that the coking industry contributes greatly to the total emission of potentially harmful elements (PHEs) in BTH, quantitative analysis on the PHE pollution caused by coking is rare. This study collected the pollutant discharge data of coking enterprises and assessed the risks of coking plants in BTH using the soil accumulation model and ecological risk index. The average contribution rate of coking emissions to the total emissions of PHEs in BTH was ~7.73%. Cross table analysis indicated that there was a close relationship between PHEs discharged by coking plants and PHEs in soil. The accumulation of PHEs in soil and their associated risks were calculated, indicating that nearly 70% of the coking plants posed a significant ecological risk. Mercury, arsenic, and cadmium were the main PHEs leading to ecological risks. Scenario analysis indicated that the percentage of coking plants with high ecological risk might rise from 8.50% to 20.00% as time progresses. Therefore, the control of PHEs discharged from coking plants in BTH should be strengthened. Furthermore, regionalized strategies should be applied to different areas due to the spatial heterogeneity of risk levels.

Keywords: accumulation; atmospheric deposition; emission; ecological risk; spatial heterogeneity

Citation: Wan, X.; Zeng, W.; Gu, G.; Wang, L.; Lei, M. Discharge Patterns of Potentially Harmful Elements (PHEs) from Coking Plants and Its Relationship with Soil PHE Contents in the Beijing–Tianjin–Hebei Region, China. *Toxics* **2022**, *10*, 240. https://doi.org/10.3390/toxics10050240

Academic Editor: Ilaria Guagliardi

Received: 14 April 2022
Accepted: 6 May 2022
Published: 10 May 2022

Publisher's Note: MDPI stays neutral with regard to jurisdictional claims in published maps and institutional affiliations.

1. Introduction

China is the largest coke producer and consumer in the world. The annual coke production in China as of 2020 has reached 430 million tons, contributing to more than 60% of the world's coke production [1]. The Beijing–Tianjin–Hebei (BTH) region, the Capital Economy Circle of China, is one of the main areas with concentrated coking plants. Coke production in BTH accounts for 12% of the whole country [1,2].

Studies have been conducted on the PHEs emission patterns from the coking industry. Copper (Cu), zinc (Zn), arsenic (As), lead (Pb), chromium (Cr), nickel (Ni), mercury (Hg), cadmium (Cd), and cobalt (Co) are the most abundant PHEs emitted from the coking industry and are commonly found in abundance in raw coal [3]. Although coking has been considered as a main contributor to PHE release in BTH [4–6], most studies concerning the coking industry focus on the emission patterns of organic contaminants [7–9], with less attention paid to PHE emissions. Until now, no official data on PHE emissions from coking plants in BTH have been available. The exact contribution of coking to the total PHEs released in BTH is still unknown, which is important for the government to establish regional policy.

Furthermore, during coking activities, PHEs are volatized, released, and then deposited on the surrounding soil [10–13]. Direct air pollutant emissions from iron and steel production plants account for 21% of the total suspended particulates emissions in the BTH region [14]. Atmospheric deposition is one of the main sources of PHEs in soil [15] and in crops [16]. Considering the intensive distribution of coking plants in BTH, coking has been regarded as one of the main sources of soil PHE pollution in this area. The apparent contribution of coking to soil pollution has also been identified in other areas of the world. Atmospheric emissions and dusting from the surface of ash dumps are essential sources for Cr, Mn, Ni, Cu, Zn, Cd, and Pb in soils around a coal-fired power station in Southern Russia [17]. The coke industry and steel metallurgy have been identified as sources of soil contamination by PHEs around the Zdzieszowice coking plant, Opole Province, Southern Poland [18].

There have been studies on the PHE pollution status in the soil around coking plants, which was found to be serious [19–21]. It has been found that PHEs in soil, especially As, Cd, and Hg, led to significant ecological risks, while Hg posed the most serious risk [22–24]. Furthermore, PHE emissions from coking plants may bring certain carcinogenic and non-carcinogenic risks to surrounding residents; and the health risks of As, Cr, and Ni are relatively high [25]. Moreover, soils with a high level of PHEs may result in the accumulation of PHEs in plants, which will reduce soil economic benefits and increase human health risks [15,26,27].

At a regional scale, coal combustion may lead to differences in the temporal and spatial distribution of PHEs, leading to the heterogeneous distribution of risks caused by coking plants [28,29]. Current risk assessments mostly focus on the soil pollution status inside a certain coking plant, with few considerations of the pollution from coking plants to the surrounding soil through atmospheric deposition of coking gas [30–32]. The lack of risk assessment at a regional scale may cause difficulties for the overall management of risks caused by coking plants in BTH. Therefore, quantifying the risks of PHE emissions from coking plants, and its spatial heterogeneity on a regional scale can provide valuable information to understand and effectively manage metal-induced risks [33,34].

This study aimed to disclose the discharge patterns of PHEs from coking plants and their relationship with soil PHEs contents in the BTH region. First, the data for PHEs discharged from coking plants in BTH were collected from the China Industrial Enterprise Pollution Emission Database [35], and POI data from Baidu map. Second, the data for PHE concentrations in soil were collected from published literature. Third, the relationship between the discharge data and soil concentration data was disclosed. Then, the ecological risks of PHEs accumulated from coking emissions were assessed. Scenario analysis was further conducted to predict the change in risk levels in the future given that PHEs could persist in soil for a very long time. The results could provide a quantified understanding of the degree of risks brought by coking plants in BTH. On this basis, possible suggestions on emission control were provided to protect the soil environment in the BTH region.

2. Materials and Methods

2.1. Study Area

The BTH region is in the northern part of the North China Plain close to the Northwest Pacific Ocean, and it is located between 113° E–119° E and 36° N–42° N (Figure 1). BTH are composed of 2 provincial-level municipalities (i.e., Beijing City and Tianjin City) and 11 prefecture-level cities, namely, Shijiazhuang (SJZ), Tangshan (TS), Qinhuangdao (QHD), Handan (HD), Hengshui (HS), Xingtai (XT), Baoding (BD), Zhangjiakou (ZJK), Chengde (CD), Cangzhou (CZ), and Langfang (LF). BTH cover an area of 218,000 km^2 and have a population of approximately 110 million (National Bureau of Statistics in China, NBSC, 2019). The BTH region has a temperate monsoon climate with an average temperature of 3.9 °C~15.3 °C.

The terrain conditions of the BTH region are complex. The overall terrain is high in the northwest and low in the southeast. From northwest to southeast are situated Bashang Plateau, Yanshan Taihang Mountain and the southeast plain. Among them, the southeast plain can be divided into five types, namely, alluvial proluvial fan, flood plain, yellow flood plain, alluvial marine plain and marine plain. The ground elevation of Bashang

Plateau, Yanshan Taihang Mountain, southeast plain, and the coastal areas is 1300–1700 m, 500–1000 m, less than 200 m, and less than 4 m, respectively.

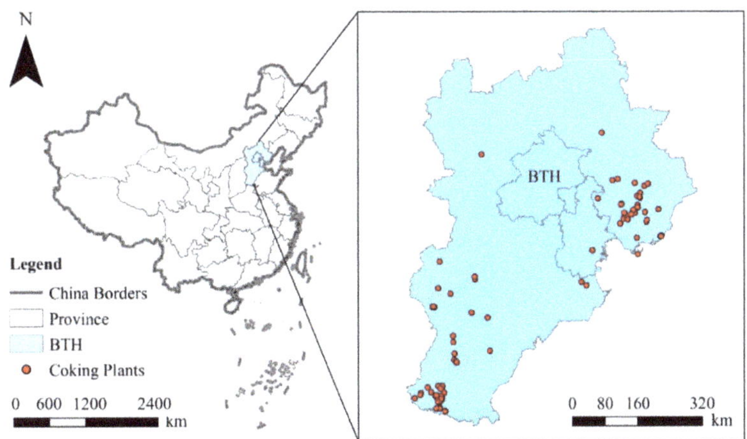

Figure 1. Location of coking plants in Beijing–Tianjin–Hebei Region, China.

The BTH region is located in the Haihe River Basin. The surface water system is well-developed and includes Yongding River, Daqing River, Ziya River, South Canal, Luanhe River and other rivers. BTH belong to the Yanliao stratigraphic division and North China Plain stratigraphic division of the Shanxi Hebei Shandong Henan stratigraphic region. The Cenozoic strata are well-developed, and most of the surface is Quaternary. The bedrock strata under the Quaternary system are basically complete, lacking a Neoproterozoic Sinian system, Paleozoic Silurian system and Devonian system. In the Cenozoic era, strong fault depressions and depressions were produced, and huge thick Cenozoic deposits were deposited.

The soil types of different geomorphic units are quite different. The main soil types in the Bashang plateau are calcium layer soil. The mountain area is mainly distributed with eluvial soil, semi eluvial soil, calcium layer soil and primary soil. The piedmont alluvial proluvial fan area is mainly semi eluvial soil. The flood plain and yellow flood plain are mainly distributed with semi hydrous soil. The marine plain is mainly developed with saline alkali soil.

2.2. Data Collection

2.2.1. Data of Coking Enterprises and Their PHE Emissions

The data of the coking enterprises were mainly from two sources: one is the list of coking companies in production in BTH, and the other is the pollution emissions by coking companies in BTH. Coking enterprises in the current study included both coking enterprises and coking section of iron and steel industry.

The list of coking plants in production in BTH was obtained through POI data from Baidu map, which is the largest map search engine in China. The searching keyword including coking plant and steelworks. To ensure the accuracy of the data, we filtered the companies that meet the access conditions of the coking industry published by the Ministry of Industry and Information Technology of China from the companies list. A total of 80 coking plants were identified in BTH.

The data of pollution emission from each coking enterprise were obtained through the China Industrial Enterprise Pollution Emission Database [35]. This database was a unique and comprehensive firm-level database, established on the basis of Statistical Report Statistics of Industrial Enterprises above Designated Size by the National Bureau of Statistics in China [1]. This database provides information on the industrial output value, energy input and pollution emission of enterprises. The emission list of databases included the locations, gross industrial output value, coal consumption, emissions of particulate matter, and exhaust gas emissions of coking plants in production. The data were independently reported by polluting enterprises, counted by environmental protection departments, and finally monitored and irregularly inspected by local environmental protection departments at the county level to ensure data quality.

PHE emissions were calculated by combining coal consumption data with specific emission factors grouped by different boiler patterns and various air pollutant control device configurations [6,11,13], as shown in Table S1 (Supplementary Sile). The combustion emission factor data of As, Cd, Cr, Cu, Hg, Ni, Pb, and Zn of the coking plants were 1.64, 0.10, 3.60, 1.71, 0.17, 0.94, 2.22, and 7.55 $\mu g \cdot kg^{-1}$ [12].

2.2.2. Coal Consumption by Enterprises in BTH

The coal consumption data of each enterprise were also obtained through the annual coal-burning data published by the China Industrial Enterprise Pollution Emission Database [35].

2.2.3. Soil Environmental Quality Data in BTH

To understand the soil environmental status of the BTH region, data of the soil PHE concentrations in this region were collected from literature published from 2010 to 2020. Two databases were searched: the China Knowledge Resource Integrated Database (the main database for Chinese research) and Web of Science. The literature selection criteria were as follows: (1) the original pollutant concentrations and sampling locations could be obtained; (2) the sampling and analysis methods of pollutants should be of international or domestic standards to facilitate the comparison of results. The studied pollutants include arsenic (As), cadmium (Cd), zinc (Zn), chromium (Cr), mercury (Hg), copper (Cu), nickel (Ni), and lead (Pb). Data were obtained from about 60 sampling locations. Representative references are listed here [36–41].

2.3. Data Analysis

2.3.1. PHE Emissions from Coal-Consuming Enterprises in BTH

PHEs emitted by coal consuming enterprises and coking industries in BTH were calculated based on the amount of coal, the content of PHEs in coal, and the release rates (Equations (1) and (2)). Proportion of PHEs discharged from coking plants in BTH was calculated by Equation (3).

$$T_{i,\,j,\,\text{BTH}} = A_{j,\,\text{BTH}} \times C_{i,\,\text{BTH}} \times r_i \tag{1}$$

$$T_{i,\,j,\,\text{Coke}} = A_{j,\,\text{Coke}} \times E_{i,\,\text{Coke}} \tag{2}$$

$$CR_{i,\,j} = T_{i,\,j,\,\text{Coke}} / T_{i,\,j,\,\text{BTH}} \times 100\% \tag{3}$$

where $T_{i,\,j,\,\text{BTH}}$ is the total emission of PHE i in city j in BTH, mg; $A_{j,\,\text{BTH}}$ is the coal consumption of city j in BTH, kg; $C_{i,\,\text{BTH}}$ is the content of PHE i in coal in BTH, $\mu g \cdot g^{-1}$; r_i is the release rate of PHE i from coal-consuming enterprises in BTH, %, shown in Table S1 (Supplementary Sile); $T_{i,\,j,\,\text{Coke}}$ is the emission of PHE i from coking plants in city j in BTH, mg; $A_{j,\,\text{Coke}}$ is the coal consumption of coking plants in city j in BTH, kg, obtained from China Industrial Enterprise Pollution Emission Database; $E_{i,\,\text{Coke}}$ is the emission factor of PHE i during the coking process, being 1.64, 0.10, 3.60, 1.71, 0.17, 0.94, 2.22, and 7.55 $\mu g \cdot kg^{-1}$ for As, Cd, Cr, Cu, Hg, Ni, Pb, and Zn, respectively (USEPA (2008); $CR_{i,\,j}$ is the contribution rate of coking to all the coal-consuming enterprises, in terms of PHE i in city j in BTH [42].

2.3.2. Relationship between PHEs Discharged from Coking Plants with the Soil PHE Concentrations in BTH

The Inverse Distance Weighted spatial interpolation tool of ArcGIS 10.2 was used to interpolate the data of soil environmental quality collected from the literature, obtaining the spatial distribution map of PHEs. Furthermore, using the value extraction to point extraction tool of ArcGIS, the interpolated PHE content corresponding to the location of those 80 coking enterprises is obtained. The correlation between the soil PHE concentrations in coking soil and the PHEs discharged from coking plants is analyzed using the cross-table analysis method of SPSS (SPSS 22.0).

2.3.3. Soil Pollutant Accumulation from Coking Emissions

The soil pollutant accumulation model (Equations (4)–(7)) was used to calculate PHEs deposited on soil surrounding the coking plants [43].

$$M_{i,\,\text{emission}} = M_{\text{coal}} \times F_i \times 10^{-3} \tag{4}$$

$$W_0 = M_{i,\,\text{emission}} / V_{\text{gas}} \tag{5}$$

$$R = W_0 \times V \times t / M \tag{6}$$

$$W_n = RK1 - K^n / (1 - K) + C \tag{7}$$

where W_n is the content of PHEs in the soil after n-year production of coking, mg·kg^{-1}; $M_{i,\,\text{emission}}$ is the total amount of PHEs i that were discharged by the coking plants, mg; M_{coal} is the coal consumption of the coking plant, kg·year^{-1}; F_i is the emission factor of PHE i during the coking process, μg·kg^{-1}; W_0 is the maximum concentration of pollutants on the ground, mg·m^{-3}; V_{gas} is the annual exhaust gas emitted from coking plants, m^3; R is the annual input content of pollutants, mg·kg^{-1}; V is the settlement rate, m·s^{-1}; t is the annual settlement time, s; M is the mass of soil per square meter, kg/m^2; n is the years of coking plants; K is the annual residual rate of PHEs in the soil, %; C is the background value of PHE content in BTH, mg·kg^{-1}, obtained from a recently published national investigation [43]. The value of soil PHEs' residue rate refers to the mentioned residue standard of 90%, the value of sedimentation rate is $0.001\ \text{m·s}^{-1}$, and the value of soil mass is 169 kg [44].

2.3.4. Calculation of Potential Ecological Risk

The potential ecological risk index method [27,45] was used to evaluate the potential ecological risks of soil PHEs obtained through sedimentation from coking waste gas in BTH (Equations (8) and (9)).

$$E_i = Tr_i \cdot C_i / S_i \tag{8}$$

$$RI = \sum_{i=1}^{n} E_i \tag{9}$$

where E_i is the potential ecological risk index of a single PHE i; Tr_i is the toxicity coefficient of PHE i, and the coefficients of As, Cd, Cr, Cu, Hg, Ni, Pb, and Zn are 10, 30, 2, 5, 40, 5, 5, and 1, respectively [45–47]; C_i is the concentration of a single PHE i, mg·kg^{-1}; S_i is a reference for a single PHE i, mg·kg^{-1}, which was the background value of PHEs in BTH; RI is the comprehensive potential ecological risk index of PHEs. Risk classification standards refer to published studies [45,48] (Table S2, Supplementary File).

2.3.5. Prediction of Ecological Risk Caused by Coking Plants

The accumulation of As, Cd, Cr, Cu, Hg, Ni, Pb, and Zn in the soil around the coking plants after 10, 20, 50, and 100 years of production was calculated [49], and compared with the thresholds of 30, 0.3, 200, 100, 2.4, 100, 120, and 250 specified by GB15618-2018 in China [50]. It was assumed that no soil remediation or discharge control measures were taken in this area, thereby remaining the same metals discharge and deposition ratio. The ecological risks posed by coking plants in BTH was calculated using Equations (8) and (9), based on the predicted PHE concentrations in the soil.

3. Results

3.1. Discharge Patterns of PHEs from Coking Plants

There are ~90 coking enterprises in the BTH region, mainly located in southeast regions, namely, Tangshan city and Handan city of Hebei Province (Figure 1). These two cities have about 30 coking enterprises in production, accounting for 67% of the coking enterprises in production in the BTH region.

The contribution rate of coking emissions of PHEs varied in different cities, and the average rate reached 7.73% (Table S3, Supplementary File). Among all the cities where coking plants were located, the order of contribution rate of the coking industry to the total PHE release was Xingtai (XT) > Tangshan (TS) > Handan (HD) > Baoding (BD) > Zhangjiakou (ZJK) > Chengde (CD) > Shijiazhuang (SJZ) > Tianjin (TJ) > Qinhuangdao (QHD). The contribution rates of PHEs in XT reached approximately 17%, and these rates were significantly higher than those of other cities. The contribution rates of coking plants to PHE emissions were higher in the southeast plain cities than in the northwest mountainous region.

Regarding the contribution rate of a single PHE in a single city, the order of contribution rate of each PHE from high to low was Hg > As > Cd > Ni > Zn > Pb > Cr > Cu (Figure 2). The contribution rate of Hg in TS and XT exceeded 50%. In addition, the contribution rates varied among different elements and in different regions. Hg, As, and Cd showed a generally higher contribution rate from coking than other elements. The contribution rates of As and Cd in TS and XT nearly exceeded 20%.

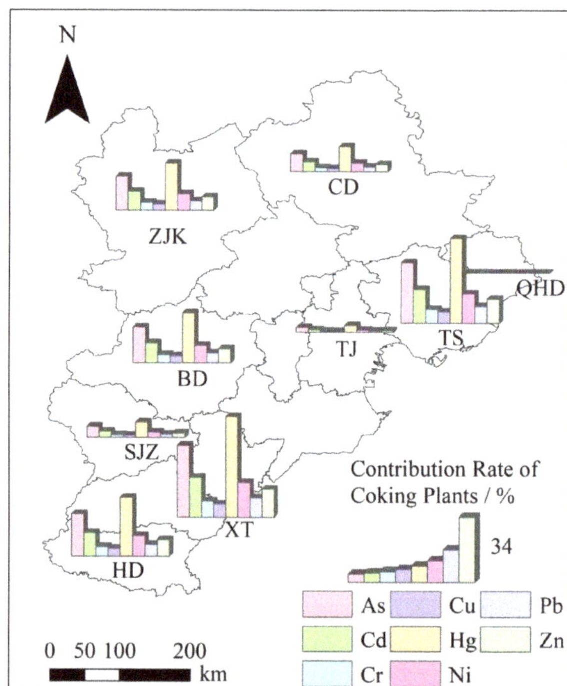

Figure 2. Contribution rate of coking plants to the PHE discharge in BTH. ZJK indicates Zhangjiakou, CD indicates Chengde, BD indicates Baoding, TJ indicates Tianjin, TS indicates Tangshan, QHD indicates Qinhuangdao, SJZ indicates Shijiazhuang, XT indicates Xingtai, and HD indicates Handan.

3.2. Relationship between Soil PHE Concentration and PHE Discharge from Coking Plants in BTH

The overall soil environmental quality in the BTH region is relatively good, only As, Cr, Hg and Zn indicated a certain degree of over-screening ratio (Table 1).

Table 1. Statistics of PHE content in the BTH under existing research (mg·kg^{-1}).

PHE	Max (mg·kg^{-1})	Min (mg·kg^{-1})	Screening Value (mg·kg^{-1}) *	Over-Screening-Value Ratio (%)
As	27.6	4.70	20	4.55
Cd	1.45	0.06	20	0.00
Cr	6929	14.6	200	1.52
Cu	85.0	12.5	2000	0.00
Hg	1.62	0.01	8	2.38
Ni	69.0	12.0	150	0.00
Pb	117	2.15	400	0.00
Zn	1670	30.0	250	6.25

* Screening value refers to Soil environmental quality—Risk control standard for soil contamination of develop land (GB 36600-2018).

The accumulation of PHEs in the southeast plain of BTH also presents certain local characteristics, in accordance with the distribution pattern of population and industry in this area. According to literature analysis, soil PHE pollution in the BTH region mainly occurs in the southeast plain, with no apparent soil PHE pollution in Yanshan Mountain in the north and Taihang Mountain in the west (Figure 3). Among them, the areas with high contents of As, Cd and Hg are mainly distributed along the coast of Bohai Bay. The contents of Cr, Zn and Cu are higher in Shijiazhuang, Xingtai and Handan in the south of Hebei Province, and the contents of Cu, Ni and Pb are higher in Tangshan and Qinhuangdao. Hg and Ni are also concentrated around Langfang city. Overall, the PHE pollution extent in the BTH region is low, showing a certain degree of local aggregation (Figure 3).

According to Table 2, the sig value of eight PHEs in soil quality corresponding to PHEs discharged by coking enterprises is greater than 0.05, and the phi value and V value are greater than 0.1, indicating that there is a close relationship between PHEs discharged by coking enterprises and the soil PHE concentration.

Table 2. Crosstabs analysis between PHEs discharged by coking and air quality.

PHE in Soil	PHE Released by Coking Plant		
	Sig.	Phi Value	V Value
As	0.252	8.367	1.000
Cd	0.364	4.243	1.000
Cr	0.252	8.367	1.000
Cu	0.367	4.123	1.000
Hg	0.367	4.123	1.000
Ni	0.252	8.367	1.000
Pb	0.283	7.141	1.000
Zn	0.367	4.123	1.000

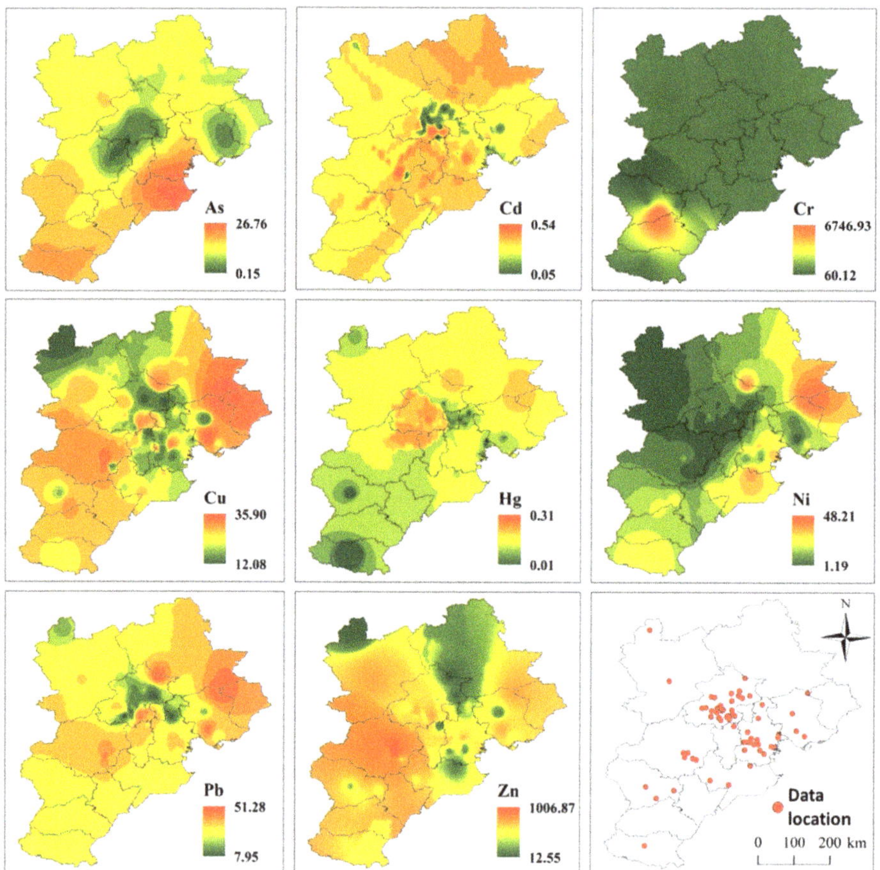

Figure 3. Distribution of PHE concentrations in BTH region (data were from published references).

3.3. PHE Accumulation in Soil from Coking Emissions and Depositions, and Its Associated Ecological Risk

Excluding other pollution sources in the soil, the scenario simulation of PHEs in the soil around coking enterprises was carried out. Through the soil accumulation model, the distribution of soil PHEs in the BTH region after the discharge and deposition of PHEs from coking enterprises were calculated and depicted on a map (Figure 4), aiming to provide a general idea of PHE discharge on soil PHE accumulation.

The spatial distribution of PHEs in coking soil due to the deposition of PHEs discharged from coking waste gas at a regional scale of BTH indicated that the area with high PHEs in coking soil is located in the southeast and northeast of the BTH area (Figure 4), in accordance with the locations of coking plants in Figure 1.

The ecological risks of soil PHEs resulting from coking were further assessed to provide suggestions for management in BTH (Figures 5 and 6). Table 3 shows the potential ecological risk index E_i of a single PHE and comprehensive ecological risk index RI of multiple PHEs after the production period. The order of coefficient of variation of E_i (CV) was Hg > Cd > As > Pb > Ni > Cu > Zn > Cr (Table 3). Strong significant spatial variation of ecological risk was observed, especially for Hg, Cd, and As (Figure 6).

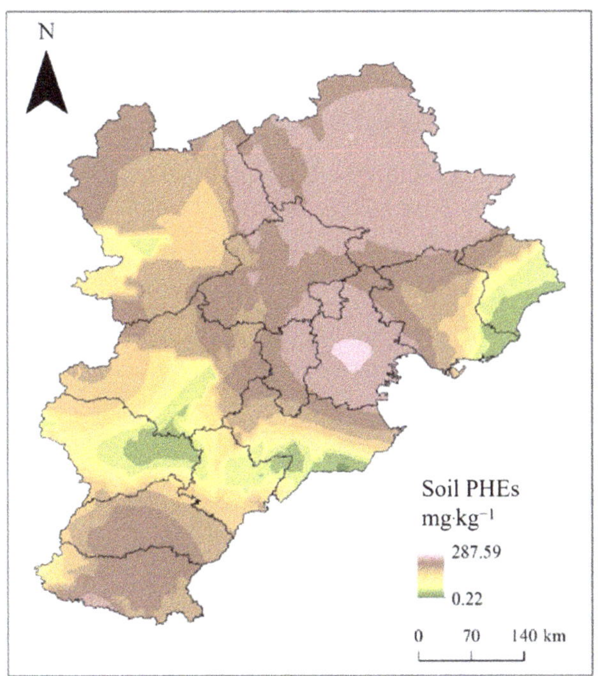

Figure 4. Sum of the eight PHEs after the discharge and deposition of PHEs from coking enterprises.

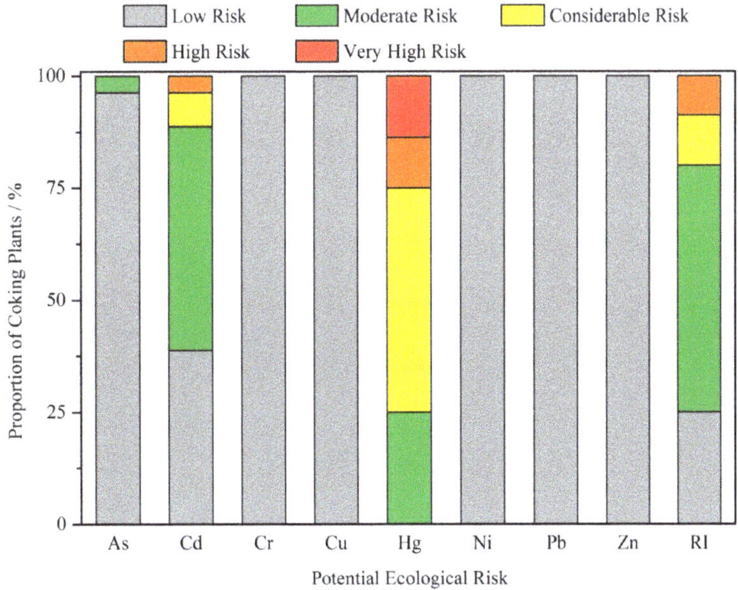

Figure 5. Proportion of coking plants causing ecological risk to the surrounding soil.

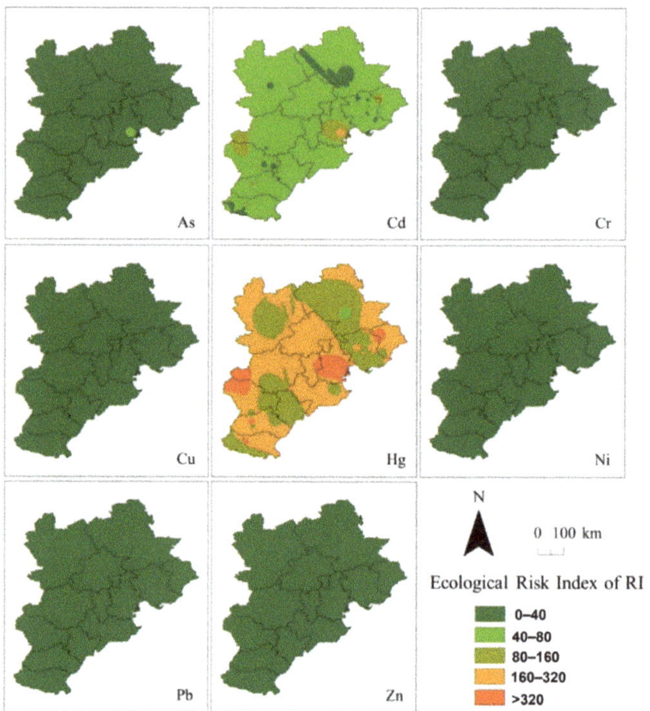

Figure 6. Distribution of RI index of coking plants causing potential ecological risk to the surrounding soil for each single PHE.

Table 3. Descriptive Statistics of Cumulative Ei and RI of Coking Plants in BTH (N = 80).

Index of Risk	Maximum	Minimum	Mean	Standard	CV *
As-Ei	52.29	10.03	15.30	7.88	51.48
Cd-Ei	209.57	30.14	52.51	33.44	63.69
Cr-Ei	2.50	2.00	2.06	0.09	4.52
Cu-Ei	11.03	5.00	5.76	1.12	19.51
Hg-Ei	1199.75	40.88	185.36	216.00	116.53
Ni-Ei	12.69	5.01	5.96	1.43	24.00
Pb-Ei	17.21	5.01	6.53	2.27	34.82
Zn-Ei	2.52	1.00	1.19	0.28	23.82
RI	1507.55	99.07	274.68	262.53	95.58

* CV is the abbreviation for coefficient of variation.

Figure 5 shows the comparative analysis results of potential ecological risk index E_i of PHEs and comprehensive ecological risk index RI of multiple PHEs. Hg, Cd, and As showed apparent ecological risk, while other PHEs had slight risk. In terms of the comprehensive potential risk, coking plants with high risk level accounted for 8.75% of the total coking plants, while coking plants with low risk level only accounted for around 25.00%. For As, 3.75% of the coking plants reached the moderate risk level, while the risk level of the remaining plants was low. For Cd, only 38.75% of the coking plants had low risk level, while 3.75% of coking plants reached high risk. For Hg, all enterprises reached a certain level of risk: 50% of the coking plants reached the moderate risk level, 11.25% of the coking plants reached the high-risk level, and 13.75% of the coking plants reached the very high-risk level.

In terms of the spatial distribution of comprehensive ecological risks (Figures 6 and 7), some cities in the northern areas were at a lower risk level, while the remaining cities affected by PHE pollution emissions from coking plants were at a moderate risk or above. The results suggested higher risks from Hg, Cd, and As than other elements. One of the reasons is that the three elements had a higher emission than other metals (Figure 2). The higher toxicity of the three elements could also contribute to their high risks.

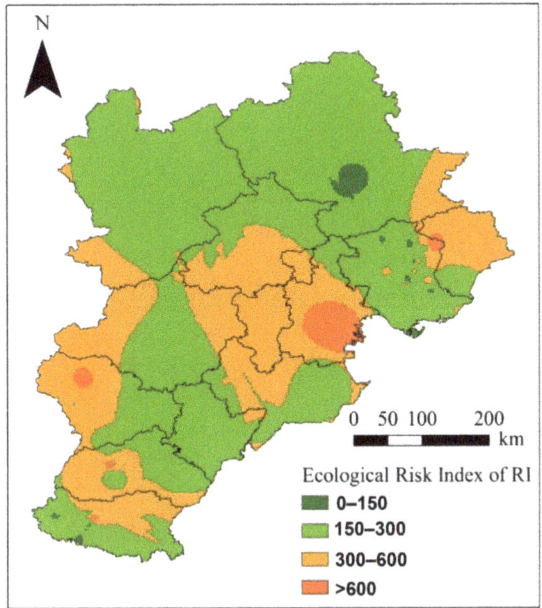

Figure 7. Distribution of RI index of coking plants causing potential ecological risk to the surrounding soil for all the PHEs.

3.4. Prediction of the Potential Ecological Risk in the Future by Scenario Analysis

Results indicated that after 10, 20, 50, and 100 years of production, some PHEs will exceed the threshold value after a certain period (Figure 8). As, Cd, and Zn will cause the surrounding soil to exceed the threshold in most coking plants. Hg, Ni, and Pb will also cause the surrounding soil to exceed the threshold in a small number of coking plants. The potential ecological risk index RI of the corresponding year and the proportion of coking plants of each risk level in different years were calculated on the basis of the simulated soil PHE content after years of coking production.

As the production time progressed, the proportion of coking plants with low and moderate risks gradually decreased, and the proportion of coking plants with considerable and high risks gradually increased. This finding indicates that the degree of risk continued to increase as production time progressed, and the trend of changes will be more serious. Notably, the role of considerable and moderate risks kept increasing as production time progressed. This result indicates that the coking plants presenting high risks should be under strict control.

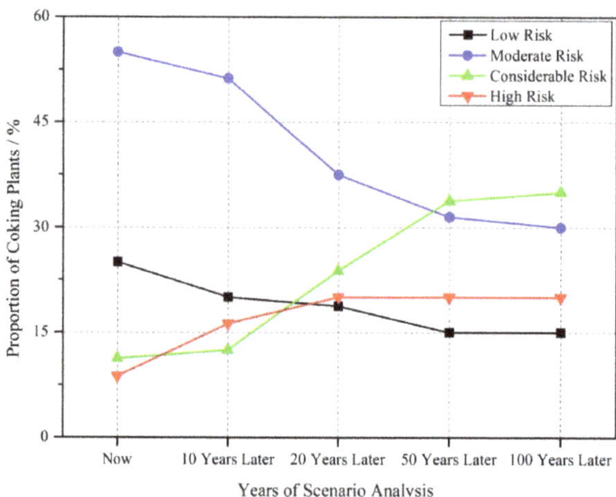

Figure 8. Scenario analysis of potential ecological risk of each coking plant in BTH.

4. Discussion

4.1. Coking Was an Important Source for the PHE Discharge in BTH

This report is the first to quantify the discharge pattern of PHEs from coking plants, and their contribution to the total PHE discharge in different districts of the BTH region. Despite the general idea that coking could be an important source for PHE discharge in BTH, there has been no quantitative analysis on the contribution rate. Through the current study, it was disclosed that coking plants also played an important role in discharging PHEs into the environment, other than discharging organic pollutants [51].

The contribution rate showed an apparent spatial heterogeneity (Figure 2), which might be related to the geomorphological characteristics of the BTH regions. The BTH region is located at the junction of coastal and inland areas in the middle latitude. The terrain is high in the northwest and low in the southeast. It transits from the northwest "Yanshan Taihang Mountain System" structure to the southeast plain. A considerably higher number of coking plants were located in the southeast region than other regions (Figure 1). Another report has also identified that pollution-intensive industries mainly concentrated in the southeast regions of BTH [52].

Another reason may come from the different categories of the main coal-consuming enterprises in different cities. The key coal consuming enterprises in BTH region included thermal power and coking and steel industries. The thermal power industries in TJ, SJZ, and TS consumed a considerably higher amount of coal than that in other cities. By contrast, the coking and steel industry in TS and HD consumed the highest amount of coal among all the cities in BTH. Due to the more intensive distribution of coking plants in XT, HD, and TS, the contribution rates of coking plants to the total PHE discharge in this area were higher.

In accordance with the current study, an investigation on the concentration of PHEs in $PM_{2.5}$ in BTH also found that high emissions of PHEs are mainly distributed in the central and southern BTH region, especially for Beijing, Tianjin, SJZ, TS, HD and BD cities [53]. This may be attributed to the significant coal consumption of industrial boilers in the iron and steel industries, which are intensive and flourishing in XT, HD, and TS.

In terms of the difference among elements, Hg, As and Cd indicated a higher discharge ratio. Hg and As are comparatively volatile, which led to a higher contribution rate [54,55]. Results were in accordance with an earlier study, which also suggested that the east and south-central regions of BTH may have higher emissions of Hg, As, and Cd than other regions [6]. An earlier study also confirmed a considerable amount of atmospheric As, Pb, and Cd in BTH [56], indicating that the highest concentration of atmospheric As at the BD site was of 59 ng m^{-3}; this value was 10 times the WHO standard (6.6 ng m^{-3}). The value was lower in other cities of BTH, but it was still 2–3 times higher than the WHO limits. It has been suggested that other than the interprovincial pathways, the contribution of local coal burning and the iron industry to the pollution of aerosol trace elements in the region should be given more attention. Consideration should especially be given to the topographic features of the BTH region; anthropogenic emissions in BTH region were trapped in the boundary layer by the surrounding mountain ranges [57], which led to a higher deposition rate of atmospheric PHEs.

4.2. Coking Contributed to the Accumuation of PHEs in Soil

The meta-analysis on the PHE concentration in soil indicated that the overall soil environmental quality in BTH is good, with only As, Cd and Hg showing an accumulation along the coast of Bohai Bay. This was in accordance with the literature, suggesting the Tianjin coastal area was one of the high-risk regions in China [52]. The coastal zone is usually a diverse and complex system subjected to diversified anthropogenic activities, which is pronounced in the enrichment of aquatic soils with a number of metals from different types of sources that likely can be of both local and whole basin scale [29].

Comparing Figures 1–3, it can be found that areas with a high content of soil PHEs, such as the coast of Bohai Bay and the south of Hebei Plain, also have more coking enterprises. This implies that the PHEs discharged by coking enterprises in Beijing, Tianjin and Hebei have a certain impact on the soil quality of Beijing, Tianjin and Hebei. The cross-table analysis further confirmed the relationship between coking emission and soil PHE concentrations.

The potential relationship between soil PHE concentration and PHE discharge from coking plants in BTH has been indirectly implied in the literature. Although also dependent on other sources such as runoff or soil additives, the PHE concentrations of topsoil was contributed significantly by atmospheric deposition [15]. In terms of the sources of PHEs in atmospheric deposition, coal burning was one of the most essential sources in BTH region [58,59].

The contribution of coking to soil PHE pollution is not only restricted to the BTH area but also exists in other areas of China, and even the world. Similar to the current study, in several steel plant areas, it has been found that PHE pollution in soil was the most serious in the coke processing plant, mostly from coke processing [60,61]. These reports have found that Hg and Pb were two of the main pollutants contributing to soil PHE pollution in this area, posing both high concentration and high bioavailability. One reference indicated that in the vicinity of coking plants, the water-soluble fraction of PHEs such as As, Cd and Sb reached as high as 50%, posing significant ecological risk and human health risks [24]. Due to the atmospheric emissions and dusting from the surface of ash dumps, Ni, Cd, and Pb were accumulated in soils around a coal-fired power station in Southern Russia and caused carcinogenic risk to human health [17]. Around a coking plant in Southern Poland, higher amounts of PHEs were also detected in soil [18].

4.3. Ecological Risk Analysis and Its Environmental Implications

The risk assessment results indicate that coking plants with moderate to high risk accounted for ~75% of the total coking plants. Scenario analysis indicates that if no measures are taken, the degree of risk might continue to increase, implying that the potential pollution will be more serious. Additionally, the high risks mainly come from Hg, Cd, and As, due to their higher emission (Figure 2) and higher toxicity [62] than other elements. Another study assessed the potential ecological risks of PHEs in topsoil of Shijiazhuang city in the BTH area, which showed that high concentrations of Cr, Cd and Ni distributed in the northeast of the urban area was related to the existence of steel and coking plants in that area [63]. They also spotted widespread contamination of Cd and Hg in the Shijiazhuang urban area, posing significant ecological risk. This similar trend between the current study and the study of Zou et al. [63] confirmed the non-negligible contribution of coking to the accumulation of PHE in soil. Furthermore, considering the potential interaction between

PHE and organic pollutants co-released by coking plants [64], the PHE contamination surrounding the coking industry in the BTH area should be paid more attention.

The North China Plain is the most populous plain in China and forms the core of the BTH economic circle. With urbanization, anthropogenic PHEs have increasingly dispersed and accumulated in urban topsoil. Despite the generally known contribution of the coking industry to PHE emissions, attention has been mainly paid to polycyclic aromatic hydrocarbons [31,65]. Through the current study, the contribution of coking plants to PHE emissions in BTH was quantified, which reminded the related governor or scientist that As and Cd from coking plants should be paid more attention, especially considering their wide distribution and persistence in soil [66]. Furthermore, diversified control strategies should be applied to areas with different risk levels considering the apparent heterogeneity of risk levels in different cities of BTH. The emission and pollution control of coking plants in the southeast and northeast regions of BTH needs to be strengthened.

4.4. Uncertainty Analysis

There are limitations of the current study. This study only considered the PHE emissions of coal-consuming enterprises above a designated size in BTH. As a result, the total PHE emissions may be lower than the actual emissions. In addition, this study selected the coking emission factors based on EPA standards, given that no emission factors for separate elements are available in China. The emission factors may have differences due to the different production processes between the two countries [3]. The accumulation of PHEs and their potential ecological risks were determined by multiple factors [16].

When evaluating the correlation between soil PHE content and PHE discharge from coking plants, we only obtained 60 detailed sampling locations from the literature, therefore the spatial interpolation results may not be very accurate. Furthermore, the interpolation results were extracted from 80 coking enterprises to obtain the soil PHE concentration of coking plants. This data calculation could lead to inaccuracy.

The risk assessment of PHEs in soil was conducted under the assumption that PHEs in coking exhaust gas were all deposited in the location soil of each coking plant, without considering the transport of PHEs through dry or wet precipitation, and the associated spatial heterogeneity of PHE distribution. This might lead to the inaccurate evaluation of the spatial distribution of ecological risks. Further analysis on the temporal or spatial heterogeneity of the long-range transport of PHEs discharged from coking plants, especially [67], should be able to further improve our research. Evaluation of pollutant emissions and their interpretations from the aspects of spatial distribution is important for the proposal of pollution control measures [6].

During the prediction of risks caused by PHE emissions from coking plants in the future, the emission rate was assumed to be kept the same, and the soil also kept the status quo without remediation or a development plan. However, the truth of this case is yet to be seen. Therefore, the prediction results in the current study might not be accurate enough.

Despite these uncertainties, this report firstly quantified the contribution of the coking industry to the total PHE discharge in BTH through the extensive collection of pollutant discharge data and data calculation via the soil accumulation model and ecological risk index. Using the cross-table analysis, a close relationship was identified between the PHEs emitted by coking and the soil PHE concentration in the BTH region. Furthermore, at a regional scale, the potential ecological risk coefficient of PHEs discharged by coking was calculated to provide some feasible suggestions for the zoning of potential ecological risks of coking enterprises and the management of waste gas discharged by coking enterprises. This research presents a generalized pattern for the PHEs discharged from coking plants in BTH and its associated risk assessment and provides information for the regional pollution control. The results could, on the one hand, provide information for the diffusion pattern of coking pollutants, and, on the other hand, aid the government in controlling the potential risks caused by coking plants in BTH.

5. Conclusions

The contribution rates of coking plants to the total discharge pattern was firstly disclosed as being ~7.73%. Furthermore, this discharge of PHEs from the coking industry was found to be closely related to the accumulation of PHEs in soil. The discharge of Hg, As, and Cd from coking in BTH were relatively higher than those of other metals, which resulted in higher ecological risks of Hg, As, and Cd in this area. Other than the difference among metals, spatial heterogeneity should also be noticed. The eastern and central-southern parts of BTH showed high coking contribution rates. Most coking plants in eastern and central southern parts of BTH were at a moderate risk level. Scenario analysis also showed that the level of ecological risks would increase rapidly with an extending production time, especially for As, Cr, Ni, and Pb. Despite some uncertainties regarding the results, the regional-scale assessment of risks caused by coking plants could produce some suggestions for the government to better control the potential pollution due to coke production in BTH.

Supplementary Materials: The following supporting information can be downloaded at: https://www.mdpi.com/article/10.3390/toxics10050240/s1, Table S1: Averaged Concentrations ($\mu g \cdot g^{-1}$) and Emission Rates (%) of PHEs from Coals in the BTH, Table S2: Classification of Potential Ecological Risk Index, Table S3: Contribution Rate of Coking Plants to the Total PHEs Emissions in BTH.

Author Contributions: X.W.: Conceptualization, Methodology, Funding acquisition. W.Z.: Investigation, Methodology. G.G.: Writing—Original draft preparation, Formal analysis, Visualization. L.W.: Draft revision. M.L.: Supervision. All authors have read and agreed to the published version of the manuscript.

Funding: This research was funded by the National Key Research and Development Program of China (Grant No. 2018YFC1800302).

Institutional Review Board Statement: Not applicable.

Informed Consent Statement: Not applicable.

Data Availability Statement: Data available on request from the authors.

Conflicts of Interest: The authors declare no conflict of interest.

References

1. National Bureau of Statistics in China (NBSC). *Chinese Statistical Yearbook*; Chinese Statistics Press: Beijing, China, 2020. Available online: http://www.stats.gov.cn/tjsj/ndsj/2020/indexch.htm (accessed on 10 April 2021).
2. National Bureau of Statistics in China (NBSC). *Chinese Statistical Yearbook*; Chinese Statistics Press: Beijing, China, 2019. Available online: http://www.stats.gov.cn/tjsj/ndsj/2019/indexch.htm (accessed on 10 April 2021).
3. Mu, L.; Peng, L.; Liu, X.F.; Bai, H.L.; Song, C.F.; Wang, Y.; Li, Z. Emission characteristics of heavy metals and their behavior during coking processes. *Environ. Sci. Technol.* **2012**, *46*, 6425–6430. [CrossRef] [PubMed]
4. Qiao, P.W.; Dong, N.; Yang, S.C.; Gou, Y.L. Quantitative analysis of the main sources of pollutants in the soils around key areas based on the positive matrix factorization method. *Environ. Pollut.* **2021**, *273*, 116518. [CrossRef] [PubMed]
5. Zhai, C.; Fan, W.Y.; Li, M.Z.; Mao, X.G.; Bao, G.D. Surface soil geochemical quality evaluation based on the space analysis of GIS. In Proceedings of the World Automation Congress (WAC), Puerto Vallarta, Mexico, 24–28 June 2012.
6. Zhu, C.Y.; Tian, H.Z.; Hao, Y.; Gao, J.J.; Hao, J.M.; Wang, Y.; Hua, S.B.; Wang, K.; Liu, H.J. A high-resolution emission inventory of anthropogenic trace elements in Beijing-Tianjin-Hebei (BTH) region of China. *Atmos. Environ.* **2018**, *191*, 452–462. [CrossRef]
7. Fan, R.F.; Li, J.N.; Chen, L.G.; Xu, Z.C.; He, D.C.; Zhou, Y.X.; Zhu, Y.Y.; Wei, F.S.; Li, J.H. Biomass fuels and coke plants are important sources of human exposure to polycyclic aromatic hydrocarbons, benzene and toluene. *Environ. Res.* **2014**, *135*, 1–8. [CrossRef] [PubMed]
8. Wang, P.; Mi, W.Y.; Xie, Z.Y.; Tang, J.H.; Apel, C.; Joerss, H.; Ebinghaus, R.; Zhang, Q.H. Overall comparison and source identification of PAHs in the sediments of European Baltic and North Seas, Chinese Bohai and Yellow Seas. *Sci. Total Environ.* **2020**, *737*, 139535. [CrossRef]
9. Chen, J.; Zhang, B.F.; Zhang, S.; Zeng, J.; Chen, P.; Liu, W.Z.; Wang, X.M. A complete atmospheric emission inventory of F, As, Se, Cd, Sb, Hg, Pb, and U from coal-fired power plants in Anhui Province, eastern China. *Environ. Geochem. Health* **2020**, *43*, 1817–1837. [CrossRef]
10. Saber, A.Z.; Zhang, H.F.; Islam, A.; Yang, M. Occurrence, fates, and carcinogenic risks of substituted polycyclic aromatic hydrocarbons in two coking wastewater treatment systems. *Sci. Total Environ.* **2021**, *789*, 147808. [CrossRef]

11. Tian, H.Z.; Zhu, C.Y.; Gao, J.J.; Cheng, K.; Hao, J.M.; Wang, K.; Hua, S.B.; Wang, Y.; Zhou, J.R. Quantitative assessment of atmospheric emissions of toxic heavy metals from anthropogenic sources in China: Historical trend, spatial distribution, uncertainties, and control policies. *Atmos. Chem. Phys.* **2015**, *15*, 12107–12166. [CrossRef]

12. USEPA. *Compilation of Air Emissions Factors*; United States Environmental Protection Agency (US, EPA): Research Triangle Park, NC, USA, 2008. Available online: https://www.epa.gov/air-emissions-factors-and-quantification/ap-42-compilation-air-emissions-factors (accessed on 15 May 2021).

13. Zhu, C.Y.; Tian, H.Z.; Cheng, K.; Liu, K.J.; Wang, K.; Hua, S.B.; Gao, J.J.; Zhou, J.R. Potentials of whole process control of heavy metals emissions from coal-fired power plants in China. *J. Clean. Prod.* **2016**, *114*, 343–351. [CrossRef]

14. China SEPA. Ministry of Ecology and Environment of the People's Republic of China. In *2013 China Statistical Yearbook China*; Environmental Science Press: Beijing, China, 2013.

15. Li, Y.; Zhou, S.L.; Jia, Z.Y.; Liu, K.; Wang, G.M. Temporal and spatial distributions and sources of heavy metals in atmospheric deposition in western Taihu Lake, China. *Environ. Pollut.* **2021**, *284*, 117465. [CrossRef]

16. Liu, H.L.; Zhou, J.; Li, M.; Hu, Y.-M.; Liu, X.; Zhou, J. Study of the bioavailability of heavy metals from atmospheric deposition on the soil-pakchoi (*Brassica chinensis* L.) system. *J. Hazard. Mater.* **2019**, *362*, 9–16. [CrossRef] [PubMed]

17. Minkina, T.; Konstantinova, E.; Bauer, T.; Mandzhieva, S.; Sushkova, S.; Chaplygin, V.; Burachevskaya, M.; Nazarenko, O.; Kizilkaya, R.; Gülser, C.; et al. Environmental and human health risk assessment of potentially toxic elements in soils around the largest coal-fired power station in Southern Russia. *Environ. Geochem. Health* **2021**, *43*, 2285–2300. [CrossRef] [PubMed]

18. Rachwał, M.; Magiera, T.; Wawer, M. Coke industry and steel metallurgy as the source of soil contamination by technogenic magnetic particles, heavy metals and polycyclic aromatic hydrocarbons. *Chemosphere* **2015**, *138*, 863–873. [CrossRef] [PubMed]

19. Dhaliwal, S.S.; Singh, J.; Taneja, P.K.; Mandal, A. Remediation techniques for removal of heavy metals from the soil contaminated through different sources: A review. *Environ. Sci. Pollut. Res.* **2020**, *27*, 1319–1333. [CrossRef] [PubMed]

20. Zhang, H.; Yao, Q.S.; Zhu, Y.M.; Fan, S.S.; He, P.J. Review of source identification methodologies for heavy metals in solid waste. *Chin. Sci. Bull.* **2013**, *58*, 162–168. [CrossRef]

21. Zhao, S.W.; Xu, W.; Zong, S.; Yao, L.Y.; Liu, J.L.; Lu, F.; Yi, Y.Q.; Shi, Y.G.; Shi, Z.Y.; Zhang, B.; et al. The Pollution Characteristics and Full-Scale Risk Assessment of Coking Plant Soil Co-Contaminated with Polycyclic Aromatic Hydrocarbons and Heavy Metals. *Environ. Eng. Sci.* **2021**, *38*, 867–876. [CrossRef]

22. He, J.J.; Wei, C.Y.; Feng, C.H.; Ren, Y.; Hu, Y.; Wu, C.F. States distribution of heavy metals in coking wastewater sludge and its potential environmental risks. *Chin. J. Environ. Eng.* **2013**, *7*, 79–85.

23. Wang, S.Y.; Kalkhajeh, Y.K.; Qin, Z.R.; Jiao, W.T. Spatial distribution and assessment of the human health risks of heavy metals in a retired petrochemical industrial area, south China. *Environ. Res.* **2020**, *188*, 109661. [CrossRef]

24. Zajusz-Zubek, E.; Radko, T.; Mainka, A. Fractionation of trace elements and human health risk of submicron particulate matter (PM$_1$) collected in the surroundings of coking plants. *Environ. Monit. Assess.* **2017**, *189*, 389. [CrossRef]

25. Hou, W.; Zhang, L.; Li, S.X.; Wu, D.; Li, X.J.; Ji, L.; Ma, X.P. Health risk assessment of heavy metals and polycyclic aromatic hydrocarbons in soil at coke oven gas plants. *Environ. Eng. Manag. J.* **2015**, *14*, 487–496.

26. Bai, J.; Zhang, W.; Liu, W.; Xiang, G.; Zheng, Y.; Zhang, X.; Yang, Z.; Sushkova, S.; Minkina, T.; Duan, R. Implications of Soil Potentially Toxic Elements Contamination, Distribution and Health Risk at Hunan's Xikuangshan Mine. *Processes* **2021**, *9*, 1532. [CrossRef]

27. Xiang, M.T.; Li, Y.; Yang, J.Y.; Lei, K.G.; Li, Y.; Li, F.; Zheng, D.F.; Fang, X.Q.; Cao, Y. Heavy metal contamination risk assessment and correlation analysis of heavy metal contents in soil and crops. *Environ. Pollut.* **2021**, *278*, 116911. [CrossRef] [PubMed]

28. Pandey, B.; Agrawal, M.; Singh, S. Assessment of air pollution around coal mining area: Emphasizing on spatial distributions, seasonal variations and heavy metals, using cluster and principal component analysis. *Atmos. Pollut. Res.* **2014**, *5*, 79–86. [CrossRef]

29. Konstantinova, E.; Minkina, T.; Nevidomskaya, D.; Mandzhieva, S.; Bauer, T.; Zamulina, I.; Burachevskaya, M.; Sushkova, S. Exchangeable form of potentially toxic elements in floodplain soils along the river-marine systems of Southern Russia. *Eurasian J. Soil Sci.* **2021**, *10*, 132–141. [CrossRef]

30. Mu, L.; Li, X.M.; Liu, X.F.; Bai, H.L.; Peng, L.; Li, Y.Y.; Tian, M.; Zheng, L.R. Characterization and emission factors of carbonaceous aerosols originating from coke production in China. *Environ. Pollut.* **2021**, *268*, 115768. [CrossRef]

31. Mu, L.; Peng, L.; Cao, J.; He, Q.; Li, F.; Zhang, J.; Liu, X.; Bai, H. Emissions of polycyclic aromatic hydrocarbons from coking industries in China. *Particuology* **2013**, *11*, 86–93. [CrossRef]

32. Qiao, P.W.; Lai, D.L.; Yang, S.C.; Zhao, Q.Y.; Wang, H.Q. Effectiveness of predicting the spatial distributions of target contaminants of a coking plant based on their related pollutants. *Environ. Sci. Pollut. Res.* **2022**. [CrossRef]

33. Ahado, S.K.; Nwaogu, C.; Sarkodie, V.Y.O.; Borůvka, L. Modeling and Assessing the Spatial and Vertical Distributions of Potentially Toxic Elements in Soil and How the Concentrations Differ. *Toxics* **2021**, *9*, 181. [CrossRef]

34. Zerizghi, T.; Guo, Q.; Tian, L.; Wei, R.; Zhao, C. An integrated approach to quantify ecological and human health risks of soil heavy metal contamination around coal mining area. *Sci. Total Environ.* **2022**, *814*, 152653. [CrossRef]

35. Lin, B.; Xu, C. Does environmental decentralization aggravate pollution emissions? Microscopic evidence from Chinese industrial enterprises. *Sci. Total Environ.* **2022**, *829*, 154640. [CrossRef]

36. Guo, H.; Yang, Z.; Li, H.; Ma, W.; Ren, J. Environmental quality and anthropogenic pollution assessment of heavy metals in topsoil of Hebei plain. *Geol. China* **2011**, *38*, 218–225.

37. Huo, X.-N.; Li, H.; Sun, D.-F.; Zhou, L.-D.; Li, B.-G. Combining geostatistics with Moran's I analysis for mapping soil heavy metals in Beijing, China. *Int. J. Environ. Res. Public Health* **2012**, *9*, 995–1017. [CrossRef] [PubMed]
38. Li, X.; Lu, A.; Wang, J.; Ma, Z.; Zhao, C. Analysis and assessment of soil environmental quality of some farmlands in Beijing. *Trans. Chin. Soc. Agric. Eng.* **2006**, *22*, 60–63.
39. Shi, R.; Zhang, Y.; Xu, M.; Zheng, X.; Zhao, Z. Pollution evaluation and source apportionment of heavy metals in soils from Tianjin suburbs, China. *J. Agro-Environ. Sci.* **2019**, *38*, 1069–1078.
40. Yang, L.Y.; Wu, S.T. Assessment of Soil Heavy Metal Cu, Zn and Cd Pollution in Beijing, China. *Adv. Mater. Res.* **2012**, *356–360*, 730–733. [CrossRef]
41. Zou, J.; Liu, X.; Dai, W.; Luan, Y. Pollution assessment of heavy metal accumulation in the farmland soils of Beijing's suburbs. *Environ. Sci. Pollut. Res.* **2018**, *25*, 27483–27492. [CrossRef]
42. Şenay, Ç.; Aykan, K.; Beyhan, P.; Savaş, A. Inventory of emissions of primary air pollutants in the city of Kocaeli, Turkey. *Environ. Monit. Assess.* **2007**, *128*, 165–175.
43. Hou, Q.Y.; Yang, Z.F.; Yu, T.; Xia, X.Q.; Cheng, H.X.; Zhou, G.H. *Soil Geochemical Parameters of China*; Geological Publishing House: Beijing, China, 2021.
44. Wang, X.R. *Environmental Chemistry*, 2nd ed.; Nanjing University Press: Nanjing, China, 1993.
45. Hakanson, L. An ecological risk index for aquatic pollution control—A sedimentological approach. *Water Res.* **1980**, *14*, 975–1001. [CrossRef]
46. Orellana, E.P.; Custodio, M.; Bastos, M.C.; Ascencion, J.C. Heavy Metals in Agriculture Soils from High Andean Zones and Potential Ecological Risk Assessment in Peru's Central Andes. *J. Ecol. Eng.* **2020**, *21*, 108–119. [CrossRef]
47. Nihal, G.; Latha, R.; Sudip, M. Occurrence, geochemical fraction, ecological and health risk assessment of cadmium, copper and nickel in soils contaminated with municipal solid wastes. *Chemosphere* **2021**, *271*, 129573.
48. Saleh, Y.S. Evaluation of sediment contamination in the Red Sea coastal area combining multiple pollution indices and multivariate statistical techniques. *Int. J. Sediment Res.* **2021**, *36*, 243–254. [CrossRef]
49. Jiang, X.; Lu, W.X.; Zhao, H.Q.; Yang, Q.C.; Yang, Z.P. Potential ecological risk assessment and prediction of soil heavy-metal pollution around coal gangue dump. *Nat. Hazards Earth Syst. Sci.* **2014**, *14*, 1599–1610. [CrossRef]
50. China SEPA. Ministry of Ecology and Environment of the People's Republic of China. In *Soil Environmental Quality, Risk Control Standard for Soil Contamination of Agricultural Land (GB15618-2018)*; China Environmental Science Press: Beijing, China, 2018.
51. Coralie, B.; Pierre, F.; Laurence, M.; Aurélie, C.; Thierry, B.; Corinne, L. Biodegradation of the organic matter in a coking plant soil and its main constituents. *Org. Geochem.* **2013**, *56*, 10–18.
52. Luo, Y.; Zhou, D.; Tian, Y.; Jiang, G. Spatial and temporal characteristics of different types of pollution-intensive industries in the Beijing-Tianjin-Hebei region in China by using land use data. *J. Clean. Prod.* **2021**, *329*, 129601. [CrossRef]
53. Liu, S.; Zhu, C.; Tian, H.; Wang, Y.; Zhang, K.; Wu, B.; Liu, X.; Hao, Y.; Liu, W.; Bai, X.; et al. Spatiotemporal Variations of Ambient Concentrations of Trace Elements in a Highly Polluted Region of China. *J. Geophys. Res. Atmos.* **2019**, *124*, 4186–4202. [CrossRef]
54. Wu, Q.R.; Gao, W.; Wang, S.X.; Hao, J.M. Updated atmospheric speciated mercury emissions from iron and steel production in China during 2000–2015. *Atmos. Chem. Phys.* **2017**, *17*, 10423–10433. [CrossRef]
55. Sun, X.F.; Zhang, L.X.; Lv, J.S. Spatial assessment models to evaluate human health risk associated to soil potentially toxic elements. *Environ. Pollut.* **2021**, *268*, 115699. [CrossRef]
56. Pan, Y.P.; Tian, S.L.; Li, X.R.; Sun, Y.; Li, Y.; Wentworth, G.R.; Wang, Y. Trace elements in particulate matter from metropolitan regions of Northern China: Sources, concentrations and size distributions. *Sci. Total Environ.* **2015**, *537*, 9–22. [CrossRef]
57. Streets, D.G.; Fu Joshua, S.; Jang Carey, J.; Hao, J.M.; He, K.B. Air quality during the 2008 Beijing Olympic Games. *Atmos. Environ.* **2008**, *41*, 480–492. [CrossRef]
58. Li, X.; Yan, C.; Wang, C.; Ma, J.; Li, W.; Liu, J.; Liu, Y. PM$_{2.5}$-bound elements in Hebei Province, China: Pollution levels, source apportionment and health risks. *Sci. Total Environ.* **2022**, *806*, 150440. [CrossRef]
59. Zhi, M.; Zhang, X.; Zhang, K.; Ussher, S.J.; Lv, W.; Li, J.; Gao, J.; Luo, Y.; Meng, F. The characteristics of atmospheric particles and metal elements during winter in Beijing: Size distribution, source analysis, and environmental risk assessment. *Ecotoxicol. Environ. Saf.* **2021**, *211*, 111937. [CrossRef] [PubMed]
60. Cui, X.W.; Geng, Y.; Sun, R.R.; Xie, M.; Feng, X.W.; Li, X.X.; Cui, Z.J. Distribution, speciation and ecological risk assessment of heavy metals in Jinan Iron & Steel Group soils from China. *J. Clean. Prod.* **2021**, *295*, 126504. [CrossRef]
61. Zhang, X.; Yang, H.H.; Sun, R.R.; Cui, M.H.; Sun, N.; Zhang, S.W. Evaluation and analysis of heavy metals in iron and steel industrial area. *Environ. Dev. Sustain.* **2021**. [CrossRef]
62. Kafka, Z.; Puncocharova, J. Toxicity of heavy metals in nature. *Chem. Listy* **2002**, *96*, 611–617.
63. Zou, J.Y.; Song, Z.F.; Cai, K.Z. Source apportionment of topsoil heavy metals and associated health and ecological risk assessments in a typical hazy city of the North China plain. *Sustainability* **2021**, *13*, 10046. [CrossRef]
64. Xie, Y.L.; Lin, T.; Yang, M.; Zhang, Z.R.; Deng, N.; Tang, M.Q.; Xiao, Y.M.; Guo, H.; Deng, Q.F. Co-exposure to polycyclic aromatic hydrocarbons and metals, four common polymorphisms in microRNA genes, and their gene-environment interactions: Influences on oxidative damage levels in Chinese coke oven workers. *Environ. Int.* **2019**, *132*, 105055. [CrossRef]
65. Zhang, Q.F.; Meng, J.; Su, G.J.; Liu, Z.L.; Shi, B.; Wang, T.Y. Source apportionment and risk assessment for polycyclic aromatic hydrocarbons in soils at a typical coking plant. *Ecotoxicol. Environ. Saf.* **2021**, *222*, 12. [CrossRef]

66. Jiang, H.; Wang, B.; Liu, X.; Zhang, W.; Li, H.; Sun, D. Early warning of heavy metals potantial risk governance in Beijing. *Acta Pedol. Sin.* **2015**, *52*, 731–746.
67. Eyrikh, S.; Shol, L.; Shinkaruk, E. Assessment of Mercury Concentrations and Fluxes Deposited from the Atmosphere on the Territory of the Yamal-Nenets Autonomous Area. *Atmosphere* **2022**, *13*, 37. [CrossRef]

Article

Influence of Urban Informal Settlements on Trace Element Accumulation in Road Dust and Their Possible Health Implications in Ekurhuleni Metropolitan Municipality, South Africa

Innocent Mugudamani [1,†], Saheed A. Oke [2,*,†] and Thandi Patricia Gumede [1,†]

[1] Department of Life Sciences, Central University of Technology, Bloemfontein 9301, Free State, South Africa; imugudamani@gmail.com (I.M.); TGumede@cut.ac.za (T.P.G.)

[2] Department of Civil Engineering, Centre for Sustainable Smart Cities, Central University of Technology, Bloemfontein 9301, Free State, South Africa

[*] Correspondence: soke@cut.ac.za or okesaheed@gmail.com

[†] These authors contributed equally to this work.

Abstract: The study was aimed at assessing the influence of urban informal settlement on trace element accumulation in road dust from the Ekurhuleni Metropolitan Municipality, South Africa, and their possible health implications. The concentration of major and trace elements was determined using the wavelength dispersive XRF method. The major elements in descending order were SiO_2 (72.76%), Al_2O_3 (6.90%), Fe_2O_3 (3.88%), CaO (2.71%), K_2O (1.56%), Na_2O (0.99%), MgO (0.94%), MnO (0.57%), TiO_2 (0.40%), and P_2O_5 (0.16%), with SiO_2 and P_2O_5 at above-average shale values. The average mean concentrations of 17 trace elements in decreasing order were Cr (637.4), Ba (625.6), Zn (231.8), Zr (190.2), Sr (120.2), V (69), Rb (66), Cu (61), Ni (49), Pb (30.8), Co (17.4), Y (14.4), Nb (8.6), As (7.2), Sc (5.8), Th (4.58), and U (2.9) mg/kg. Trace elements such as Cr, Cu, Zn, Zr, Ba, and Pb surpassed their average shale values, and only Cr surpassed the South African soil screening values. The assessment of pollution through the geo-accumulation index (Igeo) revealed that road dust was moderately to heavily contaminated by Cr, whereas all other trace elements were categorized as being uncontaminated to moderately contaminated. The contamination factor (CF) exhibited road dust to be very highly contaminated by Cr, moderately contaminated by Zn, Pb, Cu, Zr, and Ba, and lowly contaminated by Co, U, Nb, Ni, As, Y, V, Rb, Sc, Sr, and Th. The pollution load index (PLI) also affirmed that the road dust in this study was very highly polluted by trace elements. Moreover, the results of the enrichment factor (EF) categorized Cr as having a significant degree of enrichment. Zn was elucidated as being minimally enriched, whereas all other trace elements were of natural origin. The results of the non-carcinogenic risk assessment revealed a possibility of non-carcinogenic risks to both children and adults. For the carcinogenic risk, the total CR values in children and adults were above the acceptable limit, signifying a likelihood of carcinogenic risk to the local inhabitants. From the findings of this study, it can be concluded that the levels of trace elements in the road dust of this informal settlement had the possibility to contribute to both non-carcinogenic and carcinogenic risks, and that children were at a higher risk than the adult population.

Keywords: informal settlement; trace elements; road dust; health implications; carcinogenic; non-carcinogenic

Citation: Mugudamani, I.; Oke, S.A.; Gumede, T.P. Influence of Urban Informal Settlements on Trace Element Accumulation in Road Dust and Their Possible Health Implications in Ekurhuleni Metropolitan Municipality, South Africa. *Toxics* **2022**, *10*, 253. https://doi.org/10.3390/toxics10050253

Academic Editor: Ilaria Guagliardi

Received: 31 March 2022
Accepted: 13 May 2022
Published: 17 May 2022

Publisher's Note: MDPI stays neutral with regard to jurisdictional claims in published maps and institutional affiliations.

1. Introduction

The urban landscape of most developing nations has experienced a proliferation of informal settlements over many years [1], and the majority of Sub-Saharan African urban inhabitants (~55%) now reside in informal settlements [2]. Informal settlements are regarded as being neglected portions of cities where housing and living conditions are terribly poor. They vary from overcrowded, contaminated dwellings to inadvertent squatter locations with no legitimate rights, distributed at the edge of cities [3]. They are seriously characterized by man-made activities such as household heating, the combustion of coal and oil, industrial processes, unplanned construction, demolition activities, road weathering, poor waste management, the burning of waste, and dense traffic [4,5]. Such anthropogenic activities lead to the accumulation of trace elements on buildings, plants, in air, soil, water, and on road dust. Trace elements are not decomposable, and consequently, they persist for long periods of time in the environment [6]. Road dust refers to the re-suspended particulate matter found on roads, mostly in troughs. It sometimes comprises of soil and sand particles that are assorted with litter, and rubble that becomes airborne due to traffic movements [7]. As a results of traffic movements, road dust is frequently raised up, settled, and raised again, to a certain elevation, which exposes residents to any trace elements that are available in such dust [8]. Some of the trace elements connected to road dust are As (arsenic), Ba (barium), Co (cobalt), Cr (chromium), Cu (copper), Ni (nickel), Pb (lead), Zn (zinc), Zr (zirconium), [9] Nb (niobium), Rb (rubidium), Sc (scandium), Sr (strontium), Th (thorium), V (vanadium), and Y (yttrium). Exposure to road dust containing these trace elements, either through inhalation, ingestion, or dermal contact absorption [10] may ultimately lead to serious health implications on the local residents. These include medical conditions such as cancer, miscarriages, hearing and visual impairment, asthma, renal failure, high blood pressure, headaches and dizziness, problems of reproductive systems, cardiovascular disorder, writhing, ataxia, skin and eye irritation, and lung granulomas [11,12]. In South Africa, informal settlements are home-based areas to masses of households, and Ekurhuleni is a metropolitan municipality with roughly 26 percent of its inhabitants residing in informal settlements [13]. These areas are characterized by a lack of proper sanitation, dense traffic, human activities such as a high population, unplanned construction, poor waste management, unregulated waste incineration, charcoal burning, and sewage runoff onto the streets. All of these man-made activities may influence the proliferation of trace elements in road dust, and thus endanger the health of the local inhabitants. Therefore, studying the influence of urban informal settlements on trace element accumulation in road dust is not only an essential aspect of assessing the quality of urban informal settlement settings, but also of protecting the health of the local residents. Despite the available knowledge on the possible effects of trace elements in road dust on public health in South Africa and in other parts of the world, overpopulated informal settlements in South Africa are lacking scientific data on the concentration of trace elements and their associated health risks. Therefore, this study aim to fill in this lacuna by determining the concentration of trace elements, assessing their pollution levels and possible sources, evaluating the health risks, and determining possible health implications that are associated with trace element exposure in road dust. The findings of this study will provide scientific knowledge on the levels of trace elements in road dust, and create awareness about the potential health risks associated with trace element exposure for populations living in poor urban informal settlements.

2. Materials and Methods

2.1. Study Area

The study area is situated within Ekurhuleni Metropolitan Municipality in Gauteng Province, South Africa (Figure 1). The area is positioned 15 km north of Kempton Park City Centre, and approximately 39.4 km south of Pretoria. It consists of 41,581 households with a population of 91,646, and is mostly dominated by approximately 99% Black Africans over a surface area of 5.43 km^2. The area experiences some rainfall, mainly during summer. It is frequently characterized by an average mean annual rainfall of 60 mm. January is considered to be the wettest month, with an average of approximately 125 mm. The month of July is the driest month, with an average of approximately 4 mm. Furthermore, the average mean annual temperature is 16 °C, with January being the warmest, at an average of 20.1 °C, whereas June is considered to be the coldest month, at an average of 10.1 °C.

The study is characterized by human activities such as a high population, unplanned construction, poor waste management, unregulated waste incineration, charcoal combustion, firewood, and sewage waste runoff onto the streets. The settlement is decorated by barbecuing markets along the street, which bisect the settlement. Furthermore, the community transportation system functions with taxicabs, which are the most commonly used method of conveyance by inhabitants. Traffic congestion is noteworthy during the early morning and late afternoon peak hours on the main road that bisect the settlement. There is also an industrial area that is situated approximate 2 to 3 km away from the settlement, which acts as a source of employment for the residents. The area is underlain by Archean Cratonic rocks allocated to the Johannesburg Dome, also recognized as the halfway house or the Johannesburg-Pretoria Dome. The Johannesburg Dome is a dome-like window of ancient granitoid (approximately 750 km^2) positioned in the middle part of the Kaapval Craton. It consist of black reef formation which form the base of the Transvaal Supergroup, which is an outcropping to the north-eastern, northern, and north-western margin of the inlier, and un-conformably overlies the granitoids and greenstones. It also comprises of trondhjemitic and tonalitic granitic rocks intruded into mafic-ultramafic greenstones. Furthermore, it encompasses some hornblende-amphibolites dykes and dolomites of the Chuniespoort group, as shown in Figure 2 [14].

Figure 1. Study area and the location of the sampling site.

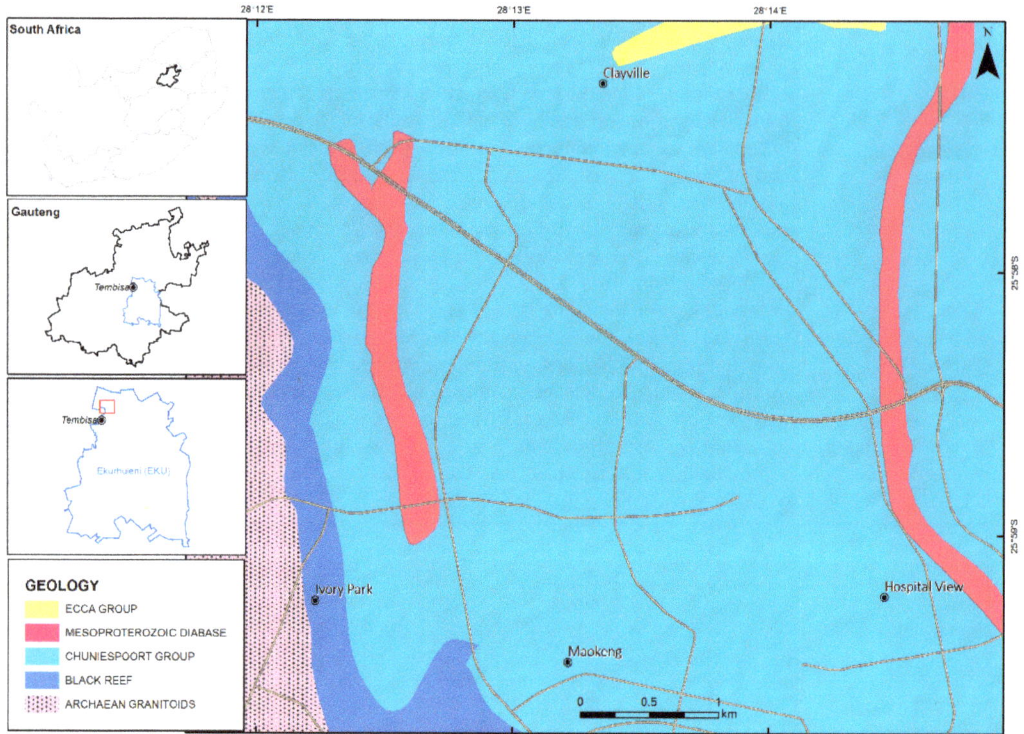

Figure 2. Geological map of the study area.

2.2. Sample Collection and Chemical Analysis

The study area is characterized by different functional areas that include commercial, residential, roadway, taxi rank, and leisure parks/playing grounds. Five (5) road dust samples were collected from these different functional areas, using a random sampling method. The samples were exactly collected at one of the major roads that bisect the settlement, near the park, at the taxi rank, next to the primary school, and at the shopping mall, in order to cover all of the functional areas around the settlement. The points were recorded using GPS, as detailed in Table 1. A sampling campaign was conducted in June 2021, which is a dry month. The road dust particles within a 5 m range of the chosen sampling point were collected by sweeping with a brush and dustpan. A randomly selected sample was collected on a paved surface, and the sampling points are presented in Figure 1. Unrelated material such as litter and debris were taken out from the samples in the course of sampling. To avoid cross-contamination, equipment was cleaned after every location. The samples were then transferred into the plastic sample bags, labelled, and conveyed to the laboratory.

A total of 10 g of sample was ground to a particle size of less than 200 mesh. The 10 g sample was then heated to 110 °C to dehydrate and devolatilize the sample, and then to 1050 °C, which breaks down minerals such as carbonates. This was conducted to determine the total loss on ignition, or the total gain on ignition. Then, a flux of 0.2445 g of La_2O_3, 0.705 g of $Li_2B_4O_7$, 0.5505 g of Li_2CO_3, and 0.02 g of $NaNO_3$ was added to 0.28 g of sample. The mixture was then heated to 1000 °C for approximately 5 min until a consistent fluid was formed within a Pt crucible. The fluid was then poured into a mold and pressed to form the disc. The application used to measure the majors was "IGS majors". It was created using the following standards: DR-N, JB-2, JF-2, JG-2, JGB-1, K8000, MA-N, MICA-

FE, MRG-1, NIM-S, SARM4, SARM5, SARM6, SARM43, SARM44, SARM47, SARM48, SARM50, SARM52, SY-2, and UB-N for quality control.

Table 1. Location and characteristics of sampling points.

Sample ID.	Description	GPS Coordinates	
		Latitude	Longitude
RD01	At one of the major roads (Madiba drive)	25°58'22.1'' S	28°13'27.1''E
RD02	At the southward drive, near the park	25°58'00.3'' S	28°12'57.6'' E
RD03	At the community taxi rank	25°58'51.4'' S	28°13'22.1'' E
RD04	Next to primary school gate	25°58'50.7'' S	28°13'30.1'' E
RD05	Within the community shopping centre or mall	25°58'33.0'' S	28°13'42.2'' E

To analyze the trace elements, 8 g of sample was added to 3 g of Hoechst wax ($C_6H_8O_3N_2$). It was then mixed for 20 min in a Turbula mixer to ensure that the sample was mixed until it was homogeneous. The mixture was then pressed to pressures of greater than 395 N/m. The calibrated application used for the trace elements analysis was 'UIC traces', and for the analysis of Na, it was 'Sodium only'. The standards used to calibrate the 'UIC traces' included: ASK-2, ASK-3, BE-N, BHVO-1, BR, GA, GH, JA-1, JB-1, JB-2, JDO-1, JG-1, JG-2, JLS-1, JP-1, JR-1, JR-2, K8000, MA-N, MICA-FE, MICA-MG, MRG-1, NIM-D, NIM-G, NIM-L, NIM-N, NIM-P, NIM-S, RGM-1, SY-2, TRABS-001, TRABS-002, TRABS-003, TRABS-004, TRACE-000, TRACE-001, TRACE-002, TRACE-003, TRACE-004, TRACE-005, TRACE-006, TRACE-007, TRACE-008, TRACE-009, TRACE-010, TRACE-011, TRACE-012, TRACE-013, TRACE-014, TRACE-015, TRACE-016, TRMAC-001, TRMAC-002, TRMAC-003, TRMAC-004, TRMAC-005, TRMAC-006, UB-N, and VS-N for the quality control. CaO, TiO_2, and Fe_2O_3 were measured to correct for line overlaps. The standards used to calibrate 'Sodium only' included: AN-G, BR, FK-N, G2, GA, GH, GS-N, GSP-1, JG-1, NIM-G, NIM-N, NIM-P, NIM-S, SARM39, and SY-2. Blank samples and duplicates were also used to determine precision and bias. The level of inconsistency was determined to be <10%. The WD-XRF machine used in this study was a Rigaku-Primus IV, with an Rh tube. The software used for the machine was ZXS, and the results are quantitative results.

2.3. Pollution Assessment of Trace Elements in Road Dust

2.3.1. Geo-Accumulation Index (Igeo)

This involves matching the level of the determined trace element to the average shale value or background levels [15,16]. This was computed using Equation (1):

$$\text{Igeo} = \log_2 (\text{Cn}/1.5\,\text{Bn}) \tag{1}$$

where Cn signifies the measured concentration in this study, and Bn represents the geochemical background value or an average shale value of an element of interest. A constant of 1.5 is presented to reduce the effect of possible variations in the background values that may be attributed to lithological differences in the sediments or soil [17]. Igeo values divide soil into different quality classes: Class 0 (Igeo ≤ 0) uncontaminated; Class 1 (0 < Igeo ≤ 1) uncontaminated to moderately contaminated; Class 2 (1 < Igeo ≤ 2) moderately contaminated; Class 3 (2 < Igeo ≤ 3) moderately to heavily contaminated; Class 4 (3 < Igeo ≤ 4) heavily contaminated; Class 5 (4 < Igeo ≤ 5) heavily to extremely contaminated; and Class 6 (Igeo > 5) extremely contaminated [15].

2.3.2. Contamination Factor (CF) and Pollution Load Index (PLI)

The contamination factor permits the evaluation of soil pollution by taking into consideration the content of trace elements in the soil and its background values or average shale value [15]. This was calculated using Equation (2):

$$\text{CF} = \text{C sample}/\text{C background} \tag{2}$$

where C sample represents the concentration of an element of interest and C background is the metal background concentration or average the shale value of an element of interest. Consistent with Addo et al. [18], the CF values were categorized into four clusters: CF < 1 (low contamination), $1 \leq CF < 3$ (moderate contamination), $3 \leq CF \leq 6$ (considerable contamination), and CF > 6 (very high contamination). The PLI was then calculated using the values of CF. This verifies how environmental conditions have deteriorated due to a rise in metal concentration [19], using Equation (3):

$$PLI = \sqrt[n]{(CF1 \times CF2 \times CF3 \times \ldots CFn)} \tag{3}$$

where n represents the number of trace elements detected in this study, and CF connotes the contamination factor computed using Equation (2). According to Kowalska et al. [15], PLI classifies site quality as: $0 < PLI \leq 1$ (unpolluted), $1 < PLI \leq 2$ (moderately to unpolluted), $2 < PLI \leq 3$ (moderately polluted), $3 < PLI \leq 4$ (moderately to highly polluted), $4 < PLI \leq 5$ (highly polluted), and $5 < PLI$ (very highly polluted).

2.3.3. Enrichment Factors (EF)

The EF was used to differentiate between elements originating from human activities and natural sources [20]. This was computed using Equation (4) by comparing the concentration of an element in a sample with its concentration in the average shale value. Scandium (Sc) was used as the reference element [21].

$$EF = (E/R) \text{ sample}/(E/R) \text{ background} \tag{4}$$

where E is the concentration of an element of interest, R is a reference element of crustal material (Sc), and (E/R) sample is the concentration ratio of E to R in the collected samples, and (E/R) background is the concentration ratio of E to R in the Earth's crust. The EF is categorized into five classes: EF < 2 (depletion to minimal enrichment), EF = 2–5 (moderate enrichment), EF = 5–20 (significant enrichment), EF = 20–40 (very high enrichment), and EF > 40 (extremely high enrichment) [20,21].

2.4. Human Health Risk Assessment of Trace Elements in Road Dust

2.4.1. Non-Cancer Risk Assessment

The average daily dose (ADD) of each analyzed trace element through ingestion, inhalation, and dermal contact was calculated using Equations (5)–(7) [22,23].

$$ADD_{ing} = C \times IngR \times CF \times EF \times ED/BW \times AT \tag{5}$$

$$ADD_{inh} = C \times InhR \times EF \times ED/BW \times AT \times PEF \tag{6}$$

$$ADD_{derm} = C \times SA \times CF \times SL \times ABS \times EF \times ED/BW \times AT \tag{7}$$

where ADD_{ing} signifies the average daily ingestion (mg/kg/day) amount for an element, ADD_{inh} indicates the average daily inhalation (mg/kg/day) amount for an element, and ADD_{derm} specifies the average daily dermal (mg/kg/day) exposure amount of metal, and their values are presented in Table 2. Non-cancer risk was then evaluated from the hazard quotient (HQ) for each trace element by dividing the ADD calculated in Equations (5)–(7) with a particular reference dose (Rfd), using Equation (8):

$$HQ = ADD/RfD \tag{8}$$

HQ > 1 suggests a possibility of health effects, while HQ < 1 shows no possibility of health effects [24]. Furthermore, the hazard index (HI) was then calculated by adding the HQ of the three exposure pathways for a corresponding element [25], using Equation (9):

$$HI = (HQ) \text{ ing} + (HQ) \text{ inh} + (HQ) \text{ derm} \tag{9}$$

A HI value < 1 describes a very low risk, a HI value between 1 and 4 shows that the risk effects are possible, and HI values > 4 describe a high risk [25].

Table 2. Exposure factors for dose models.

Items	Parameter	Meaning	Unit	Value		References
				Children	Adult	
Basic parameter	C	Concentration of a metal	mg/kg	This study	This study	[9,23,26]
	D	Daily dose	mg/kg			[9,23,26]
	CF	Conversion factor	kg/mg	1×10^{-6}	1×10^{-6}	[9,23,26]
Exposure behavioral parameter	ED	Exposure duration	years	6	24	[9,23,26]
	BW	Body weight	kg	15	55.9	[9,23,26]
	EF	Exposure frequency	days/year	350	350	[9,23,26]
		Average time (carcinogen)	days	365×70	365×70	[9,23,26]
	AT	Average time (non-carcinogen)	days	$365 \times ED$	$365 \times ED$	[9,23,26]
Digestive tract/inhalation	InhR	Inhalation rate	m^3/kg	5	20	[9,23,26]
	IngR	Ingestion rate	mg/kg	200	100	[9,23,26]
	PEF	Particle emission factor	m^3/kg	1.32×10^9	1.32×10^9	[9,23,26]
Skin contact	SL	Skin adherence factor	mg/cm^2	1	1	[9,23,26]
	SA	Skin surface area	cm^2	1800	5000	[9,23,26]
	ABS	Dermal absorption	-	0.001	0.001	[9,23,26]

2.4.2. Cancer Risk Assessment

The lifetime average daily dose (LADD) of each of the analyzed elements was also calculated for ingestion, inhalation, and dermal exposure pathways, using Equations (10)–(12) [9].

$$LADD_{ing} = C \times CF \times EF/AT \times (IngR_{child} \times ED_{child}/BW_{child} + IngR_{adult} \times ED_{adult}/BW_{adult}) \tag{10}$$

$$LADD_{inh} = C \times EF/AT \times PEF \times (InhR_{child} \times ED_{child}/BW_{child} + InhR_{adult} \times ED_{adult}/BW_{adult}) \tag{11}$$

$$LADD_{derm} = C \times CF \times EF \times SL \times ABS/AT \times (SA_{child} \times ED_{child}/BW_{child} + SA_{adult} \times ED_{adult}/BW_{adult}) \tag{12}$$

where, $LADD_{ing}$ connotes the lifetime average daily ingestion (mg/kg/day) amount of a metal, $LADD_{inh}$ implies the lifetime average daily inhalation (mg/kg/day) amount of an element, and $LADD_{derm}$ indicates the lifetime average daily dermal (mg/kg/day) exposure amount of a metal, and their values are presented in Table 2 [9,23,26–29]. After calculating the LADD of each exposure pathway, a lifetime cancer risk (CR) was then computed by multiplying the LADD with an equivalent slope factor (SF) using Equation (13). The permissible risk usually ranged from 10^{-6} to 10^{-4}: very low ($<10^{-6}$), low (10^{-6}–10^{-5}), medium (10^{-5}–10^{-4}), high (10^{-4}–10^{-3}), and very high ($>10^{-3}$) [30].

$$CR = LADD \times SF \tag{13}$$

2.5. Statistical Methods

The laboratory results were analyzed using Statistical Package for Social Sciences (SPSS) from Microsoft Excel. The data were presented as the minimum, maximum, average mean, and standard deviation. Furthermore, to determine the possible relationship between the elemental concentration and the possible source of origin, Pearson's correlation coefficient and a one way analysis of variance (ANOVA) were adopted.

3. Results and Discussion

3.1. Concentration of Major Elements in Road Dust

The descriptive statistics of major elements is summarized in Table 3, with the average shale values (ASV). The average mean concentrations of SiO_2 and P_2O_5 were higher than their average shale values. A higher silica content may expose the community to serious health implications, such as damaged lung tissue [20]. Its higher concentration in road dust may be associated with its hardness, which makes it difficult to undergo physical weathering [31]. Additionally, the dominance of quartz in road dust may also be linked to the geographical location of the study area, which falls under the Johannesburg dome, which is underlain with sedimentary rocks of the Witwatersrand and Venterdorp Supergroup.

The concentration of P_2O_5, which was also higher than its average shale value, might have been influenced by poor waste management practices in this informal settlement, particularly waste containing phosphate [31]. Runoff from gardens, illegal dumping, and nearby roadside soil polluted by phosphate used in different agricultural activities might have exacerbated the concentration of P_2O_5 in road sediments. Exposure to dust particles containing high concentrations of P_2O_5 may lead to the risk of respiratory distress, and problems of the liver, kidneys, and brain [32].

Furthermore, the average mean concentration of major elements such as Al_2O_3, K_2O, MgO, MnO, and CaO were below their average shale values, suggesting that they are of natural origin and were comparable to the findings of Li et al. [33] in Xi'an city, China. Although these major elements were below their average shale values, their concentrations may possible rise in the near future, due to daily increases in the population and uncontrolled waste generation in this poor informal settlement.

Table 3. Composition of major elements in road dust samples (Wt. %).

Elements	RD01	RD02	RD03	RD04	RD05	Min-Max	Mean	±SD	ASV
Al_2O_3	7.22	8.43	6.21	5.9	6.72	5.9–8.43	6.9	0.99	15.4
CaO	4.23	3.62	1.97	1.09	2.66	1.09–4.23	2.71	1.26	3.1
Fe_2O_3	3.41	3.52	3.68	4.11	4.67	3.41–4.67	3.88	0.52	4.02
K_2O	1.43	2.44	1.27	1.45	1.22	1.22–2.44	1.56	0.5	3.24
MgO	1.18	1.15	0.72	0.32	1.33	0.32–1.33	0.94	0.41	2.44
MnO	0.42	0.24	0.66	1.03	0.51	0.24–1.03	0.57	0.3	trace
Na_2O	0.98	1.89	0.71	0.76	0.62	0.62–1.89	0.99	0.52	1.3
P_2O_5	0.12	0.37	0.11	0.08	0.12	0.08–0.37	0.16	0.12	0.14
SiO_2	69.17	65.37	75.07	79.92	74.25	65.37–79.92	72.76	5.62	58.11
TiO_2	0.36	0.35	0.46	0.35	0.48	0.35–0.48	0.4	0.06	0.65
LOI	11.3	10.52	7.47	4.98	6.61	4.98–11.3	8.18	2.67	-

Notation: LOI = loss on ignition; SD = Standard deviation; ASV = Average shale value; Min = Minimum; Max = Maximum; Average shale value [34].

3.2. Concentration of Trace Elements in Road Dust

The concentration of trace elements presented in Table 4 were ranging as Cr > Ba > Zn > Zr > Sr > V > Rb > Cu > Ni > Pb > Co > Y > Nb > As > Sc > Th > U. Cr, Cu, Zn, Zr, Ba, and Pb were above their average shale values [35], which support the findings of Cai and Li [8] in the street dust of Shijiazhuang, China. When compared with the South African soil screening values [36] for metals in informal settlements only Cr was above this value. This outcome corresponds to the findings of the study conducted by Kamunda et al. [37] in soils from the Witwatersrand Gold Mining Basin, South Africa. The high levels of these trace elements might have been influenced by man-made activities. Dense traffic, which is mostly seen during the early hours and late hours of the day, mostly in tar roads that bisect the settlement, is one of the factors that influence the accumulation of trace elements. Vehicle exhaust is associated with Zr [38], while tire rubber, break wear re-suspended particles, and fuel combustion are sources of Zn [31]. The accumulation of Cu in street dust is associated with brake pad wear [31], while Pb mostly emanated from brake friction, batteries, and gasoline [39].

Table 4. Statistical analysis of trace elements concentration (mg/kg) in road dust samples with average shale values (mg/kg) and South African soil screening values (mg/kg).

Elements	LOD	RD01	RD02	RD03	RD04	RD05	Min-Max	Mean	±SD	ASV	SASSV
As	1	9	10	4	4	9	4–10	7.2	2.9	13	23
Ba	17	563	636	546	729	654	546–729	625.6	73.9	580	n.a
Co	3	16	15	17	20	19	15–20	17.4	2.1	19	300
Cr	4	493	283	694	1088	629	283–1088	637.4	297	90	6.5
Cu	4	125	42	45	43	50	42–125	61	35.9	45	1100
Nb	1	9	8	9	8	9	8–9	8.6	0.5	11	n.a
Ni	4	37	32	53	80	43	32–80	49	19	68	620
Pb	2	32	17	30	41	34	17–41	30.8	8.8	20	110
Rb	2	61	82	59	72	56	56–82	66	10.8	140	n.a
Sc	3	7	8	4	5	5	4–8	5.8	1.6	13	n.a
Sr	2	156	153	99	81	112	81–156	120.2	33.2	300	n.a
Th	3	6	6	5	2.9	3	2.9–6	4.58	1.5	12	n.a
U	3	2.9	2.9	2.9	2.9	2.9	2.9–2.9	2.9	0	3.7	n.a
V	5	57	55	76	75	82	55–82	69	12.2	130	150
Y	1	15	14	14	14	15	14–15	14.4	0.5	26	n.a
Zn	2	700	131	147	67	114	67–700	231.8	263.4	95	9200
Zr	1	164	167	221	172	227	164–227	190.2	31.1	160	n.a

Notation: LOD = limit of detection; RD = road dust; n.a = not available; SD = Standard deviation; ASV = Average shale value; SASSV = South African Soil Screening Value; Min = Minimum; Max = Maximum. Average shale values [35], SASSV [36].

According to Moryani et al. [40], Cr may be released from the combustion of lubricants and fuel. Charcoals is used most of the time in this poor informal settlement for household warming and cooking, and street barbecuing may also distribute Cr around the settlement through fly ash. According to Cui et al. [41], volatile condensing elements such as Cr are enriched in fine fly ash. Unplanned construction in this informal settlement, which occurs most of the time, may also release dust containing Co, possibly from materials containing cobalt, such as alloys and paints [42]. Furthermore, the dumping of waste in any available space or adjacent to the streets may release elements such as Ba. Waste materials such as ceramics, glass, or plastics are considered to be possible sources of Ba [43]. Tires and brakes are also sources of Ba.

When trace elements in road dust were compared with other cities around the world, as shown in Table 5, there were variations in their concentrations. This variation may be attributed to various factors such as a high population, unregulated waste burning, unplanned construction, charcoal burning and the use of firewood, emissions from nearby industrial areas, sewage waste, and poor waste management practices in the settlement. Salah et al. [44] agrees that the accumulation of trace elements in different regions may be influenced by factors such as the type of man-made activities. According to Shi et al. [42], the population level and stages of development may also influence differences in trace element concentration.

3.3. Pollution Assessment and Identification of Sources of Trace Elements in Road Dust

The inclusive Igeo values as presented in Table 6 were as follows: Cr > Zn > Pb > Cu > Zr > Ba > Co > U > Nb > Ni > Y > As > V > Sc > Rb > Sr > Th. Cr was the leading element, indicating possible anthropogenic sources (Figure 3). The possible anthropogenic sources of Cr in this study may be flying ashes from the combustion of charcoal used for indoor warming, cooking, and street barbecuing. Furthermore, the corrosion of vehicular parts, and the combustion of lubricants and fuel are also considered to be sources of Cr [40]. On the basis of the Igeo average values, Cr in this study is an element of concern, and its exposure at high concentrations may trigger serious health implications for the local inhabitants, particularly vulnerable groups such as children, elders, and pregnant women.

Table 5. Comparison of trace elements (mg/kg) in road dust with other global cities.

Trace Elements (mg/kg)	City and Country									
	Ekurhuleni (South Africa)	Dhaka City (Bangladesh)	Tyumen (Russia)	Viana do Castelo (Portugal)	Barbican Downtown (China)	Luanda (Angola)	Ahvaz (Iran)	Seoul (Korea)	Xian (China)	Villavicencio (Columbia)
As	7.2	-	5.7	35	11.7	5	6	24.9	-	-
Ba	625.6	-	317.1	390	748.2	351	-	570	-	-
Co	17.4	-	39.6	-	13.7	2.9	13	17.9	34.1	-
Cr	637.4	-	507.9	210	175.2	26	57	130	175.3	9.4
Cu	61	59.3	57.4	260	50.9	42	45	351	48.9	126.3
Nb	8.6	-	6.6	-	11.9	131	-	-	-	-
Ni	49	-	632.1	16	21	10	58	62	28.3	5.3
Pb	30.8	59.6	33.9	86	93.5	1.7	86	214	97.6	87.5
Rb	66	88.1	27.1	240	44.2	-	-	-	-	-
Sc	5.8	-	10.1	-	-	1.3	-	-	-	-
Sr	120.2	289.9	147.3	190	186.5	172	-	-	-	-
Th	4.6	-	1.9	-	-	1	-	-	-	-
U	2.9	-	1.1	3.6	-	-	-	-	-	-
V	69	-	66.4	15	69.3	20	184	35	55.8	-
Y	14.4	-	7.3	-	18.6	-	-	-	-	-
Zn	231.8	189	160.8	1180	272	317	999	1476	164.9	133.3
Zr	190.2	165.6	60.7	360	120.1	-	-	-	-	-
References	Current study	[45]	[46]	[31]	[33]	[27]	[47]	[48]	[42]	[49]

Table 6. Geo-accumulation (Igeo) values of trace element contamination in road dust.

Igeo Values of Heavy Metals in Road Samples				
Elements	Min	Max	Mean	Classification
Cr	0.63	2.42	1.42	Moderately contaminated
Zn	0.14	1.47	0.49	Uncontaminated to moderately contaminated
Pb	0.17	0.41	0.31	Uncontaminated to moderately contaminated
Cu	0.19	0.55	0.27	Uncontaminated to moderately contaminated
Zr	0.12	0.28	0.24	Uncontaminated to moderately contaminated
Ba	0.19	0.24	0.21	Uncontaminated to moderately contaminated
Co	0.16	0.21	0.2	Uncontaminated to moderately contaminated
U	0.16	0.16	0.16	Uncontaminated to moderately contaminated
Nb	0.14	0.16	0.16	Uncontaminated to moderately contaminated
Ni	0.09	0.23	0.14	Uncontaminated to moderately contaminated
Y	0.1	0.11	0.11	Uncontaminated to moderately contaminated
As	0.06	0.15	0.11	Uncontaminated to moderately contaminated
V	0.08	0.13	0.1	Uncontaminated to moderately contaminated
Sc	0.06	0.12	0.09	Uncontaminated to moderately contaminated
Rb	0.08	0.12	0.09	Uncontaminated to moderately contaminated
Sr	0.05	0.1	0.08	Uncontaminated to moderately contaminated
Th	0.05	0.09	0.08	Uncontaminated to moderately contaminated

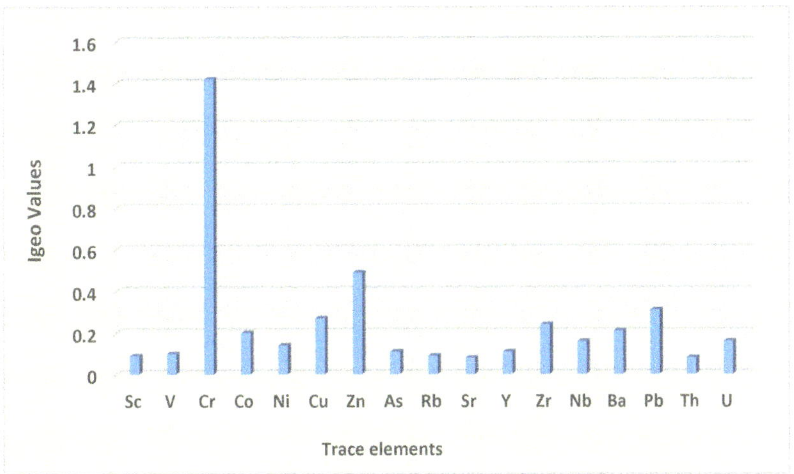

Figure 3. Trace element contamination levels in road dust based on Igeo values.

3.3.1. Contamination Factor (CF) and Pollution Load Index (PLI)

The CF and PLI were adopted to determine the degree of pollution, and to verify how environmental conditions have deteriorated due to the rise in trace element concentrations. The CF average mean value for Cr was above 6, indicating a very high rate of contamination from human activities. This outcome surpasses the findings of Dat et al. [51] for street dust in a metropolitan area of Southern Vietnam, which recorded considerable contamination by Cr. Moderate contamination was noted for elements such as Zn, Pb, Cu, Zr, and Ba, signifying a natural origin with the moderate influence of anthropogenic activities. Similarly, Al-Dabbas et al. [52] also witnessed the moderate contamination of Pb and Zn in the street dust of Diwaniya, Iraq. The majority of trace elements such as Co, U, Nb, Ni, As, Y, V, Rb, Sc, Sr, and Th were classified as having low levels of contamination resembling a natural origin, and this agrees with a study conducted in Bolgatanga Municipality, Ghana [10], which observed a low level of contamination by Co, Ni, and As in road dust, and in the street dust of Diwaniya, Iraq, which was contaminated with a low level of V [52].

As summarized in Table 7, the overall contamination factor values descended in the order of Cr > Zn > Pb > Cu > Zr > Ba > Co > U > Nb > Ni > As > Y > V > Rb > Sc > Sr > Th. As depicted in Figure 4, Cr was the leading pollution contributor amongst trace elements, possibly originating from traffic and charcoal burning, which is practiced daily for household purposes such as indoor warming and cooking. Furthermore, the use of charcoal by street vendors, and emissions from nearby industrial areas, may also be a possible source of Cr. The outcomes of the PLI results showed that road dust was very highly polluted, a similar outcome to the study conducted in the street dust of Ho Chi Minh City, Vietnam [51]. This outcome of PLI may have been influenced by factors such as daily heavy traffic within the study area [31,40,43]. Furthermore, runoff from sewage and uncollected waste materials that lie adjacent to the streets, unregulated waste incineration, and coal fly ashes might also be a contributor to the level of trace elements in road dust. The results of the contamination factor and pollution load index agree with the results of the Igeo accumulation index showing that Cr is an element of public concern in this study, and that remediation and regular monitoring are highly advocated.

Table 7. Contamination factor (CF) and pollution load index (PLI) values of trace elements in road dust.

CFs and PLI Values of Trace Elements in Road Dust Samples			
Elements	**Min**	**Max**	**Mean**	**Classification**
Cr	3.14	12.08	7.08	Very high contamination
Zn	0.7	7.37	2.44	Moderate contamination
Pb	0.85	2.05	1.54	Moderate contamination
Cu	0.93	2.78	1.35	Moderate contamination
Zr	1.02	1.42	1.2	Moderate contamination
Ba	0.94	1.26	1.08	Moderate contamination
Co	0.79	1.05	0.91	Low contamination
U	0.78	0.78	0.8	Low contamination
Nb	0.72	0.81	0.8	Low contamination
Ni	0.47	1.18	0.72	Low contamination
As	0.31	0.77	0.55	Low contamination
Y	0.54	0.58	0.55	Low contamination
V	0.42	0.63	0.53	Low contamination
Rb	0.4	0.58	0.47	Low contamination
Sc	0.31	0.61	0.45	Low contamination
Sr	0.27	0.52	0.4	Low contamination
Th	0.02	0.5	0.38	Low contamination
PLI	0.68	72.9	5.37	Very highly polluted

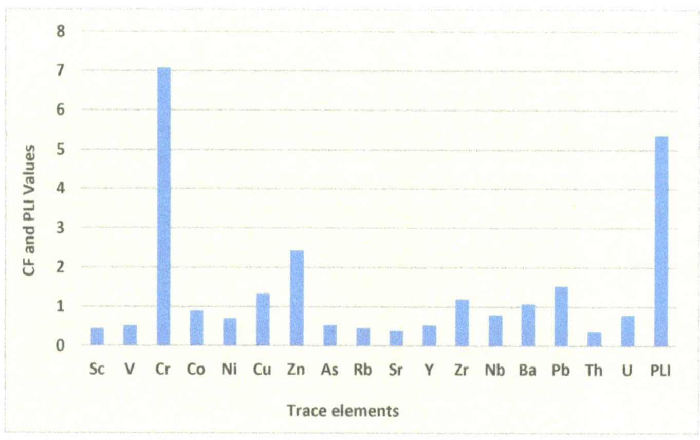

Figure 4. Contamination factor and pollution load index of individual trace elements in road dust.

3.3.2. Enrichment Factor (EF)

To compute the EF values, the average shale values were used as the background concentration, and scandium (Sc) was chosen as a reference metal. The results of EF presented in Table 8 showed Cr to be of significant enrichment, possibly from human activities that matched the findings of the study conducted in the road dust of Katowice and Wroclaw, Poland [53], and they surpassed the findings observed by Cai and Li [8] in the street dust of Shijiazhuang, China, who reported Cr to be of minimal enrichment. Moderate enrichment was reported for Zn, indicating anthropogenic sources. These outcomes are better than the study conducted in Lagos metropolis, Nigeria [54], which observed a very high enrichment of Zn. Other trace elements, Pb, Cu, Zr, Ba, Sc, Co, U, Nb, Ni, Y, As, V, Rb, Sr, and Th were classified as having minimal enrichment, signifying a natural origin, in agreement with the study conducted on the road dust of Dhaka city, Bangladesh [55], which reported Pb, and As to be of minimal enrichment. In the road dust of Bolgatanga Municipality, Ghana, Cu, Zn, Pb, Ni, As, and Co were classified as having minimal enrichment [10]. The geology of the area, the development of the area, the selection of reference materials in calculating EF, and the selection of an element of reference may have an influence on the results of EF [56].

Table 8. Enrichment factor (EF) values of trace elements in road dust.

Elements	EF Average Mean Values	Enrichment Category
Cr	7.08	Significant enrichment
Zn	2.44	Moderate enrichment
Pb	1.54	Minimal enrichment
Cu	1.35	Minimal enrichment
Zr	1.2	Minimal enrichment
Ba	1.08	Minimal enrichment
Sc	1	Minimal enrichment
Co	0.92	Minimal enrichment
U	0.78	Minimal enrichment
Nb	0.78	Minimal enrichment
Ni	0.72	Minimal enrichment
Y	0.55	Minimal enrichment
As	0.55	Minimal enrichment
V	0.53	Minimal enrichment
Rb	0.5	Minimal enrichment
Sr	0.4	Minimal enrichment
Th	0.38	Minimal enrichment

The values of EF were as follows: Cr > Zn > Pb > Cu > Zr > Ba > Sc > Co > U > Nb > Ni > Y > As > V > Rb > Sr > Th. Cr was the major contributor to pollution, followed by Zn, which shows an influence from anthropogenic activities (Figure 5). Human activities such as waste incineration, heavy traffic, poor waste management, coal fly ashes, sewage waste run-off onto the streets, and emissions from nearby industrial areas might be possible anthropogenic sources in this informal settlement. Other researchers also agree that sewage and the incineration of plastics waste may release Zn into urban areas [57], whereas the corrosion of vehicular parts may be the source of Cr [10]. From the results of EF, it can be concluded that the high level of Cr in road dust needs serious attention. Other trace elements were of crustal origin. The natural sources of this trace element may be attributed to the geology of the area, precipitation, or wind-borne soil particles [58].

3.3.3. Pearson Correlation Coefficient Analysis

Pearson's correlation coefficient was performed to establish trace element relationships and to determine their common sources of origin. A correlation matrix of trace elements in road dust samples generated a diverse relationship between the elements. From the correlation analysis in Table 9, a sufficiently high degree of correlation, a moderate degree, and no positive correlation results were observed. A strong correlation was witnessed between pairs of Ni–Cr (0.97), Zn–Cu (r = 0.99), Co–V (r = 0.83), Co–Cr (r = 0.89), Ni–Co (r = 0.81), As–Sc (r = 0.78), Sr–Sc (r = 0.88), Sr–As (r = 0.88), Zr–V (r = 0.81), Pb–Cr (r = 0.89), Pb–Co (r = 0.87), and Pb–Ni (0.77). According to Weissmannova et al. [59], high levels of correlation among trace elements indicates the same sources of pollution, or anthropogenic sources. Therefore, the high degree of correlation between these trace elements in this study is suggestive of the same source of pollution, potentially from anthropogenic activities such as unplanned construction, coal fly ashes, poor waste management, the burning of waste, and dense traffic. Elements such as Zn, Ni, and Cr may be attributed to the burning of fuel [31,40]. Cu and Pb may be associated with brake pad wear and brake friction [31]. Coal burning releases As and Sr [43,45]. Trace elements such as Zr potentially emanate from vehicle exhaust [38], Co from construction materials such as paints [42], and V from oil combustion [43]. Furthermore, the contribution of the concentration of Sc in road dust is understood to be from soil re-suspension, as they are crustal elements [60].

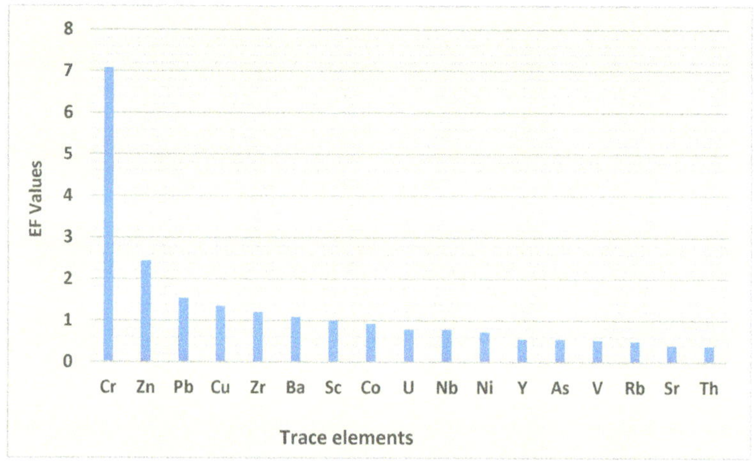

Figure 5. Enrichment factor values of trace elements in road dust.

Table 9. Correlation matrix of trace elements in road dust.

Elements	Sc	V	Cr	Co	Ni	Cu	Zn	As	Rb	Sr	Y	Zr	Nb	Ba	Pb	Th	U
Sc	1																
V	−0.91	1															
Cr	−0.72	**0.67**	1														
Co	−0.70	**0.83**	**0.89**	1													
Ni	−0.63	0.56	**0.97**	**0.81**	1												
Cu	0.37	−0.49	−0.27	−0.34	−0.37	1											
Zn	0.42	−0.57	−0.35	−0.46	−0.43	**0.99**	1										
As	**0.78**	−0.55	−0.82	−0.55	−0.86	0.36	0.37	1									
Rb	**0.63**	−0.58	−0.20	−0.32	0.01	−0.33	−0.28	0.16	1								
Sr	**0.88**	−0.85	−0.90	−0.85	−0.89	0.59	**0.65**	**0.85**	0.24	1							
Y	0.11	0.04	−0.23	0.04	−0.43	**0.67**	**0.61**	0.56	−0.63	0.38	1						
Zr	−0.75	**0.81**	0.14	0.36	0.02	−0.40	−0.42	−0.24	−0.70	−0.47	0.16	1					
Nb	−0.39	0.30	−0.15	−0.04	−0.34	0.47	0.46	0.06	−0.93	0.09	**0.67**	**0.61**	1				
Ba	−0.03	0.29	0.54	**0.67**	0.59	−0.48	−0.56	−0.12	0.43	−0.46	−0.21	−0.21	−0.70	1			
Pb	−0.66	**0.65**	**0.89**	**0.87**	**0.77**	0.11	0.00	−0.59	−0.51	−0.70	0.23	0.18	0.19	0.38	1		
Th	**0.66**	−0.86	−0.78	−0.97	−0.70	0.47	0.58	0.42	0.28	0.82	−0.05	−0.44	0.08	−0.71	−0.75	1	
U	0.00	0.00	0.00	0.00	0.00	0.00	0.00	0.00	0.00	0.00	0.00	0.00	0.00	0.00	0.00	0.00	1

Coefficients above 0.6 are in bold.

3.3.4. Statistical Analysis of Trace Elements

A one-way ANOVA was performed to test the difference between the concentrations of elements in road dust samples. The analysis of variance revealed the p-value to be 5.09×10^{-20}, as shown in Table 10. The p-value was less than the alpha level of 0.05, demonstrating significant differences ($p < 0.05$). This outcome signifies that the trace element pollutants were not from common anthropogenic sources, which is comparable with the study conducted in Yola, Nigeria [61]. In this informal settlement vehicle emissions, re-suspended dust, construction dust, coal fly ashes, emissions from nearby industrial areas, sewage waste, waste burning, or poor waste management may be considered as being possible sources of these trace elements.

Table 10. Single factor analysis of variance (ANOVA) of trace element concentrations in road dust.

Analysis of Variance in Road Dust						
Source of Variation	**SS**	**df**	**MS**	**F**	***p*-Value**	**F Crit**
Between Groups	3,245,171	16	202,823	21	5.09×10^{-20}	1.79
Within Groups	668,540	68	9832			
Total	3,913,712	84				

3.4. Non-Carcinogenic and Carcinogenic Health Risk Assessments of Trace Elements in Road Dust

The health risk assessment of trace element contaminants through various exposure pathways was evaluated by calculating both the non-cancer and the cancer risks for children and adults. The non-cancer risk assessment was calculated with the use of Equations (5)–(9), while cancer risk assessment was computed by using Equations (10)–(13). The non-cancer risk values computed for road dust were based on the reference doses (RfD) in Table 11 and the average daily dose (ADD) values summarized in Table 12. Furthermore, the slope factors in Table 11 and the average daily doses (ADD) in Table 13 were used to assess the lifetime carcinogenic risks of Cr, Ni, As, Pb, and Co in road dust. They were calculated from the average contribution of the individual trace elements in road dust for all of the exposure pathways. The overall health risk assessment of the trace elements in this study does not consider the size of the road dust particles; thus, it uses the dust content as the inhalation amount to determine the general estimation of the health risk assessment. Therefore, the findings of this study signify the utmost likely occurrence of health risks, which also provide data that are valuable for risk cautioning, monitoring, and evading pollution.

Table 11. Reference doses for non-cancer risks and slope factors for cancer risk assessment.

Elements	RfD (mg/kg/d)			SF (mg/kg/d)			References
	RfDing	**RfDinh**	**RfDderm**	**SFing**	**SFinh**	**SFderm**	
As	3.00×10^{-4}	3.00×10^{-4}	1.20×10^{-4}	1.50×10^{0}	1.51×10^{1}	3.66×10^{0}	[50,62]
Ba	2.0×10^{-1}	1.43×10^{-4}	4.90×10^{-3}	-	-	-	[27,63]
Co	3.0×10^{-4}	2.00×10^{-2}	5.40×10^{-3}	-	8.40×10^{-1}	-	[23,62–64]
Cr(VI)	3.00×10^{-3}	2.86×10^{-6}	6.00×10^{-5}	5.00×1^{-1}	4.10×10^{-1}	-	[23,50,62]
Cr(III)	1.5	-	-	-	-	-	[63]
Cu	4.00×10^{-2}	4.00×10^{-2}	1.20×10^{-2}	-	-	-	[23,62,65]
Nb	-	-	-	-	-	-	-
Ni	1.1×10^{-2}	6.00×10^{-6}	1.60×10^{-2}	1.70×10^{0}	9.80×10^{0}	-	[23,62,63,65]
Pb	3.50×10^{-3}	3.50×10^{-3}	5.25×10^{-4}	8.50×10^{-3}	4.20×10^{-2}	-	[23,62,63,65]
Rb	-	-	-	-	-	-	-
Sc	-	-	-	-	-	-	-
Sr	6.00×10^{-1}	-	1.20×10^{-1}	-	-	-	[27,63]
Th	-	-	-	-	-	-	-
U	2.0×10^{-4}	-	5.10×10^{-4}	-	-	-	[27,63]
V	5.0×10^{-3}	7.00×10^{-3}	7.00×10^{-3}	-	-	-	[50,62,63]
Y	-	-	-	-	-	-	-
Zn	3.00×10^{-1}	3.00×10^{-1}	6.00×10^{-2}	-	-	-	[23,63,65]
Zr	-	-	-	-	-	-	-

Table 12. Average daily dose (ADD), HQ, and HI values for trace elements in road dust via ingestion, inhalation, and dermal exposure pathways for children and adults.

Pathways	Average Daily Dose (ADD) for Trace Elements in Road Dust (mg/kg/day)																		Total ADD
	Sc	V	CrVI	CrIII	Co	Ni	Cu	Zn	As	Rb	Sr	Y	Zr	Nb	Ba	Pb	Th	U	
Children																			
ADDing	7.43×10^{-5}	8.80×10^{-4}	8.10×10^{-3}	8.10×10^{-3}	2.20×10^{-4}	6.30×10^{-4}	7.80×10^{-4}	2.90×10^{-3}	9.20×10^{-5}	8.43×10^{-4}	1.53×10^{-3}	1.84×10^{-4}	2.43×10^{-3}	1.11×10^{-4}	8.00×10^{-3}	3.90×10^{-4}	5.85×10^{-5}	3.70×10^{-5}	3.54×10^{-2}
ADDinh	1.40×10^{-9}	1.70×10^{-8}	1.54×10^{-7}	1.54×10^{-7}	4.21×10^{-9}	1.20×10^{-8}	1.50×10^{-8}	5.61×10^{-8}	1.74×10^{-9}	1.60×10^{-8}	2.91×10^{-8}	3.50×10^{-9}	4.60×10^{-8}	2.08×10^{-9}	1.51×10^{-7}	7.46×10^{-9}	1.10×10^{-9}	7.02×10^{-10}	6.72×10^{-7}
ADDderm	6.67×10^{-9}	7.94×10^{-6}	7.33×10^{-5}	7.33×10^{-5}	2.00×10^{-6}	5.64×10^{-6}	7.02×10^{-6}	2.67×10^{-5}	8.30×10^{-7}	7.59×10^{-6}	1.40×10^{-5}	1.66×10^{-6}	2.20×10^{-6}	9.90×10^{-7}	7.20×10^{-5}	3.54×10^{-6}	5.27×10^{-7}	3.34×10^{-7}	3.00×10^{-4}
ADD	7.43×10^{-5}	8.88×10^{-4}	8.17×10^{-3}	8.17×10^{-3}	2.22×10^{-4}	6.36×10^{-4}	7.87×10^{-4}	2.93×10^{-3}	9.28×10^{-5}	8.51×10^{-4}	1.54×10^{-3}	1.86×10^{-4}	2.43×10^{-3}	1.12×10^{-4}	8.07×10^{-3}	3.94×10^{-4}	5.90×10^{-5}	3.73×10^{-5}	3.57×10^{-2}
Adults																			
ADDing	9.94×10^{-6}	1.20×10^{-4}	1.10×10^{-3}	1.10×10^{-3}	2.98×10^{-5}	8.40×10^{-5}	1.00×10^{-4}	4.00×10^{-4}	1.23×10^{-5}	1.13×10^{-4}	2.06×10^{-4}	2.47×10^{-5}	3.26×10^{-4}	1.47×10^{-5}	1.07×10^{-3}	5.28×10^{-5}	7.86×10^{-6}	4.97×10^{-6}	4.78×10^{-3}
ADDinh	1.50×10^{-9}	1.80×10^{-8}	1.66×10^{-7}	1.66×10^{-7}	4.52×10^{-9}	1.30×10^{-8}	1.60×10^{-8}	6.02×10^{-8}	1.90×10^{-9}	1.71×10^{-8}	3.12×10^{-8}	3.74×10^{-9}	4.94×10^{-8}	2.23×10^{-9}	1.62×10^{-7}	8.00×10^{-9}	1.20×10^{-9}	7.54×10^{-10}	7.23×10^{-7}
ADDderm	4.97×10^{-7}	5.92×10^{-6}	5.50×10^{-5}	5.50×10^{-5}	1.49×10^{-6}	4.20×10^{-6}	5.23×10^{-6}	1.99×10^{-5}	6.17×10^{-7}	5.66×10^{-6}	1.03×10^{-5}	1.23×10^{-6}	1.63×10^{-5}	7.40×10^{-7}	5.40×10^{-5}	2.64×10^{-6}	3.93×10^{-7}	2.49×10^{-7}	2.39×10^{-4}
ADD	1.04×10^{-5}	1.26×10^{-4}	1.16×10^{-3}	1.16×10^{-3}	3.13×10^{-5}	8.82×10^{-5}	1.05×10^{-4}	4.20×10^{-4}	1.29×10^{-5}	1.19×10^{-4}	2.16×10^{-4}	2.59×10^{-5}	3.42×10^{-4}	1.54×10^{-5}	1.12×10^{-3}	5.54×10^{-5}	8.25×10^{-6}	5.22×10^{-6}	5.02×10^{-3}

Pathways	Non−cancer risk values for trace elements in road dust (mg/kg/day)																		Total HI
	Sc	V	CrVI	CrIII	Co	Ni	Cu	Zn	As	Rb	Sr	Y	Zr	Nb	Ba	Pb	Th	U	
Children																			
HQing	-	1.80×10^{-1}	2.70×10^{0}	5.40×10^{-3}	7.30×10^{-1}	6.00×10^{-3}	1.95×10^{-2}	9.67×10^{-3}	3.06×10^{-1}	-	2.55×10^{-3}	-	-	-	4.00×10^{-2}	1.11×10^{-1}	-	1.90×10^{-1}	4.30×10^{0}
HQinh	-	2.43×10^{-6}	5.38×10^{-2}	-	2.10×10^{-7}	2.00×10^{-3}	3.75×10^{-7}	1.87×10^{-7}	5.80×10^{-6}	-	-	-	-	-	1.05×10^{-3}	2.13×10^{-6}	-	-	5.69×10^{-2}
HQderm	-	1.13×10^{-3}	1.22×10^{0}	-	3.70×10^{-4}	3.52×10^{-4}	5.85×10^{-4}	4.45×10^{-4}	6.92×10^{-3}	-	-	-	-	-	1.47×10^{-2}	6.74×10^{-3}	-	6.55×10^{-4}	1.25×10^{0}
HI	-	1.81×10^{-1}	3.97×10^{0}	5.40×10^{-3}	7.30×10^{-1}	8.35×10^{-3}	2.01×10^{-2}	1.01×10^{-2}	3.13×10^{-1}	-	2.55×10^{-3}	-	-	-	5.58×10^{-2}	1.18×10^{-1}	-	1.91×10^{-1}	5.61×10^{0}
Adults																			
HQing	-	2.40×10^{-2}	3.67×10^{-1}	7.30×10^{-4}	1.49×10^{-3}	8.00×10^{-4}	2.50×10^{-3}	1.33×10^{-3}	4.10×10^{-2}	-	3.43×10^{-3}	-	-	-	2.00×10^{-2}	1.51×10^{-2}	-	2.50×10^{-2}	5.02×10^{-1}
HQinh	-	2.57×10^{-6}	5.80×10^{-2}	-	2.26×10^{-7}	2.17×10^{-3}	4.00×10^{-7}	2.01×10^{-7}	6.33×10^{-6}	-	-	-	-	-	1.13×10^{-3}	2.28×10^{-6}	-	-	6.13×10^{-2}
HQderm	-	8.46×10^{-4}	9.17×10^{-1}	-	2.76×10^{-4}	2.62×10^{-4}	4.36×10^{-4}	3.32×10^{-4}	5.14×10^{-3}	-	8.58×10^{-5}	-	-	-	1.10×10^{-2}	5.03×10^{-3}	-	4.88×10^{-4}	9.41×10^{-1}
HI	-	2.48×10^{-2}	1.34×10^{0}	7.30×10^{-4}	1.77×10^{-3}	3.23×10^{-3}	2.94×10^{-3}	1.66×10^{-3}	4.61×10^{-2}	-	3.52×10^{-3}	-	-	-	3.21×10^{-2}	2.01×10^{-2}	-	2.55×10^{-2}	1.50×10^{0}

89

Table 13. Carcinogenic average daily dose (ADD) for trace elements in road dust via ingestion, inhalation, and dermal exposure pathways for children and adults.

Pathways	Cancer Average Daily Dose (ADD) for Trace Elements in Road Dust (mg/kg/day)																		Total ADD
	Sc	V	CrVI	CrIII	Co	Ni	Cu	Zn	As	Rb	Sr	Y	Zr	Nb	Ba	Pb	Th	U	
Children																			
ADDing	7.43×10^{-5}	8.80×10^{-4}	8.10×10^{-3}	8.10×10^{-3}	2.20×10^{-4}	6.30×10^{-4}	7.80×10^{-4}	2.90×10^{-3}	9.20×10^{-5}	8.43×10^{-4}	1.53×10^{-3}	1.84×10^{-4}	2.43×10^{-3}	1.11×10^{-4}	8.00×10^{-3}	3.90×10^{-4}	5.85×10^{-5}	3.70×10^{-5}	3.54×10^{-2}
ADDinh	1.40×10^{-9}	1.70×10^{-8}	1.54×10^{-7}	1.54×10^{-7}	4.21×10^{-9}	1.20×10^{-8}	1.50×10^{-8}	5.61×10^{-8}	1.74×10^{-9}	1.60×10^{-8}	2.91×10^{-8}	3.50×10^{-9}	4.60×10^{-8}	2.09×10^{-9}	1.51×10^{-7}	7.46×10^{-9}	1.10×10^{-9}	7.02×10^{-10}	6.72×10^{-7}
ADDderm	6.67×10^{-9}	7.94×10^{-6}	7.33×10^{-5}	7.33×10^{-5}	2.00×10^{-6}	5.64×10^{-6}	7.02×10^{-6}	2.67×10^{-5}	8.30×10^{-7}	7.59×10^{-6}	1.40×10^{-5}	1.66×10^{-6}	2.20×10^{-6}	9.90×10^{-7}	7.20×10^{-6}	3.54×10^{-6}	5.27×10^{-7}	3.34×10^{-7}	3.00×10^{-4}
Total	7.43×10^{-5}	8.88×10^{-4}	8.17×10^{-3}	8.17×10^{-3}	2.22×10^{-4}	6.36×10^{-4}	7.87×10^{-4}	2.93×10^{-3}	9.28×10^{-5}	8.51×10^{-4}	1.54×10^{-3}	1.86×10^{-4}	2.43×10^{-3}	1.12×10^{-4}	8.07×10^{-3}	3.94×10^{-4}	5.90×10^{-5}	3.73×10^{-5}	3.57×10^{-2}
Adults																			
ADDing	9.94×10^{-6}	1.20×10^{-4}	1.10×10^{-3}	1.10×10^{-3}	2.98×10^{-5}	8.40×10^{-5}	1.00×10^{-4}	4.00×10^{-4}	1.23×10^{-5}	1.13×10^{-4}	2.06×10^{-4}	2.47×10^{-5}	3.26×10^{-4}	1.47×10^{-5}	1.07×10^{-3}	5.28×10^{-5}	7.86×10^{-6}	4.97×10^{-6}	4.78×10^{-3}
ADDinh	1.50×10^{-9}	1.80×10^{-8}	1.66×10^{-7}	1.66×10^{-7}	4.52×10^{-9}	1.30×10^{-8}	1.60×10^{-8}	6.02×10^{-8}	1.90×10^{-9}	1.71×10^{-8}	3.12×10^{-8}	3.74×10^{-9}	4.94×10^{-8}	2.23×10^{-9}	1.62×10^{-7}	8.00×10^{-9}	1.20×10^{-9}	7.54×10^{-10}	7.23×10^{-7}
ADDderm	4.97×10^{-7}	5.92×10^{-6}	5.50×10^{-5}	5.50×10^{-5}	1.49×10^{-6}	4.20×10^{-6}	5.23×10^{-6}	1.99×10^{-5}	6.17×10^{-7}	5.66×10^{-6}	1.03×10^{-5}	1.23×10^{-6}	1.63×10^{-5}	7.40×10^{-7}	5.40×10^{-5}	2.64×10^{-6}	3.93×10^{-7}	2.49×10^{-7}	2.39×10^{-4}
Total	1.04×10^{-5}	1.26×10^{-4}	1.16×10^{-3}	1.16×10^{-3}	3.13×10^{-5}	8.82×10^{-5}	1.05×10^{-4}	4.20×10^{-4}	1.29×10^{-5}	1.19×10^{-4}	2.16×10^{-4}	2.59×10^{-5}	3.42×10^{-4}	1.54×10^{-5}	1.12×10^{-3}	5.54×10^{-5}	8.25×10^{-6}	5.22×10^{-6}	5.02×10^{-3}
LADD	8.47×10^{-5}	1.01×10^{-3}	9.33×10^{-3}	9.33×10^{-3}	2.53×10^{-4}	7.24×10^{-4}	8.92×10^{-4}	3.35×10^{-3}	1.06×10^{-4}	9.69×10^{-4}	1.76×10^{-3}	2.12×10^{-4}	2.77×10^{-3}	1.27×10^{-4}	9.20×10^{-3}	4.49×10^{-4}	6.73×10^{-5}	4.26×10^{-5}	4.07×10^{-2}

3.4.1. Non-Carcinogenic Risk Assessment

The total HI value in the population of children exhibited a possibility for non-carcinogenic risk (Table 12). This was driven greatly by ingestion and dermal pathways with the likelihood of non-carcinogenic risk. As shown in Figure 6, the total HI values through various exposure pathways in descending order were ingestion > dermal > inhalation, while the contribution of individual elements to the total HI was, in order, Cr(VI) > Co > As > U > V > Pb > Ba > Cu > Zn > Ni > Cr(III) > Sr (Figure 7). The discoveries of this study are similar to the findings of the studies conducted in the road dust of Viano do Castelo, Portugal [31], in Wroclaw and Katowice, Poland [53], the urbanized cities of Pakistan [9], and in Lagos metropolis, Nigeria [54].

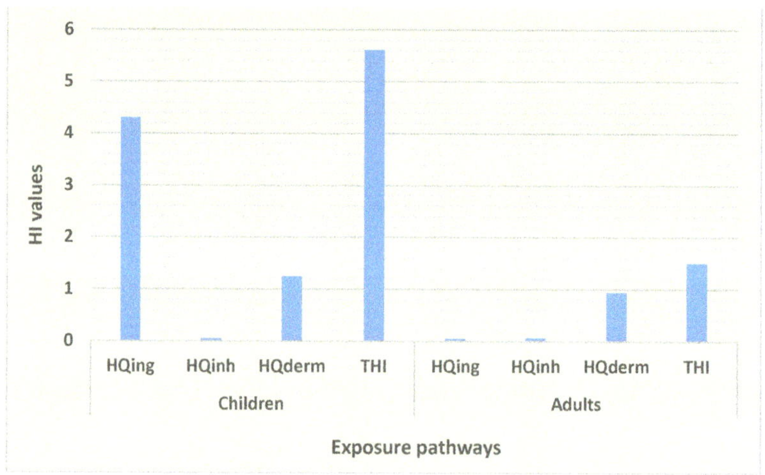

Figure 6. Hazard index for exposure to trace elements via various pathways.

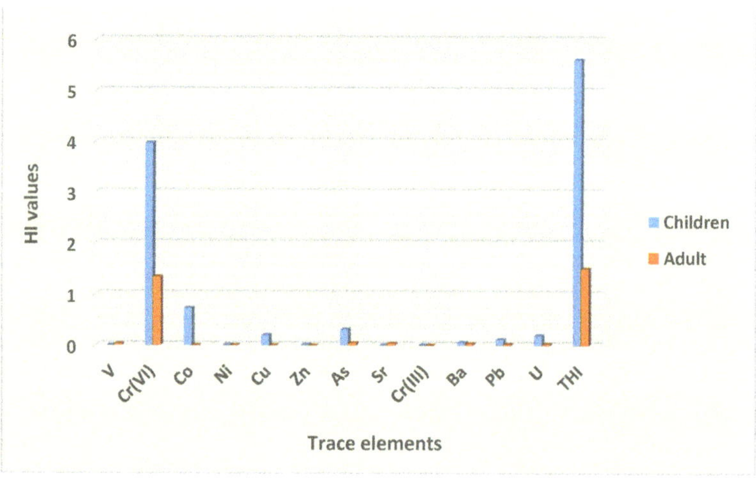

Figure 7. Contribution of trace elements to non-carcinogenic risk.

Chromium (Cr VI) was the only element that presented a probability for non-cancer risk. Taiwo et al. [54] in Lagos metropolis, Nigeria also reported Cr as being the principal contributor to non-cancer risk in road dust, which supports the findings of this study. The outcomes of this study suggest that the children group in this informal settlement are at risk of non-carcinogenic cancer, mainly through ingestion and dermal exposure, which is a concern, considering that they have a custom of playing in the dust and sucking their fingers or hands while playing. Children may be exposed to Cr mainly through inhalation and dermal contact, as it is the only element with the possibility of non-cancer risk. Other elements and inhalation pathways showed no possibility of non-cancer risk.

In the adult population, the total HI value showed a likelihood of causing non-carcinogenic risks. All exposure pathways had no chance for causing non-carcinogenic risks, and their trends were in the order of dermal > ingestion > inhalation (Figure 6). There was also no possibility of non-cancer risk from all of the trace elements, except from Cr (VI). Their contributions to the total hazard index were as follows: Cr(VI) > As > Ba > U > V > Pb > Sr > Ni > Cu > Co > Zn > Cr(III) (Figure 7). In adults, dermal contact was the major contributor to non-cancer risk, which makes re-suspended particles a serious concern. Exposure to Cr through dermal contact may lead to the possibility of non-cancer risks among adults in this poor informal settlement. The overall level of trace elements in road dust in this informal settlement showed the possibility of non-cancer risks to the local inhabitants. Additionally, the total HI value in the children population was higher than the total HI value in the adult population, evidencing the high possibility for heavy metals to cause non-cancer risks to children compared to the adult population (Figure 6). Similar outcomes were reported by Qadeer et al. [9] in urbanized cities of Pakistan, and by Candeias et al. [31] in Viano do Castelo, Portugal.

3.4.2. Carcinogenic Risk (CR) Assessment

The results of the cancer risk assessment, as summarized in Table 14, revealed that the total CR values in children and adults were between 10^{-4} and 10^{-3}. Children were at a higher risk than adults. Furthermore, the lifetime cancer risk value for the entire population was also between 10^{-4} and 10^{-3}. This value is considered to be a high risk, suggesting a concern for the residents regarding the possible CR of trace elements in road dust. Only the ingestion pathway exhibited the probability for cancer risk to both children and adults. The total cancer risks through various exposure pathways were as follows: ingestion > dermal > inhalation and ingestion > dermal > inhalation, for both children and adults, respectively (Figure 8).

Table 14. Cancer risk assessment for trace elements in road dust via ingestion, inhalation, and dermal exposure pathways for children and adults.

Pathways	Cancer Risk Values for Trace Elements in Road Dust (mg/kg/day)					Total CR
	Cr	Co	Ni	As	Pb	
Children						
CRing	3.54×10^{-4}	-	3.60×10^{-5}	1.18×10^{-5}	2.86×10^{-7}	4.02×10^{-4}
CRinh	5.41×10^{-9}	3.03×10^{-10}	9.99×10^{-9}	2.26×10^{-10}	2.68×10^{-11}	1.60×10^{-8}
CRderm	-	-	-	2.59×10^{-7}	-	2.59×10^{-7}
CR	3.54×10^{-4}	3.03×10^{-10}	3.60×10^{-5}	1.21×10^{-5}	2.86×10^{-7}	4.02×10^{-4}
Adults						
CRing	1.87×10^{-4}	-	1.90×10^{-5}	6.34×10^{-6}	1.54×10^{-7}	2.12×10^{-4}
CRinh	2.33×10^{-8}	1.30×10^{-9}	4.28×10^{-8}	9.69×10^{-10}	1.15×10^{-10}	6.85×10^{-8}
CRderm	-	-	-	7.76×10^{-7}	-	7.76×10^{-7}
CR	1.87×10^{-4}	1.30×10^{-9}	1.90×10^{-5}	7.12×10^{-6}	1.54×10^{-7}	2.13×10^{-4}
LCR	5.41×10^{-4}	1.60×10^{-9}	5.51×10^{-5}	1.92×10^{-5}	4.40×10^{-7}	6.16×10^{-4}

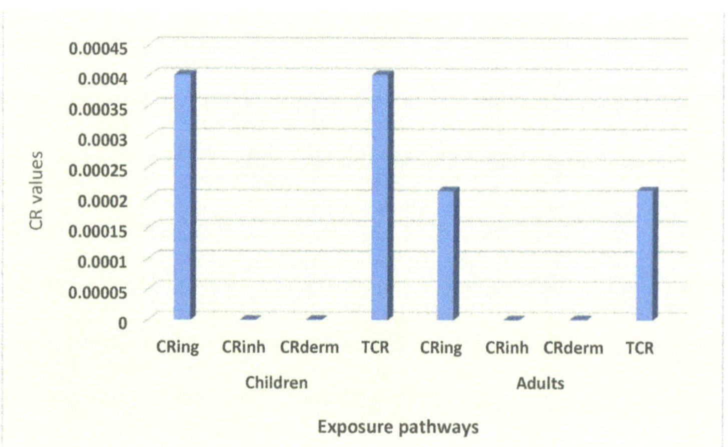

Figure 8. Contribution of exposure pathway carcinogenic risks.

Furthermore, all of the trace element cancer risks values were within the acceptable limit in both the children and adult groups, except for the Cr risk value. The CR values of all of the elements were as follows: Cr > Ni > As > Pb > Co and Cr > Ni > As > Pb > Co for children and adults, respectively (Figure 9. According to various researchers, including Qadeer et al. [9] in the urbanized cities of Pakistan, and Candeias et al. [31] in Viano do Castelo, Portugal, the ingestion of road dust in the children group was the leading contributor to cancer risk, which is comparable to the outcomes of this study. Similar findings confirming Cr to be the main driver of carcinogenic risk were also reported by Dat et al. [51] in the street dust of a metropolitan area in Southern Vietnam. Similarly, Dat et al. [51], for the street dust of a metropolitan area in Southern Vietnam, and Taiwo et al. [54], for Lagos metropolis, Nigeria, elucidated the ingestion pathway and determined Cr as being the leading contributor to the total cancer risk among adults. From the results of the cancer risk assessment, it was clear that trace elements in road dust had the possibility for contributing to lifetime carcinogenic risks for the entire population, and Cr was the leading contributor (Figure 10). Children were at a high risk of cancer compared to the adult group (Figure 9). The community should be educated on the importance of the environment, pollution, waste management, and health. Remediation, cleaning, and the regular monitoring of heavy metals is desirable for the safety of the inhabitants and the sustainability of the settlement.

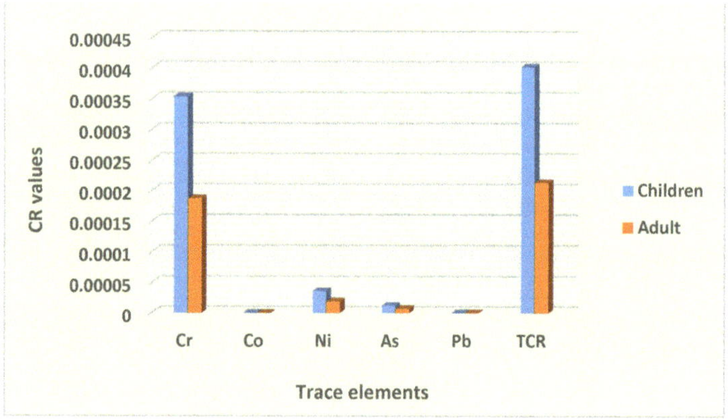

Figure 9. Contribution of trace elements to cancer risks.

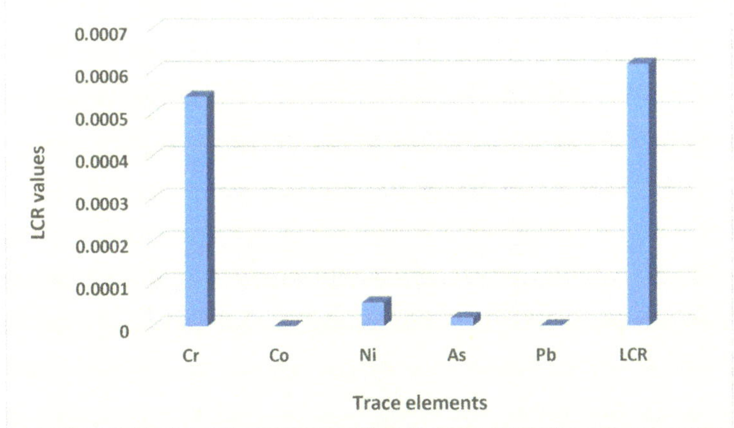

Figure 10. Contribution of trace elements to lifetime carcinogenic risks.

3.5. Health Implications Associated with Trace Elements Exposure in Road Dust

Among the examined elements, only six trace elements were above their average shale values, which is a health concern for the local population, particularly children. As reported by Jin et al. [66], outdoor play areas represent important exposure situations for children in many urban settlement areas. Hassaan et al. [57], stated that trace elements at high concentrations are toxic. The toxicity of trace elements in this study can be listed as Cr > Ba > Zn > Zr > Cu > Pb, and their exposure may lead to serious health implications to the local inhabitants. Cr may trigger liver disorder and irritation [11]. Barium (Ba) may lead to renal failure and lung sclerosis [67]. Exposure to zinc (Zn) may cause risks of prostate cancer, and exhaustion [12]. Zirconium (Zr) may lead to pulmonary effects and hoarseness [68]. Copper (Cu) is associated with dizziness and stomach cramps [12]. Furthermore, exposure to lead (Pb) may cause hormonal changes and reduced potency in males [69].

4. Conclusions

This study was aimed at determining the influence of urban informal settlements on trace element accumulation in road dust from Ekurhuleni Metropolitan Municipality, South Africa, and their health implications. The outcomes of the study have revealed that informal settlement activities have considerable influences on the accumulation of trace elements in road dust during the winter season. In this poor informal settlement, major elements (SiO_2 and P_2O_5) and trace elements (Cr, Cu, Zn, Zr, Ba, and Pb) were particular health concerns as they were above their corresponding average shale values. In particular, Cr was a major health concern to the inhabitants, as its level of accumulation in the road dust was high. Furthermore, the level of trace elements in road dust exhibited a possibility for non-cancer risks and cancer risks to the entire population, and Cr was the major driver for both non-carcinogenic and carcinogenic risks for the entire population. Cleaning and the regular monitoring of heavy metals in poor urban informal settlements is desirable for the safety of the inhabitants and the sustainability of the settlement. The study will provide valuable scientific data on urban informal settlement geochemistry and health risks, which will be used as a reference value and for remediation measures. The overall health risk assessment of trace elements in this study did not take into account the size of the road dust particles; thus, future work should cover the variation in particle size distribution and morphology over different seasons of the year.

Author Contributions: I.M., S.A.O. and T.P.G. conceived and designed the study. Samples were collected by I.M. Data was analyzed by all authors. All authors have read and agreed to the published version of the manuscript.

Funding: This research was financed by research development and postgraduate studies at the Central University of Technology, Free State. The APC was also funded by the Central University of Technology.

Institutional Review Board Statement: Not applicable.

Informed Consent Statement: Not applicable.

Data Availability Statement: The raw data utilized in this study are obtainable upon request from the corresponding author.

Acknowledgments: The authors gratefully acknowledge financial support from the Central University of Technology, Free State. Moreover, the Department of Geology at the University of Free State for laboratory analysis is acknowledged.

Conflicts of Interest: The authors have not declared any conflict of interest.

References

1. Msimang, Z. A Study of the Negative Impacts of Informal Settlements on the Environment. A Case Study of Jika Joe, Pietermaritzburg. Master's Thesis, University of KwaZulu-Natal, Berea, Durban, South Africa, 2017.
2. Weimann, A.; Oni, T. A Systematised Review of the Health Impact of Urban Informal Settlements and Implications for Upgrading Interventions in South Africa, a Rapidly Urbanising Middle-Income Country. *Int. J. Environ. Res. Public Health* **2019**, *16*, 3608. [CrossRef] [PubMed]
3. Menshawya, A.; Alya, S.S.; Salmana, A.M. Sustainable upgrading of informal settlements in the developing world, case study: Ezzbet Abd El Meniem Riyadh, Alexandria, Egypt. *Procedia Eng.* **2011**, *21*, 168–177. [CrossRef]
4. Charlesworth, S.M.; De Miguel, E.; Ordóñez, A. A review of the distribution of particulate trace elements in urban terrestrial environments and its application to considerations of risk. *Environ. Geochem. Health* **2011**, *33*, 103–123. [CrossRef]
5. Odat, O. Calculating Pollution Indices of Heavy Metal along Irbid/Zarqa Highway-Jordan. *Int. J. Appl. Sci. Technol.* **2013**, *3*, 72–76.
6. Yang, J.; Teng, Y.; Song, L.; Zuo, R. Tracing Sources and Contamination Assessments of Heavy Metals in Road and Foliar Dusts in a Typical Mining City, China. *PLoS ONE* **2016**, *11*, e0168528. [CrossRef]
7. Sampson, C. Trace Elements in Christchurch Road Dust. Master's Thesis, University of Canterbury, Christchurch, New Zealand, 2017.
8. Cai, K.; Li, C. Street Dust Heavy Metal Pollution Source Apportionment and Sustainable Management in A Typical City—Shijiazhuang, China. *Int. J. Environ. Res. Public Health* **2019**, *16*, 2625. [CrossRef]
9. Qadeera, A.; Saqibb, Z.A.; Ajmalb, Z.; Xinga, C.; Khalila, S.K.; Usman, M.; Huanga, Y.; Bashir, S.; Ahmad, Z.; Ahmede, S.; et al. Concentrations, pollution indices and health risk assessment of heavy metals in road dust from two urbanized cities of Pakistan: Comparing two sampling methods for heavy metals concentration. *Sustain. Cities Soc.* **2020**, *52*, 101953. [CrossRef]
10. Victoria, A.; Cobbina, S.J.; Dampare, S.B.; Duwiejuah, A.B. Heavy Metals Concentration in Road Dust in the Bolgatanga Municipality, Ghana. *J. Environ. Pollut. Hum. Health* **2014**, *2*, 74–80.
11. Ray, R.R. Adverse hematological effects of hexavalent chromium: An overview. *Interdiscip. Toxicol.* **2016**, *9*, 55–65. [CrossRef]
12. Okereafor, U.; Makhatha, M.; Mekuto, L.; Uche-Okereafor, N.; Sebola, T.; Mavumengwana, V. Toxic Metal Implications on Agricultural Soils, Plants, Animals, Aquatic life and Human Health. *Int. J. Environ. Res. Public Health* **2020**, *17*, 2204. [CrossRef]
13. Tissington, K. Legislation, Policy, Programmes and Practice: A Resource Guide to Housing in South Africa 1994–2010. Socio-Economic Rights Institute (SERI), South Africa. 2011. Available online: http:www.urbanlandmark.org.za/downloads/SERI_Housing_Resource_Guide_Feb11.pdf (accessed on 28 March 2022).
14. RoadLab. Detailed Design of the Midrand K109 Road Located between K27 and Dale Road, Midrand Area, Gauteng Province. 2014. Available online: https://nalisustainabilitysolutions.co.za (accessed on 3 January 2022).
15. Kowalska, J.B.; Mazurek, R.; Gasiorek, M.; Zaleski, T. Pollution indices as useful tools for the comprehensive evaluation of the degree of soil contamination—A review. *Environ. Geochem. Health* **2018**, *40*, 2395–2420. [CrossRef] [PubMed]
16. Timothy, N.; Williams, E.T. Environmental Pollution by Heavy Metal: An Overview. *Int. J. Environ. Chem.* **2019**, *3*, 72–82. [CrossRef]
17. Ma, J.; Singhirunnusorn, W. Distribution and Health Risk Assessment ofHeavy Metals in Surface Dusts of Maha Sarakham Municipality. *Procedia Soc. Behav. Sci.* **2012**, *50*, 280–293. [CrossRef]
18. Addo, M.A.; Darko, E.O.; Gordon, C.; Nyarko, B.J.B.; Gbadago, J.K. Heavy Metal Concentrations in Road Deposited Dust at Ketu-South District, Ghana. *Int. J. Sci. Technol.* **2012**, *2*, 28–39.
19. Taofeek, Y.A.; Tolulope, T.O. Evaluation of some Heavy Metals in Soils along a Major Road in Ogbomoso, South West Nigeria. *J. Environ. Earth* **2012**, *2*, 71–79.
20. Sebaiwa, M.M. Characterization of Dust Fallout around the City of Tshwane (Cot), Gauteng, South Africa. Master's Thesis, University of South Africa, Pretoria, South Africa, 2016.

21. Ghanavati, N.; Nazarpour, A.; Watts, M.J. Status, source, ecological and health risk assessment of toxic metals and polycyclic aromatic hydrocarbons (PAHs) in street dust of Abadan, Iran. *Catena* **2019**, *177*, 246–259. [CrossRef]

22. USEPA (United States Environmental Protection Agency). Soil Screening Guidance: Technical Background Document. Office of Soild Waste and Emergency Response. Washington, DC. 1996. Available online: https:www.epa.gov/superfund/superfund-soil-screening-guidance#:~[]:text=The%20Soil%20Screening%2 (accessed on 23 June 2021).

23. Gabarrón, M.; Faz, A.; Martínez-Martínez, S.; Zornoza, R.; Acosta, J.A. Assessment of metals behaviour in industrial soil using sequential extraction, multivariable analysis and a geostatistical approach. *J. Geochem. Explor.* **2017**, *172*, 174–183. [CrossRef]

24. Chonokhuu, S.; Batbold, C.; Chuluunpurev, B.; Battsengel, E.; Dorjsuren, B.; Byambaa, B. Contamination and health risk assessment of heavy metals in soil of major cities in Mongolia. *Int. J. Environ. Res. Public Health* **2019**, *16*, 2552. [CrossRef]

25. Zgłobicki, W.; Telecka, M. Heavy Metals in Urban Street Dust: Health Risk Assessment (Lublin City, E Poland). *Appl. Sci.* **2021**, *11*, 4092. [CrossRef]

26. USEPA (United States Environmental Protection Agency). Supplemental Guidance for Developing Soil Screening Levels for Superfund Sites. Office of Emergency and Remedial Response. Washington DC. 2002. Available online: https:epa.gov/superfund/superfund-soil-screening-guidance (accessed on 25 June 2021).

27. Ferreira-Baptista, L.; De Miguel, E. Geochemistry and risk assessment of street dust in Luanda, Angola: A tropical urban environment. *Atmos. Environ.* **2005**, *39*, 4501–4512. [CrossRef]

28. Li, H.; Qian, X.; Hu, W.; Wang, Y.; Gao, H. Chemical speciation and human health risk of trace metals in urban street dusts from a metropolitan city, Nanjing, SE China. *Sci. Total Environ.* **2013**, *456*, 212–221. [CrossRef] [PubMed]

29. Lu, X.; Wu, X.; Wang, Y.; Chen, H.; Gao, P.; Fu, Y. Risk assessment of toxic metals in street dust from a medium-sized industrial city of China. *Ecotoxicol. Environ. Saf.* **2014**, *106*, 154–163. [CrossRef]

30. USEPA (United States Environmental Protection Agency). Role of the Baseline Risk Assessment in Superfund Remedy Selection Decisions. April 22—Memorandum. Office of Solid Waste and Emergency Response. OSWER Directive. 1991. Available online: https://www.epa.gov/risk/role-baseline-risk-assessment-superfund-remedy-selection-descisions (accessed on 22 June 2021).

31. Candeias, C.; Vicente, E.; Tomé, M.; Rocha, F.; Ávila, P.; Célia, A. Geochemical, Mineralogical and Morphological Characterisations of Road Dust and Associated Health Risks. *Int. J. Environ. Res. Public Health* **2020**, *17*, 1563. [CrossRef] [PubMed]

32. USEPA (United State Environmental Agency). *Summary Review of Health Effects Associated with Elemental and Inorganic Phosphorus Compounds: Health Issue Assessment*; Office of Research and Development: USA Environmental Protection Agency: Washington, DC, USA, 1990. Available online: https://nepis.epa/EPA/html/DLwait.htm?url=/Exe/ZPDF.cgi/30001IQN.PDF?Dockey=30001IQN.PDF (accessed on 23 October 2021).

33. Li, X.; Liu, B.; Zhang, Y.; Wang, J.; Ullah, H.; Zhou, M.; Peng, L.; He, A.; Zhang, X.; Yan, X.; et al. Spatial Distributions, Sources, Potential Risks of Multi-Trace Metal/Metalloids in Street Dusts from Barbican Downtown Embracing by Xi'an Ancient City Wall (NW, China). *Int. J. Environ. Res. Public Health* **2019**, *16*, 2992. [CrossRef]

34. Clarke, F.W.; Washington, H.S. *The Composition of the Earth's Crust*; USA Government Printing Office: Washington, DC, USA, 1924; Volume 127, p. 117.

35. Turekian, K.K.; Wedepohl, K.H. Distribution of the elements in some major units of the earth crust. *Geol. Soc. Am. Bull.* **1961**, *72*, 175–192. [CrossRef]

36. Department of Environmental Affairs (DEA). The Framework for the Management of Contaminated Land, South Africa. 2010. Available online: http://sawic.environment.gov.za/documents/562.pdf (accessed on 11 April 2022).

37. Kamunda, C.; Mathuthu, M.; Madhuku, M. Health Risk Assessment of Heavy Metals in Soils from Witwatersrand Gold Mining Basin, South Africa. *Int. J. Environ. Res. Public Health* **2016**, *13*, 663. [CrossRef]

38. Chen, H.Y.; Chang, H.L.R. Development of low temperature three-way catalysts for future fuel efficient vehicles. *Johns. Matthey Technol. Rev.* **2015**, *59*, 64–67. [CrossRef]

39. Adamiec, E. Chemical fractionation and mobility of traffic-related elements in road environments. *Environ. Geochem. Health* **2017**, *39*, 1457–1468. [CrossRef]

40. Moryani, H.T.; Kong, S.; Du, J.; Bao, J. Health Risk Assessment of Heavy Metals Accumulated on PM2.5 Fractioned Road Dust from Two Cities of Pakistan. *Int. J. Environ. Res. Public Health* **2020**, *17*, 7124. [CrossRef]

41. Cui, W.; Meng, Q.; Feng, Q.; Zhou, L.; Cui, Y.; Li, W. Occurrence and release of cadmium, chromium and lead from stone coal combustion. *Int. J. Coal Sci. Technol.* **2019**, *6*, 586–594. [CrossRef]

42. Shi, D.; Lu, X.; Wang, Q. Evaluating Health Hazards of Harmful Metals in Roadway Dust Particles Finer than 100 μm. *Pol. J. Environ. Stud.* **2017**, *27*, 2729–2737.

43. Sager, M. Urban Soils and Road Dust—Civilization Effects and Metal Pollution—A Review. *Environments* **2020**, *7*, 98. [CrossRef]

44. Salah, E.; Turki, A.; Noori, S. Heavy Metals Concentration in Urban Soils of Fallujah City, Iraq. *J. Environ. Earth Sci.* **2013**, *3*, 100–113.

45. Rahman, M.S.; Khan, M.D.H.; Jolly, Y.N.; Kabir, J.; Akter, S.; Salam, A. Assessing risk to human health for heavy metal contamination through street dust in the Southeast Asian Megacity: Dhaka, Bangladesh. *Sci. Total Environ.* **2019**, *660*, 1610–1622. [CrossRef] [PubMed]

46. Moskovchenko, D.; Pozhitkov, R.; Ukarkhanova, D. Geochemistry of Street Dust in Tyumen, Russia: Influence of Traffic Load. *Environ. Sci. Pollut. Res. Int.* **2022**, *4*, 223. [CrossRef]

47. Kianpor, M.; Payandeh, K.; Ghanavati, N. Environmental Assessment of Some Heavy Metals Pollution in Street Dust in the Industrial Areas of Ahvaz. *Jundishapur J. Health Sci.* **2019**, *1*, 87212. [CrossRef]
48. Sager, M.; Chon, H.T.; Marton, L. Spatial variation of contaminant elements of roadside dust samples from Budapest (Hungary) and Seoul (Republic of Korea), including Pt, Pd and Ir. *Environ. Geochem. Health* **2015**, *37*, 181–193. [CrossRef] [PubMed]
49. Trujillo-González, J.M.; Torres-Mora, M.A.; Keesstra, S.; Brevik, E.C.; Jiménez-Ballesta, R. Heavy metal accumulation related to population density in road dust samples taken from urban sites under different land uses. *Sci. Total Environ.* **2016**, *553*, 636–642. [CrossRef]
50. Tan, B.; Wang, H.; Wang, X.; Ma, C.; Zhou, J.; Dai, X. Health Risks and Source Analysis of Heavy Metal Pollution from Dust in Tianshui, China. *Minerals* **2021**, *11*, 502. [CrossRef]
51. Dat, N.D.; Nguyen, V.; Vo, T.; Bui, X.; Bui, M.; Nguyen, L.S.P.; Nguyen, X.; Tran, A.T.; Nguyen, T.; Ju, Y.; et al. Contamination, source attribution, and potential health risks of heavy metals in street dust of a metropolitan area in Southern Vietnam. *Environ. Sci. Pollut. Res.* **2021**, *28*, 50405–50419. [CrossRef]
52. Al-Dabbas, M.A.; Mahdi, K.H.; Al-Khafaji, R.; Obayes, K.H. Heavy metals characteristics of settled particles of streets dust from Diwaniyah City-Qadisiyah Governorate-Southern Iraq. *IOP Conf. Ser. J. Phys.* **2018**, *1003*, 012023. [CrossRef]
53. Rybak, J.; Wróbel, M.; Bihałowicz, J.S.; Rogula-Kozłowska, W. Selected Metals in Urban Road Dust: Upper and Lower Silesia Case Study. *Atmosphere* **2020**, *11*, 290. [CrossRef]
54. Taiwo, A.M.; Musa, M.O.; Oguntoke, O.; Afolabi, T.A.; Sadiq, A.Y.; Akanji, M.A.; Shehu, M.R. Spatial distribution, pollution index, receptor modelling and health risk assessment of metals in road dust from Lagos metropolis, Southwestern Nigeria. *Environ. Adv.* **2020**, *2*, 100012. [CrossRef]
55. Khan, M.D.H.; Talukder, A.; Rahman, M.S. Spatial distribution and contamination assessment of heavy metals in urban road dusts from Dhaka city, Bangladesh. *IOSR J. Appl. Chem.* **2018**, *11*, 90–99.
56. Mmolawa, K.B.; Likuku, A.S.; Gaboutloeloe, G.K. Assesment of heavy meatal pollution in soil along a major roadside areas in Botswana. *Afr. J. Environ. Sci. Technol.* **2011**, *5*, 186–196.
57. Hassaan, M.A.; El Nemr, A.; Madkour, F.F. Environmental Assessment of Heavy Metal Pollution and Human Health Risk. *Am. J. Water Sci. Eng.* **2016**, *2*, 14–19.
58. Maeaba, W.; Prasad, S.; Chandra, S. First assessment of metals contamination in road dust and roadside soil of Suva City, Fiji. *Arch. Environ. Contamin. Toxicol.* **2019**, *77*, 249–262. [CrossRef]
59. Weissmannova, H.D.; Mihocova, S.; Chovanec, P.; Pavlovsky, J. Potential ecological risk and human health risk assessment of heavy metal pollution in industrial affected soils by coal mining and metallurgy in Ostrava, Czech Republic. *Int. J. Environ. Res. Public Health* **2019**, *16*, 4495. [CrossRef]
60. Yalala, B.N. Characterization, Bioavailability and Health Risk Assessment of Mercury in Dust Impacted by Gold Mining. Ph.D. Thesis, University of the Witwatersrand, Johannesburg, South Africa, 2015.
61. Shinggu, D.Y.; Ogugbuaja, V.O.; Toma, I.; Barminas, J.T. Determination of heavy metal pollutants in street dust of Yola, Adamawa State, Nigeria. *Afr. J. Pure Appl. Chem.* **2009**, *4*, 17–21.
62. United State Environmental Protection Agency (USEPA). Assessment. Risk Assessment Guidance for Superfund (Rags), Volume I: Human Health Evaluation Manual (Part E, Supplemental Guidance for Dermal Risk Assessment) Interim. 2009. Available online: https://www.epa.gov/sites/default/files/2015-09/documents/rags_a.pdf (accessed on 10 April 2022).
63. United State Environmental Protection Agency (USEPA). Regional Screening Levels (RSLs). 2021. Available online: https://www.epa.gov/risk/regional-screening-levels-rsls-generic-tables (accessed on 10 April 2022).
64. United State Environmental Protection Agency (USEPA). Framework for Determining a Mutagenic Mode of Action for Carcinogenicity: Review Draft. 2007. Available online: http://www.epa.gov/osa/mmoaframework/pdfs/MMOA-ERD-FINAL-83007.pdf (accessed on 12 April 2022).
65. RAIS-The Risk Assessment Information System. 2016. Available online: https://rais.ornl.gov/tools/tox_profiles.html (accessed on 10 April 2022).
66. Jin, Y.; O'Connora, D.; Ok, Y.S.; Tsang, D.C.W.; Liu, A.; Houa, D. Assessment of sources of heavy metals in soil and dust at children's playgrounds in Beijing using GIS and multivariate statistical analysis. *Environ. Int.* **2019**, *124*, 320–328. [CrossRef]
67. Kravchenko, J.; Darrah, T.H.; Miller, R.K.; Lyerly, H.K.; Vengosh, A. A review of the health impacts of barium from natural and anthropogenic exposure. *Environ. Geochem. Health* **2014**, *36*, 797–814. [CrossRef] [PubMed]
68. Liippo, K.K.; Anttila, S.L.; Taikina-Aho, O.; Ruokonen, E.L.; Toivonen, S.T.; Tuomi, T. Hypersensitivity pneumonitis and exposure to Zirconium silicate in a young ceramic tile worker. *Am. Rev. Respir. Dis.* **1993**, *148*, 1089–1092. [CrossRef] [PubMed]
69. Aremu, M.O. *Environment, Health and Nutrition: Global Perspective*; Basu, S.K., Datta, B., Eds.; A.P.H. Publishing Inc.: New Delhi, India, 2008; pp. 79–96.

Article

Water-Rock Interaction Processes: A Local Scale Study on Arsenic Sources and Release Mechanisms from a Volcanic Rock Matrix

Daniele Parrone *, Stefano Ghergo, Elisabetta Preziosi and Barbara Casentini

Water Research Institute—National Research Council, IRSA-CNR, Via Salaria km 29.300, PB 10, 00015 Rome, Italy; stefano.ghergo@irsa.cnr.it (S.G.); elisabetta.preziosi@irsa.cnr.it (E.P.); barbara.casentini@irsa.cnr.it (B.C.)
* Correspondence: daniele.parrone@irsa.cnr.it

Abstract: Arsenic is a potentially toxic element (PTE) that is widely present in groundwater, with concentrations often exceeding the WHO drinking water guideline value (10.0 µg/L), entailing a prominent risk to human health due to long-term exposure. We investigated its origin in groundwater in a study area located north of Rome (Italy) in a volcanic-sedimentary aquifer. Some possible mineralogical sources and main mechanisms governing As mobilization from a representative volcanic tuff have been investigated via laboratory experiments, such as selective sequential extraction and dissolution tests mimicking different release conditions. Arsenic in groundwater ranges from 0.2 to 50.6 µg/L. It does not exhibit a defined spatial distribution, and it shows positive correlations with other PTEs typical of a volcanic environment, such as F, U, and V. Various potential As-bearing phases, such as zeolites, iron oxyhydroxides, calcite, and pyrite are present in the tuff samples. Arsenic in the rocks shows concentrations in the range of 17–41 mg/kg and is mostly associated with a minor fraction of the rock constituted by FeOOH, in particular, low crystalline, containing up to 70% of total As. Secondary fractions include specifically adsorbed As, As-coprecipitated or bound to calcite and linked to sulfides. Results show that As in groundwater mainly originates from water-rock interaction processes. The release of As into groundwater most likely occurs through desorption phenomena in the presence of specific exchangers and, although locally, via the reductive dissolution of Fe oxy-hydroxides.

Keywords: potentially toxic elements; aquifer; volcanic tuff; sequential extraction; Fe oxyhydroxides; sorption

Citation: Parrone, D.; Ghergo, S.; Preziosi, E.; Casentini, B. Water-Rock Interaction Processes: A Local Scale Study on Arsenic Sources and Release Mechanisms from a Volcanic Rock Matrix. *Toxics* **2022**, *10*, 288. https://doi.org/10.3390/toxics10060288

Academic Editor: Ilaria Guagliardi

Received: 29 April 2022
Accepted: 25 May 2022
Published: 27 May 2022

Publisher's Note: MDPI stays neutral with regard to jurisdictional claims in published maps and institutional affiliations.

1. Introduction

Arsenic is a natural element easily found throughout the environment. Its presence in drinking water represents a hazard to human health [1]. Long-term exposure to As-rich waters can cause serious diseases, from skin lesions, cardiovascular diseases, and type II diabetes to bladder, lung, and skin cancers [1–3]. In groundwater, this element frequently shows concentrations exceeding 10.0 µg/L, which is the guideline value suggested by WHO [4] for drinking water and was implemented in Europe as standard by 98/83/EC [5], entailing a prominent risk to human health.

The presence of this potentially toxic element (PTE) in groundwater depends on the hydrogeological setting as well as on various natural processes, including climate, biological activity, and volcanic emissions. Since the 1990s, wide occurrences of As in well water in Bangladesh have attracted the attention of the scientific community, which seeks to deepen its knowledge of the origin and fate of arsenic in groundwater [6–9]. Water-rock interactions under favorable biogeochemical conditions are considered to be the most important mechanism for the release of As in aquifers by far [10]. Arsenic concentrations in natural waters may range from 0.5 to > 5000 µg/L [9]. Arsenic-enriched groundwater has been reported in many parts of the world, mainly in Asia (Bangladesh, China, India,

Vietnam, Nepal), the Americas (USA, Mexico, Argentina, Chile, etc.), and Europe (Italy, Greece, the Pannonian Basin, etc.) [11]. Arsenic release in groundwater has been mainly attributed to three possible processes: the reductive dissolution of Fe/Mn hydroxides, ion exchange with different competitive exchangers (e.g., phosphate, bicarbonate, and silicate), and the oxidation of arsenic-bearing minerals [9,12,13]. Precipitation/dissolution, sorption/desorption, biotic and abiotic oxidation/reduction [14,15], and complex formation are the main reactions controlling the distribution of arsenic between the aqueous and solid phases.

The scientific literature is abundant regarding the presence of As and the main processes leading to its release in groundwater in alluvial environments [16–20], though less so in volcanic contexts. Alarcón-Herrera et al. [21] found a co-occurrence of As and F in oxidized and alkaline environments due to interactions with volcanic glass and, to a lesser extent, with hydrothermal minerals. Further, they identified Fe, Mn, and Al (hydr-)oxides as secondary sources due to their great adsorbing properties. In Argentina, Francisca and Carro Perez [22] reported the presence of As, F, and V in groundwater due to volcanic glass leaching. In a volcanic aquifer in Mexico, under oxidative conditions and high temperature, Morales et al. [23] observed an As mobilization from metallic sulfides, which was favored by thermal water rising through faults and fractures. Ren et al. [24] identify Tertiary volcanic tuffs as the main source of the high As content in groundwater, hypothesizing an origin linked to the leaching of Arsenogoyazite-like arsenic minerals. In a geothermally active area of northern Greece, Winkel et al. [25] analyzed different travertines containing high levels of arsenic (up to 913 mg/kg) and found that it mainly coprecipitated with calcite. In this regard, Di Benedetto et al. [26] claimed that As incorporation in calcite may be an effective limit of As mobility under conditions where immobilization through sorption by Fe and/or Mn oxy-hydroxides is not acting.

In the volcanic aquifers of northern Latium (central Italy), the presence of As and other PTEs typical of the volcanic environment is well known [27,28]. However, the main origin of the As contamination, as well as whether a common process associates arsenic and the other PTEs, is still unclear. Vivona et al. [29] claimed volcanic rock leaching in cold groundwater, while Angelone et al. [30] pointed to a major influence of localized thermal fluids rising along fractures. In the same region, Armiento et al. [31] and Cinti et al. [32] ascribed the diffused As concentration in groundwater to water-rock interaction processes, locally enhanced by thermalism and volcanic gas emissions. Casentini et al. [33] performed batch leaching tests to investigate the release of As induced by interactions of volcanic rocks (lava and tuffs) with inorganic anions and found a positive influence of F and HCO_3, likely due to anion exchange processes.

The main laboratory experiments aiming to understand the mechanisms of arsenic release include column tests [34–36] and batch tests [37–39]. Several laboratory studies showed that As is released from sediments under anaerobic conditions [40–42], particularly following the reductive dissolution of Fe and Mn oxyhydroxides [43]. Batch tests are often used to study the effects of oxidation-reduction potential (ORP) variations [19], which also affect pH [44] and could naturally occur as a consequence of the mixing of waters with different chemical compositions. Frohne et al. [38] found that dissolved As concentration decreases significantly with increasing ORP and concluded that low ORP levels promote As mobility. Batch tests have also been used to investigate As sorption and dissolution kinetics linked to various mineral phases [45,46].

Arsenic is relatively scarce in the Earth's continental crust (1.5–2 mg/kg) and can be associated with different As-bearing phases [47]. In this regard, Selective Sequential Extraction (SSE) is a valuable procedure to quantify the distribution of As between different solid fractions that constitute the rocks [48–51], potentially providing important information about the adsorption-bearing phases and related processes occurring at the soil/water interface [52]. The literature does not lack works on sequential extraction methods aimed at arsenic fractionation [53–59], and additional procedures have been proposed to improve the extraction of specific As-bearing phases [60,61]. SSE has been used to characterize As distribution and mobility both in contaminated areas [62–65] and in pristine aquifers.

This work aims to study the main mechanisms that govern As behavior in the water-rock interaction processes in a volcanic context where aquifer rocks are dominated by tuffs. The research was organized in the following three phases:

- Sampling and analysis of water samples from 69 private wells and springs tapping a volcanic aquifer located just north of Rome for the geochemical characterization of groundwater.
- Selective Sequential Extraction operated on selected tuff samples extracted from three quarries within the study area for defining the As distribution among the different solid phases that potentially constitute the aquifer matrix.
- Two batch tests realized on one of the samples analyzed by SSE in order to simulate water-rock interaction processes in different conditions.

Lastly, field and laboratory data have been analyzed in order to evaluate the significant connections between aquifer mineralogy and groundwater geochemistry, with particular reference to As, and also to deepen its relationships with other PTEs in groundwater.

2. Materials and Methods

2.1. Study Area

The study area (Figure 1) lies on the eastern flank of the Sabatini volcanic apparatus.

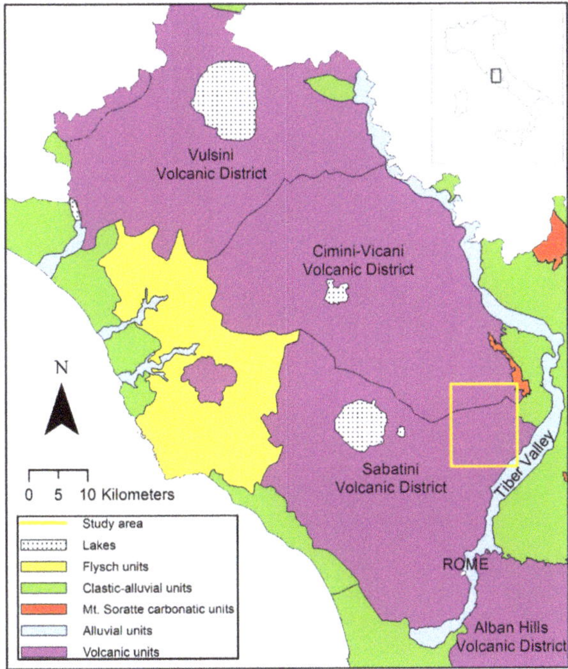

Figure 1. Hydrogeological scheme of Latium volcanic domain, from Parrone et al. [66]. The yellow box indicates the investigated area.

Its activity developed from 0.6 to 0.04 M.a. b.p. through five main phases, during which ultrapotassic volcanic rocks ranging from trachybasalts to trachytes to phonolites were erupted. During the Sabatini volcanic district formation, the Bracciano lake volcano-tectonic depression and several calderas developed [67–69]. The volcanic activity developed through several vents distributed over a large area, the main one being the Sacrofano system in the east and the Bracciano system in the west.

Pleistocene volcanic products, which are usually characterized by an overall medium permeability [70], overlap, in angular unconformity, Pliocene and Pleistocene sedimentary deposits of marine and continental origin along with clays and silts that form the low permeability bottom of the groundwater body.

The volcanic deposits outcropping in the study area include mainly pyroclastic deposits: "Via Tiberina Yellow Tuff" (VTYT), "Sacrofano Tuff", and "La Storta Tuff" (Figure 2). Among these, the VTYT is the most important formation in the investigated aquifer. Cineritic levels and paleosoils determine local decreases of the vertical permeability resulting in a multilayer aquifer with semiconfined horizons [71].

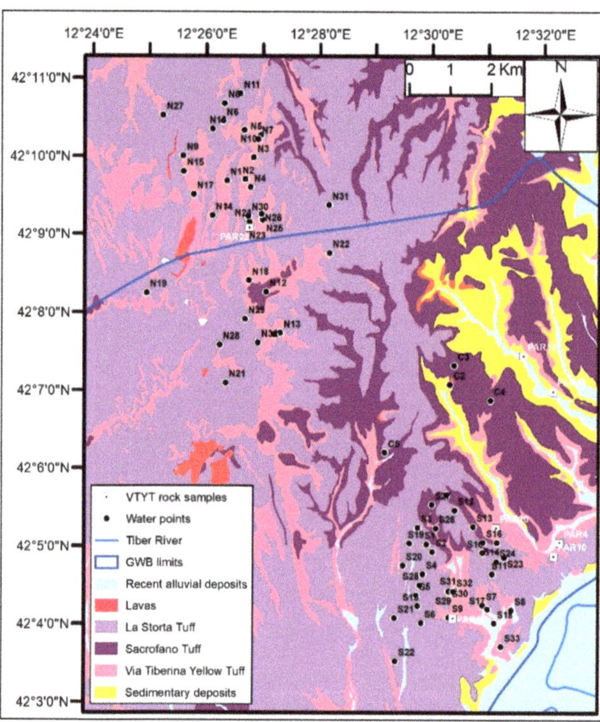

Figure 2. Simplified geological scheme with the location of the rock sampling points of the "Via Tiberina Yellow Tuff" (VTYT) and the groundwater sampling points.

The VTYT is a pyroclastic flow (ignimbrite) produced by the activity of the Sacrofano center, located about 15 km west of the Tiber Valley, which has been set in place directly on the Plio-Pleistocene sedimentary basement. The VTYT has several overlapping units with different lithological characters. Nappi et al. [72] identified a stratigraphically upper lithology with components of smaller dimensions and, between the abundant zeolites, chabazite prevailing over phillipsite; the other, which can be found in the lower levels of the formation, has coarse grain size, has more abundant phillipsite, and shows alterations (clays) due to the action of groundwater. De Rita et al. [73] distinguish three units, spaced by pyroclastite layers of different nature, while Campobasso et al. [74] recognize at least seven depositional units.

The formation has been extensively described from the petrographic and chemical points of view by Lombardi and Marra [75]. The VTYT appears as a massive aggregate, mainly lithoid, of pumices of centimetric sizes linked by a very fine matrix, with colors ranging from yellow to light gray. The average porosity stands at 48% [75]. As reported by Jackson et al. [76], glassy matrix represents 42–54% of the rock, lithic fragments and crystals 6–14%, and secondary minerals such as zeolites and calcite 37–48%. Zeolitization processes of the vitreous mass are evident, with the formation of chabazite and phillipsite aggregates that are primarily responsible for the high degree of lithification of the rock. Another widespread secondary mineral is calcite, resulting from precipitation of the water circulating in diagenetic phases and present not only in the common carbonate fragments but also in the cementing matrix [75].

2.2. Groundwater Sampling and Analysis

The first phase of the work consisted of sampling and analysis of groundwater samples from the investigated area. This phase is aimed at the geochemical characterization of circulating groundwater, with a particular focus on the As content and distribution in the two study sectors.

A total of 69 groundwater samples (Figure 2) were collected from wells (64) and springs (5), of which 32 were in the northern sector and 37 were in the southern one. These are mainly private wells for domestic/irrigation use.

At each site, pH, EC, Eh, DO, and water T were measured with probes (Hach-Lange). The sampling was carried out after a purging of the wells until the physical-chemical parameters stabilized (usually 20–30 min). All samples were filtered in the field with 0.45 μm membrane filters under N_2 and stored in HNO_3 1% rinsed polyethylene bottles. One fraction was immediately acidified by HNO_3 (Suprapure, Merck) for major cations and trace metal determination. Bicarbonates were determined in laboratory by HCl (Suprapure, Merck) titration on 50 mL of sample within 24 h from sampling. Field blanks with ultrapure water were periodically collected, checking for possible environmental contaminations. The samples were analyzed for anions by ion chromatography (IC, Dionex DX-120), for major cations by ICP-OES (Perkin Elmer P400) and trace elements by inductively coupled plasma mass spectrometry (ICP-MS, Agilent 7500c); certified materials (NIST 1640a, trace elements in natural waters) were used to check accuracy of the laboratory measurements. The electro-neutrality (EN%) evaluated as the percent difference for major cations and anions, including F and NO_3, ranges between −3.96% and +5.10%. For statistical analysis, the concentrations below detection limit (LOD) were assumed to be equal to half of LOD itself.

2.3. Statistical Analysis of Groundwater Chemical Data

A univariate statistical analysis of groundwater chemical data was carried out. Groundwater analytical data were processed by means of descriptive basic statistics, and the concentrations of some parameters were represented via scatterplots and distribution maps.

Normality or lognormality of As distribution was verified using the Shapiro-Wilk goodness of fit test [77]. In addition, Rosner's parametric test [78] and Huber's non-parametric test [79] for detection of data outliers were applied.

Finally, a bivariate correlation analysis in the two study sectors was carried out using both Pearson's linear correlation index, which is useful in case of normal data distributions, and Spearman's rank correlation index, which is more robust and suitable for skewed distributions or in presence of data outliers.

Data processing was performed through different software: ProUCL 5.1 [80] for the normality/lognormality and outlier detection statistical tests, Past 4.05 [81] for the bivariate correlation analysis, and ArcGIS 10.2.2 for the As concentration mapping.

2.4. Outcropping Rock Sampling and Mineralogical Analysis

Seven rock samples from the "Via Tiberina Yellow Tuff" (VTYT) were collected for mineralogical analysis (Figure 2). Four of these samples (PAR04, PAR10, PAR11, and PAR22) were extracted from excavation fronts of VTYF tuff quarries. The superficial portion of the samples was removed in order to eliminate the part of the rock that has undergone exogenous alteration processes. The samples were then transported to the laboratory for the analysis.

Tuff samples were analyzed by X-Ray Powder Diffraction (XRPD) and Scanning Electron Microscopy (SEM), in order to have a first mineralogical characterization of the aquifer matrix. The XRPD patterns were collected with a Siemens D5000 diffractometer operating with Bragg-Brentano $\theta/2\theta$ geometry, $CuK\alpha = 1.518$ Å, 40 kV, and 40 mA. Each XRPD pattern was collected from 4 to $80°$ of 2θ, with a step scan of $0.02°$. Identified Bragg reflections were assigned to the corresponding crystalline standards contained in the inorganic crystal structure database (ICSD). Powdered samples were prepared by hand grinding using an agate mortar and pestle, then sieved with a sieve of 0.5 mm mesh. Powders were further ground prior to the analysis to achieve a particle size < 50 μm.

SEM observations were performed with the Philips XL30 Analytical Scanning Electron Micro-scope equipped with secondary (2 nm imaging resolution), backscattered (0.1 AZ elemental resolution) electron detectors, and probe for Energy Dispersive X-ray Analysis (EDAX 134 eV) for the execution of punctual elemental analysis of the mineralogical phases and spectrum representation. SEM investigations were carried out on ordinary 30 μm thin sections after graphite sputter-coating of the samples.

Preliminary observations of the thin sections at the optical polarizing microscope (Nikon Eclipse E400 Pol) were also performed.

2.5. Selective Sequential Extraction

For three selected VTYT samples (PAR10, PAR11, and PAR22) we performed a Selective Sequential Extraction (SSE) in order to determine the distribution of As between the different solid bearing phases forming the rocks and to better understand the possible processes governing As mobility. Several protocols are available in literature to perform SSE analysis [82,83]. For the extraction procedure, we chose the methodology proposed by Wenzel [57] and modified it in order to consider a specific step for calcite from Costagliola et al. [61] and one for sulfides from Torres and Auleda [84].

The final procedure is able to distinguish seven fractions that can potentially bind As: (1) easily exchangeable As, (2) specifically adsorbed As, (3) As bound with calcite, (4) low crystalline Fe(III)-oxyhydroxides, (5) crystalline Fe(III)-oxides, (6) sulfides, and (7) residual fraction. The seven different steps that constitute the procedure are presented in Table 1.

Table 1. Selected sequential extraction procedure for As fractioning.

Step	Extractant	Time/Temperature	As Fraction	Reference
1	$(NH_4)_2SO_4$ 0.05M	4 h/25 °C	Easily exchangeable As	[57]
2	$(NH_4)H_2PO_4$ 0.05M	16 h/20 °C	Specifically adsorbed As	[57]
3	Acetic buffer 1M (pH 5)	12 h/25 °C	As bound with calcite	[61]
4	NH_4-oxalate buffer 0.2M (pH 3.2)	4 h/20 °C + 10 min (wash)	Low crystalline Fe(III)-oxyhydroxides	[57]
5	NH_4-oxalate buffer 0.2M + ascorbic acid 0.1M (pH 3.2)	30 min/96 °C + 10 min (wash)	Crystalline Fe(III)-oxides	[57]
6	HNO_3 8M	3 h/80 °C	Sulfides	[84]
7	Microwave digestion ($HNO_3 + H_2O_2$)	30 min/180 °C	Residual fraction	[57]

For the SSE, we started from 1 g of tuff powder. After each step, the suspension was centrifuged at 5000 rpm for 10 min, supernatant separated, diluted with MilliQ (18.2 MΩ cm at 25 °C), acidified with 1% HNO_3, and stored at 4 °C until analysis conducted by ICP-MS (Agilent 7500c). Two replicates were performed for each rock sample.

2.6. Batch Tests

In order to investigate the water-rock interaction processes and evaluate the release of arsenic and other elements under different conditions, two batch tests were performed on a previously selected Yellow Tuff sample (PAR11). In polyethylene bottles, two different extracting solutions (40 mL) were added to 2 g of pulverized rock (solid/liquid ratio 1/20) using:

- Synthetic water consisting of MilliQ (18.2 MΩ cm at 25 °C) to which KNO_3 was added until obtaining a low ionic strength solution (0.5 mM) to simulate contact with rainwater.
- As-free groundwater with ionic strength and physical-chemical characteristics similar to those of groundwater circulating in the study area, simulating the water-rock interaction processes.

The chemical composition of the two extracting solutions is reported in Table S1.

The bottles were placed in a horizontal shaker at 400 rpm at room temperature, and aliquots (20 mL) were taken at fixed intervals (3, 6, 12, 24, 48, 120, 720 h). Samples were centrifuged at 5000 rpm for 10 min and supernatant separated, diluted with MilliQ (18.2 MΩ cm at 25 °C), acidified 1% with HNO_3, and stored at 4 °C until analysis. Before each sampling step, pH value of the samples was measured with probes. After each sampling, 20 mL of fresh solution was added to the bottles, inducing new dissolution at each step.

A second test was carried out with an experimental apparatus consisting of a 500 mL polyethylene bottle filled with the solid matrix (PAR11) and the synthetic rainwater. Oxygen, redox, and pH variations in the system were continuously monitored by DO, Eh, and pH electrodes. The system was also equipped to be N_2 fluxed, and water samples were collected at fixed intervals. In addition, a solution of sulfide (25 mg/L S^{2-}) was injected into the batch solution to simulate sulfide-rich hypoxic conditions. A total of 20 g of the pulverized tuff were placed in 400 mL of synthetic solution and stirred for the whole duration of the test using a magnetic stirrer. The test was divided into different steps:

- STEP 1 (dissolution under aerobic conditions): The sample was in solution under oxygenated conditions. We carried out water sampling after 0, 4, 24, 48, 72, and 96 h.
- STEP 2a (induced anaerobic conditions): subsequently, the test continued by blowing N_2 within the system in order to eliminate the oxygen present. In this phase, we collected water samples after 24, 48, and 72 h.
- STEP 2b (induced anaerobic condition with sulfide presence): After repeating STEP 1 (4 days in oxygenated conditions with no sampling), the test continued under N_2 flux and adding sulfide to reach 0.5 mg/L final concentration. We then collected aliquots at 0, 3, 6, 24, 48, and 72 h.

For each sample, 15 mL of water was collected with a syringe, filtered with 0.4 μm (polycarbonate filters), diluted with MilliQ (18.2 MΩ cm at 25 °C), acidified by 1% with HNO_3, and stored at 4 °C until analysis. Samples of the two batch tests were analyzed for anions (only filtered aliquots) by IC (Dionex DX-120), for Fe by UV/Vis spectrophotometry (Hach Lange DR2800), and for trace elements by ICP-MS (Agilent 7500c).

3. Results and Discussion

3.1. Groundwater Geochemistry

As observed by Parrone et al. [66], piezometric levels decrease according to the main flow direction from NNW-SSE towards the Tiber River. Groundwater shows the dominance of a bicarbonate-earth-alkaline facies, with samples ranging from Ca-HCO_3 types to Na+K-HCO_3 types.

Water samples clearly show different pH values, passing from acidic (median = 6.1) to alkaline (median = 7.7) conditions from the northern to the southern sector (Figure 3). Low pH values in the north are probably related to a widespread circulation of CO_2 that can be hypothesized to be in the area of the hydrogeological watershed near the ancient Sacrofano caldera, as evidenced by the bubbling observed during the sampling of numerous wells. Some parameters typical of volcanic environments (e.g., PO_4, V), together with Al, show higher concentrations in the northern area, while an enrichment in HCO_3 and Na can be observed in the south (Table S2). This suggests a greater influence of the sedimentary deposits on the groundwater chemistry in the southern sector. As shown in Figure 3, in support of this hypothesis, the Na/K ratio increases from the north (median = 1.05) to the south (median = 1.55). The enrichment in K represents a typical feature of the waters circulating in the potassium-alkaline volcanites of Central Italy [29,85].

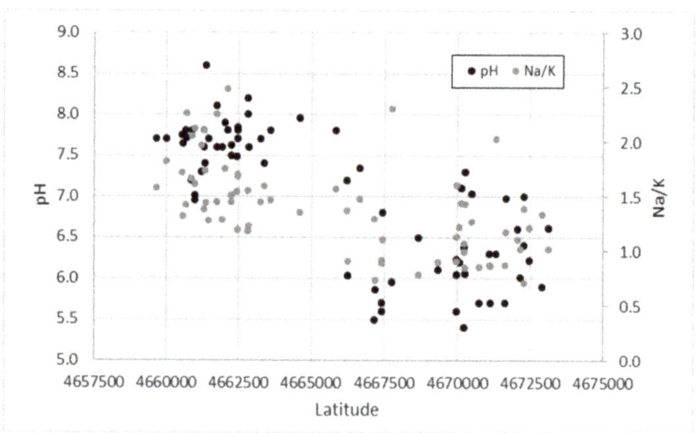

Figure 3. Latitude trend of the pH values (left ordinate scale) and the Na/K ratio (right ordinate scale).

Arsenic does not present a clear spatial distribution (Figure S1), showing slightly higher values on average in the southern area and some localized peaks. In most of the samples (91.3%), As exceeds the drinking water standard (10.0 µg/L), without particular differences between the northern (maximum value 50.6 µg/L and 90.6% of exceedances) and the southern sector (maximum value 50.2 µg/L and 91.9% of exceedances).

In both sectors, As shows lognormal distributions (Shapiro-Wilk test, $\alpha = 0.05$) and no data outliers (Rosner's parametric test, Huber's non-parametric test).

In Table 2, As' correlation with the chemical and physical-chemical parameters for the two sectors is shown. In the northern area, As shows direct correlations, both parametric and non-parametric, with F, U, and V. In the southern sector, the strong positive correlation with F remains along with Na, K, Li, and B, while a significant inverse correlation with oxygen can be observed.

Table 2. Pearson's (r) and Spearman's (r_s) correlation coefficients among As and other chemical and physical-chemical parameters. Positive correlations are highlighted on a green scale, negative correlations on a red scale. Statistically significant values in bold ($\alpha = 0.05$).

Parameter	North		South	
	r	r_s	r	r_s
Eh	0.083	−0.175	−0.255	**−0.338**
T	0.185	0.109	0.291	0.318
pH	0.291	**0.379**	0.242	0.302
DO	−0.066	−0.040	**−0.618**	**−0.619**
Cond	−0.066	0.183	0.156	0.121
F	**0.658**	**0.690**	**0.741**	**0.805**
Cl	−0.157	−0.121	−0.167	−0.180
PO_4	−0.107	0.025	−0.223	−0.301
SO_4	−0.327	−0.264	0.288	**0.449**
HCO_3	0.165	0.261	0.161	0.161

Table 2. *Cont.*

Parameter	North		South	
	r	r_s	r	r_s
Na	0.216	0.280	**0.630**	**0.488**
K	0.055	0.194	**0.541**	**0.542**
Mg	−0.250	−0.172	−0.356	**−0.449**
Ca	−0.076	0.241	−0.105	−0.225
Al	0.002	0.036	−0.070	0.213
Mn	−0.189	0.028	0.034	0.244
Fe	0.114	−0.040	0.156	−0.013
U	**0.613**	**0.536**	0.230	**0.399**
Li	0.230	0.326	**0.711**	**0.686**
B	**0.366**	0.111	**0.365**	**0.497**
V	**0.533**	**0.574**	−0.326	−0.309
Ni	−0.217	−0.041	−0.116	0.011
Cu	−0.192	−0.324	0.083	0.029
Zn	**0.397**	0.060	−0.219	−0.195
Ba	0.045	0.003	−0.161	−0.177

3.2. Mineralogical Characterization of the Rocks

The VTYT looks generally massive and coherent, with a yellow-gray color. The formation is usually affected by significant subvertical fractures and is often mineralized. Macroscopically, the tuff appears as an aggregate of pumice embedded in an ashy matrix. At the microscopic level, the VTYT samples show different fragments of rocks and minerals, such as sanidine, plagioclase, pyroxene, leucite (often analcimized), biotite, garnet, and apatite. The tuff skeleton also presents fragments of carbonate rocks, presumably ripped from the basement during the eruption.

XRPD highlighted the constant and important presence of zeolites (chabazite/hershcelite), a result consistent with data previously reported in the literature about this type of tuffs [75]. Other minerals identified within the rock with this technique include calcite, quartz, K-feldspar, mica (biotite), plagioclase, kaolinite, and chlorite.

SEM-EDX elemental analysis allowed us to identify other minerals within the rock: pyrite (Figure 4), magnetite, secondary calcite, titanite, rutile, zircon, monazite, other Fe, and rare earth oxides.

One of the objectives was the research of possible mineral phases containing particularly relevant As concentration; however, the qualitative analysis of the investigated crystalline forms showed no traces of elevated As, suggesting its possible presence as homogeneously dispersed arsenic at a level below the LOD of this technique.

3.3. As Fractioning within the Selected Tuffs

We performed a Selective Sequential Extraction on three selected VTYT samples (PAR10, PAR11, PAR22), with the purpose of quantifying the total As content and its distribution among the different solid phases that potentially constitute the aquifer matrix. The results allowed us to hypothesize the As-bearing phases that are more easily mobile from circulating waters and, accordingly, are the most likely geochemical mechanisms that can cause arsenic release from the investigated volcanic matrix.

The total arsenic in the sampled rocks is in the range of 17.5–40.9 mg/kg, which is data that is consistent with the findings of Armiento et al. [31] regarding the deposits of the Cimino-Vicano volcanic district. The sequential extraction procedure (Figure 5) shows As distribution among different solid fractions.

The As found in easily exchangeable fraction is negligible (0.22–0.41 mg/kg, corresponding to 0.9–1.4% of the total As). Conversely, the arsenic specifically adsorbed onto mineral surfaces ranges from 1.31 to 4.53 mg/kg (6.9–16.7%). Competitive ions in solution (e.g., phosphate) can affect the mobilization of this fraction, which can thus play an important role in the As contamination of groundwater in the study area.

Figure 4. SEM images: zeolites and pyrite (right) within the Via Tiberina Yellow Tuff. For pyrite the elemental analysis is also shown.

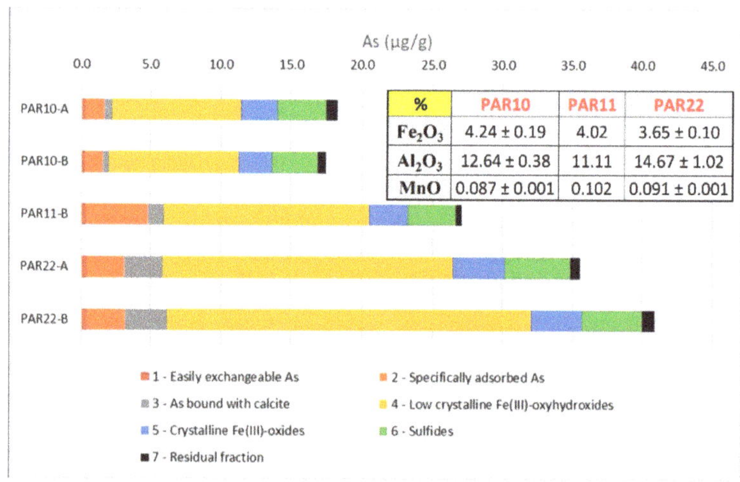

%	PAR10	PAR11	PAR22
Fe_2O_3	4.24 ± 0.19	4.02	3.65 ± 0.10
Al_2O_3	12.64 ± 0.38	11.11	14.67 ± 1.02
MnO	0.087 ± 0.001	0.102	0.091 ± 0.001

Figure 5. As fractioning resulting from the sequential extraction (two replicates for PAR10 and PAR22). The table (right) shows the weight percentage of interesting oxides in the rocks (mean ± std. dev. for PAR10 and PAR22).

Arsenic association with calcite can be significant (0.52–2.97 mg/kg, 3.4–7.8%), and the calcimetry measurements showed a total content of $CaCO_3$ between 5.5% and 12.0%. As stated by Alexandratos et al. [86], the mechanisms of arsenate sorption on the calcite surface and of its incorporation in the calcite lattice are very similar to the point where surface sorption can be considered to be a precursor of incorporation in the crystal lattice; therefore, it may be difficult to distinguish between adsorbed and incorporated As(V) complexes at the calcite surface. As a result, at least part of the specifically adsorbed As extracted in step 2 is likely derived from As incorporated in the lattice at the calcite surface.

In this step, the acetic buffer is used to selectively dissolve the carbonate, but Fe-oxyhydroxides may re-adsorb the As oxyanions liberated by the calcite dissolution. At pH 5 (acetate buffer), the capacity of Fe-oxyhydroxides to adsorb As-oxyanions is high, thus leading us to underestimate the amount of As bound to calcite, and conversely, to overestimate As bound to Fe-oxyhydroxides in the subsequent step. This effect cannot be quantitatively assessed, and therefore, the amount of As bound to calcite recovered by the specific step must be considered as a minimum estimation [61]. A release of As linked to calcite in groundwater of this area cannot be excluded, particularly where favorable conditions for dissolution exist (e.g., acidic conditions in the northern sector near the hydrogeological watershed).

Most of the As (11.69–29.51 mg/kg, 63.8–72.1%) that was extracted during the procedure is linked to Fe oxy-hydroxides, especially low crystalline (50.2–63.2%), which nevertheless constitutes only about 4% of the solid matrix. Commonly, amorphous phases present a larger specific surface area [87] than crystalline structures and tend to also adsorb As within their loose and hydrated structures, not only on their outer surfaces [88]. Furthermore, the conversion of amorphous ferrihydrite to crystalline Fe-oxide phases, which may gradually occur over time [89], can reduce the density of As sorption sites [90–92], leading to the desorption of adsorbed As [93]. These considerations are relevant in the study of As release for the dissolution of amorphous and crystalline phases in presence of reducing conditions of the aquifers [94]. Although oxidizing conditions are largely dominant in the study area, local conditions potentially favorable to the reductive dissolution of iron oxyhydroxides can also occur, with the consequent possible release of associated As.

Arsenic's association with sulfide fractions (3.17–4.64 mg/kg, 10.7–18.4%) confirms the important presence of sulfide minerals, such as pyrite, which was already observed during the SEM observations. Smedley and Kinniburgh [9] reported this association, emphasizing the chemical closeness of As and S and the occurrence of the highest As concentrations in sulfide minerals, among which pyrite is generally abundant in volcanic and geothermal contexts. It is, however, an As fraction hardly mobilized in the normal conditions of groundwater of the study area, which are always strongly undersaturated with respect to As sulfides.

3.4. Batch Experiments

Two batch tests were realized on one of the samples (PAR11) characterized with SSE. The first experiment, simulated rainwater-rock and groundwater-rock interaction processes to evaluate the As release from the tuff in contact with solutions of different ionic strength. The second experiment provided information about the influence that pH and redox potential can exert on the As-release dynamics and allowed us to investigate processes that may naturally occur in the aquifer.

In Figure 6a, the pH monitoring during the first batch test on the three selected tuffs is shown. The pH values are always higher (8.7–9.3) for the synthetic water than the As-free groundwater, which always shows values closer to neutrality (7.4–7.8) due to buffer capacity. As and V are strongly affected by high pH values and, to a lesser extent, by the presence of other exchanging anions, as observed for U (Figure 6b). Arsenic values are always higher for the synthetic water, reaching a concentration of 202 μg/L (against 77 μg/L for the As-free groundwater), corresponding to 14.9% of As released. The greater

As and V release at high pH, compared to near-neutral conditions, is promoted by the alkaline desorption processes.

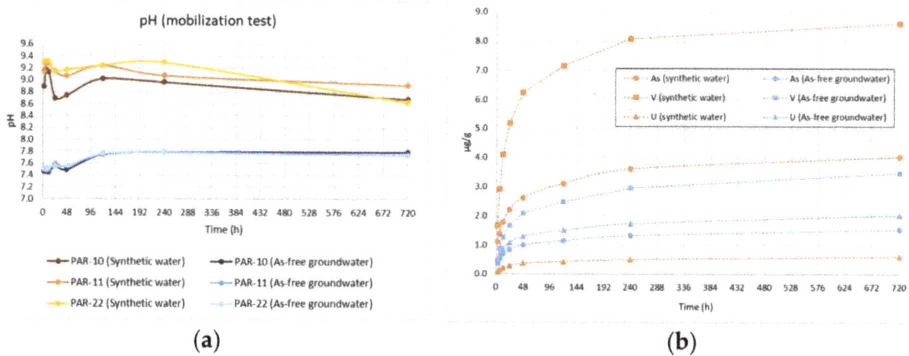

Figure 6. Batch test results. (**a**)—pH monitoring (3 rock samples, mean of two replicates); (**b**)—As, V and U released during the test (sample PAR-11, mean of two replicates).

A batch desorption study by Kim et al. [18] reported that hydrolysis of Fe (hydr-) oxides released As and F under reducing conditions; however, the same study also found that under oxidizing conditions, an increase in pH was the main mode of As and F mobilization. At pH above 8, a significant displacement of arsenate is induced by OH- ions through an anion exchange mechanism.

The second test was divided into different steps, simulating the water-rock interaction processes in oxygenated conditions (STEP 1) under anaerobic conditions and weakly positive redox potential (STEP 2a) and in anoxic and strongly reducing conditions (STEP 2b), with the latter being locally observed in the study area.

Figure 7a shows the trend of pH and Eh values during the second batch test. DO showed stable values around 7.8 mg/L during the aerobic phase, while it suddenly reduces during the second step (complete anaerobic conditions are established 15 min after fluxing with N_2).

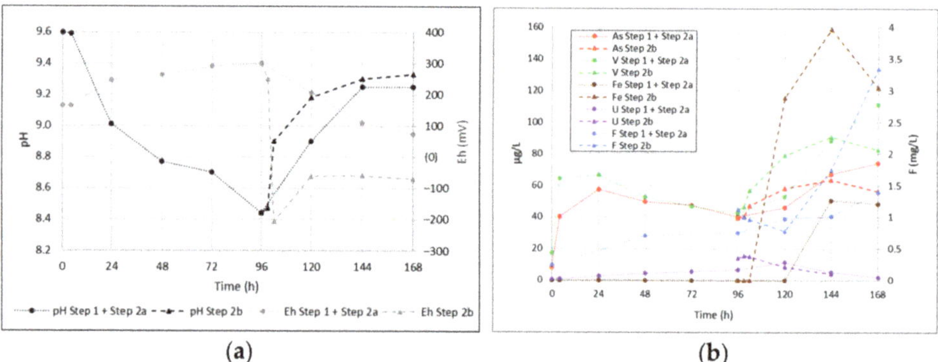

Figure 7. Behavior of different parameters during the second batch test. (**a**)—Eh and pH values; (**b**)—As, V, Fe, U and F concentrations.

Eh showed a slight and progressive increase in STEP 1, followed by a decrease to slightly positive values in STEP 2a (about +70 mV), while a sharp decrease due to the addition of a reducing species (S^{2-}) can be observed in STEP 2b before it stabilized at slightly negative values (about −60 mV). The pH after the water-rock interaction always remains at rather high values (8.4–9.6) and tends to progressively decrease during the aerobic phase and increase in the anaerobic ones, due to N_2 blowing.

Arsenic and vanadium (Figure 7b) follow a similar trend during the test. The two elements show an increase in the first 24 h during the aerobic phase and then a slight decrease towards the end of this step, probably due to the re-sorption of the oxyanions. The desorption/resorption processes from zeolites cannot be excluded. In the anaerobic step, the concentration of As and V gradually increased and may also have been affected by the dissolution of the iron oxides in an anoxic environment, as shown by the increase in Fe content, which was up to 150 ug/L.

The uranium trend appears to be strongly influenced by pH variations, with an increase in the aerobic phase and then a decrease to 1.1 µg/L at the end of the anaerobic step, suggesting a probable re-adsorption, given the affinity of uranyl for Fe oxide phases [95]. In contrast, F shows a gradual increase of the concentration both in the aerobic and anaerobic steps (only 3 samples in STEP1), thus appearing to be linked to dissolution processes and not influenced by the variable pH-Eh conditions affecting the test.

Regarding STEP2b (reducing conditions), As and V again show a very similar trend, with an increase of concentrations for a large part of the step and a final slight decrease, probably due to the partial precipitation of sulfides. Here, Fe, undetectable for a large part of the test, shows appreciable concentrations and apparently similar behavior. Uranium still appears constrained by the pH increase and tends to the gradual reduction of the concentration towards the end of the test. Fluoride, on the other hand, shows a distinct trend almost opposite to that of As and V, with a slight concentration decrease in the first 24 h and then a marked increase in the following 48. Therefore, the element is strongly mobilized under alkaline and reducing conditions.

3.5. Arsenic and Other PTEs: Linking Solid Matrix Analysis with Groundwater Geochemistry

As observed in Table 2, arsenic in groundwater in the northern sector (more representative of the volcanic domain) is well correlated with other PTEs, suggesting common processes. The link between these elements and the release from the volcanic tuff under different experimental conditions was therefore analyzed, investigating possible connections with groundwater data.

Uranium in the two batch tests shows a different behavior in regard to As, appearing strongly linked to pH variations in the frankly alkaline environment. The lack of correspondence between field and laboratory data could therefore be attributed to the different existing conditions, with lower pH values in the study area (especially in the northern sector) than those measured during the tests. However, it should be emphasized that the correlation with As is observable at the low concentrations (median = 0.8 µg/L; max = 13.9 µg/L) found in the northern sector. In the southern area, U, although showing more important concentrations (median = 6.6 µg/L; max = 34.6 µg/L), poorly correlates with As.

The strong As-F correlation is constant in groundwater in both sectors. A correlation is also partially observable in the second batch test, but only in the aerobic/anaerobic steps (1 and 2a). Beyond the common origin of the two elements due to water-rock interaction processes, in the study area, As and F remain strongly linked in aqueous solution, and processes capable of breaking this common mobility do not intervene. On a regional scale, Parrone et al. [96] attributed this stable co-presence to the widespread geochemical background of cold groundwater circulating volcanic formations, also highlighting areas where the good correlation is lost due to localized peculiarities (e.g., in the proximity of fractures/faults systems or mineral deposits).

In contrast, during the batch test, As and F showed almost opposite behaviors in the reducing step (2b). The F concentration shows a noticeable increase, probably due to an exchange with OH at high pH, while As is probably released by reductive processes and partially re-precipitated as sulfide. Reducing conditions are uncommon in the study area but can locally affect the As-F correlation. For example, the well N32, one of the few points having strongly reducing conditions (Eh = −323 mV), shows unusually low concentrations of As and V (0.2 and 2.2 µg/L) and high values of F (1.9 mg/L). This could be due to the mixing of groundwater with reducing deep fluids, leading to the subsequent precipitation of As and V as sulfides and the loss of the As-F correlation. Once eliminated, these local geochemical anomalies, usually characterized by physical-chemical parameters' abnormal values (negative Eh, low DO, high EC), cause the As-F correlation to further increase, up to a correlation of 0.8. Additionally, Kim et al. [18] identify a good correlation between As and F under oxidizing conditions and increasing pH, associated with the desorption from Fe (hydr)oxides, while a poorer correlation between the two elements verifies in reducing aquifers due to sulfate reduction and consequent As precipitation as sulfide, a process not affecting F concentration.

The good As-V correlation observed in groundwater of the northern sector is confirmed by the results of the two batch tests performed on the solid matrix. The two elements show completely overlapping trends in all conditions and therefore can be considered representative of the water-VTYT interaction process. The good correlation in groundwater is lost in the southern sector, where the importance of sedimentary deposits increases. The presence of clays could affect the As-V correlation through sorption/desorption processes. Zhu et al. [97] observed the consistent pH-dependent sorption of V onto kaolinite and montmorillonite, with relatively high sorption capacity occurring at the pH range of 4–10. However, the electrostatic repulsion between the V anions and the negatively charged surface determines the very low maximum adsorption capacities of the two clay minerals (<1.0 mg/g) compared to the result of V adsorption onto soil [98] and soil colloids [99]. This is due to the high surface area of soil colloids and the presence of Fe/Al (hydr)oxides, which have a high tendency for V adsorption [100]. Recently Gonzalez-Rodriguez and Fernandez-Marcos [101] observed high sorption of vanadate and arsenate onto non-crystalline iron and aluminum species. Furthermore, they found that phosphate can induce the desorption of arsenate, while it can hardly displace adsorbed vanadate. This different tendency to displace As and V may be a plausible explanation for the loss of positive correlation in the southern sector (Figure 8a), where PO_4 indeed shows very low concentrations and may have undergone sorption processes (Figure 8b). This could explain also why As in the southern area remains in solution and maintains its strong correlation with F.

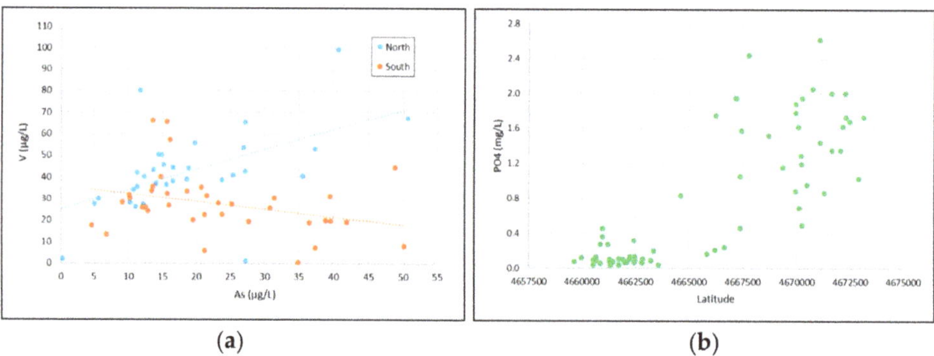

Figure 8. (**a**) scatterplot As-V for the northern and southern sector; (**b**) Latitude trend of the PO_4 concentrations.

4. Conclusions

The analysis of the distribution of PTEs in the different matrices (groundwater, rocks) and the study of the geochemical processes responsible of their mobilization can provide a comprehensive framework on the origin and evolution of geogenic contaminants in groundwater.

The joint analysis of field data and laboratory observations suggests the existence of a diffused As geochemical background due to the water-rock interactions. In many hydrogeological settings, as alluvial aquifers, the release of As has been proved to be triggered by reductive dissolution, while in geothermal contexts, it is often explained by upraising geothermal fluids, mobilization from volcanic rocks due to acidity or high temperature, and local processes as a release from carbonate trapping. Our study pointed out that in volcanic areas, although the reductive dissolution of Fe oxyhydroxides could cause an enhanced localized release of As into groundwater, the anion exchange induced by specific exchangers (e.g., phosphates) with As adsorbed on the surface seems to be the most widely diffused process, which is often not taken into consideration in other studies.

The strong and stable As-F correlation in groundwater already observed at the regional scale has been confirmed, while the As-V correlation is lost during the transition towards the sedimentary deposits, thus representing a good geochemical marker of the volcanic hydrogeological domain.

Investigating PTEs adsorption/desorption behavior at variable pH and Eh conditions could give useful hints for the comprehension of the phenomena that release arsenic in groundwater exploited for human consumption, hence suggesting suitable technology and management options for the distribution of As-free waters in affected areas.

Supplementary Materials: The following supporting information can be downloaded at: https://www.mdpi.com/article/10.3390/toxics10060288/s1, Table S1: Chemical analysis of the two solutions used for the first experiment; Table S2: Comparison of the main statistics for all the analyzed parameters, distinguished between northern and southern sectors; Figure S1: Distribution of As concentrations in groundwater of the study area.

Author Contributions: Conceptualization, D.P., S.G., B.C. and E.P.; methodology, D.P., B.C. and S.G.; validation, D.P., B.C. and S.G.; formal analysis, D.P.; investigation, D.P., B.C. and S.G.; resources, D.P., B.C. and S.G.; writing—original draft preparation, D.P.; writing—review and editing, D.P., S.G., B.C. and E.P.; visualization, D.P.; supervision, S.G. and E.P.; project administration, S.G. and E.P. All authors have read and agreed to the published version of the manuscript.

Funding: This research received no external funding.

Institutional Review Board Statement: Not applicable.

Informed Consent Statement: Not applicable.

Data Availability Statement: The data presented in this study are available on request from the corresponding author.

Acknowledgments: We wish to thank Paola Tuccimei for the coordination and supervision of the activities carried out at the Roma Tre University, Fabio Bellatreccia and Sergio Lo Mastro of the Roma Tre University for the mineralogical observations at the optical microscope and SEM analysis, Gianluca Iezzi of the Chieti University for the XRPD analysis, Jasper Griffioen of the Utrecht University for his suggestions on the laboratory tests, David Rossi of the CNR-IRSA for his support in the groundwater sampling activities and Domenico Mastroianni of the CNR-IRSA for the chemical analysis of water samples.

Conflicts of Interest: The authors declare no conflict of interest.

References

1. Cubadda, F.; D'Amato, M.; Mancini, F.R.; Aureli, F.; Raggi, A.; Busani, L.; Mantovani, A. Assessing Human Exposure to Inorganic Arsenic in High-Arsenic Areas of Latium: A Biomonitoring Study Integrated with Indicators of Dietary Intake. *Annu. Ig. Med. Prev. Comun.* **2015**, *27*, 7458. [CrossRef]
2. Karim, M.M. Arsenic in Groundwater and Health Problems in Bangladesh. *Water Res.* **2000**, *34*, 128. [CrossRef]
3. Rossman, T.G.; Uddin, A.N.; Burns, F.J. Evidence That Arsenite Acts as a Cocarcinogen in Skin Cancer. *Toxicol. Appl. Pharmacol.* **2004**, *198*, 457. [CrossRef] [PubMed]
4. WHO. *Guidelines for Drinking-Water Quality*, 4th ed.; Incorporating the First Addendum; World Health Organization: Geneva, Switzerland, 2017; p. 4.
5. European Union. 98/83/EC. In *Quality of Water Intented for Human Consumption*; European Union: Brussels, Belgium, 1998; p. 3.

6. Anawar, H.M.; Akai, J.; Komaki, K.; Terao, H.; Yoshioka, T.; Ishizuka, T.; Safiullah, S.; Kato, K. Geochemical Occurrence of Arsenic in Groundwater of Bangladesh: Sources and Mobilization Processes. *J. Geochem. Explor.* **2003**, *77*, 273. [CrossRef]

7. McArthur, J.M.; Banerjee, D.M.; Hudson-Edwards, K.A.; Mishra, R.; Purohit, R.; Ravenscroft, P.; Cronin, A.; Howarth, R.J.; Chatterjee, A.; Talukder, T.; et al. Natural Organic Matter in Sedimentary Basins and Its Relation to Arsenic in Anoxic Ground Water: The Example of West Bengal and Its Worldwide Implications. *Appl. Geochem.* **2004**, *19*, 56. [CrossRef]

8. Nickson, R.T.; Mcarthur, J.M.; Ravenscroft, P.; Burgess, W.G.; Ahmed, K.M. Mechanism of Arsenic Release to Groundwater, Bangladesh and West Bengal. *Appl. Geochem.* **2000**, *15*, 86–90. [CrossRef]

9. Smedley, P.L.; Kinniburgh, D.G. A Review of the Source, Behaviour and Distribution of Arsenic in Natural Waters. *Appl. Geochem.* **2002**, *17*, 754. [CrossRef]

10. Mukherjee, A.; Verma, S.; Gupta, S.; Henke, K.R.; Bhattacharya, P. Influence of Tectonics, Sedimentation and Aqueous Flow Cycles on the Origin of Global Groundwater Arsenic: Paradigms from Three Continents. *J. Hydrol.* **2014**, *518*, 44. [CrossRef]

11. BGS. Arsenic Contamination of Groundwater. Available online: https://www.bgs.ac.uk/research/groundwater/health/arsenic/home.html (accessed on 12 April 2022).

12. Cullen, W.R.; Reimer, K.J. Arsenic Speciation in the Environment. *Chem. Rev.* **1989**, *89*, 94. [CrossRef]

13. Meng, X.; Bang, S.; Korfiatis, G.P. Effects of Silicate, Sulfate, and Carbonate on Arsenic Removal by Ferric Chloride. *Water Res.* **2000**, *34*, 272–283. [CrossRef]

14. Nicholas, D.R.; Ramamoorthy, S.; Palace, V.; Spring, S.; Moore, J.N.; Rosenzweig, R.F. Biogeochemical Transformations of Arsenic in Circumneutral Freshwater Sediments. *Biodegradation* **2003**, *14*, 695. [CrossRef] [PubMed]

15. Wang, S.; Mulligan, C.N. Effect of Natural Organic Matter on Arsenic Release from Soils and Sediments into Groundwater. *Environ. Geochem. Health* **2006**, *28*, 45. [CrossRef] [PubMed]

16. Bhattacharya, P.; Claesson, M.; Bundschuh, J.; Sracek, O.; Fagerberg, J.; Jacks, G.; Martin, R.A.; Storniolo, A.D.R.; Thir, J.M. Distribution and Mobility of Arsenic in the Río Dulce Alluvial Aquifers in Santiago Del Estero Province, Argentina. *Sci. Total Environ.* **2006**, *358*, 48. [CrossRef] [PubMed]

17. Currell, M.; Cartwright, I.; Raveggi, M.; Han, D. Controls on Elevated Fluoride and Arsenic Concentrations in Groundwater from the Yuncheng Basin, China. *Appl. Geochem.* **2011**, *26*, 12. [CrossRef]

18. Kim, S.H.; Kim, K.; Ko, K.S.; Kim, Y.; Lee, K.S. Co-Contamination of Arsenic and Fluoride in the Groundwater of Unconsolidated Aquifers under Reducing Environments. *Chemosphere* **2012**, *87*, 25. [CrossRef]

19. Molinari, A.; Guadagnini, L.; Marcaccio, M.; Straface, S.; Sanchez-Vila, X.; Guadagnini, A. Arsenic Release from Deep Natural Solid Matrices under Experimentally Controlled Redox Conditions. *Sci. Total Environ.* **2013**, *444*, 93. [CrossRef]

20. Nicolli, H.B.; García, J.W.; Falcón, C.M.; Smedley, P.L. Mobilization of Arsenic and Other Trace Elements of Health Concern in Groundwater from the Salí River Basin, Tucumán Province, Argentina. *Environ. Geochem. Health* **2012**, *34*, 9429. [CrossRef]

21. Alarcón-Herrera, M.T.; Bundschuh, J.; Nath, B.; Nicolli, H.B.; Gutierrez, M.; Reyes-Gomez, V.M.; Nuñez, D.; Martín-Dominguez, I.R.; Sracek, O. Co-Occurrence of Arsenic and Fluoride in Groundwater of Semi-Arid Regions in Latin America: Genesis, Mobility and Remediation. *J. Hazard. Mater.* **2013**, *262*, 5. [CrossRef]

22. Francisca, F.M.; Carro Perez, M.E. Assessment of Natural Arsenic in Groundwater in Cordoba Province, Argentina. *Environ. Geochem. Health* **2009**, *31*, 9245. [CrossRef]

23. Morales, I.; Villanueva-Estrada, R.E.; Rodríguez, R.; Armienta, M.A. Geological, Hydrogeological, and Geothermal Factors Associated to the Origin of Arsenic, Fluoride, and Groundwater Temperature in a Volcanic Environment "El Bajío Guanajuatense", Mexico. *Environ. Earth Sci.* **2015**, *74*, 4554–4563. [CrossRef]

24. Ren, M.; Rodríguez-Pineda, J.A.; Goodell, P. Arsenic Mineral in Volcanic Tuff, a Source of Arsenic Anomaly in Groundwater: City of Chihuahua, Mexico. *Geosciences* **2022**, *12*, 69. [CrossRef]

25. Winkel, L.H.E.; Casentini, B.; Bardelli, F.; Voegelin, A.; Nikolaidis, N.P.; Charlet, L. Speciation of Arsenic in Greek Travertines: Co-Precipitation of Arsenate with Calcite. *Geochim. Cosmochim. Acta* **2013**, *106*, 49. [CrossRef]

26. Di Benedetto, F.; Costagliola, P.; Benvenuti, M.; Lattanzi, P.; Romanelli, M.; Tanelli, G. Arsenic Incorporation in Natural Calcite Lattice: Evidence from Electron Spin Echo Spectroscopy. *Earth Planet. Sci. Lett.* **2006**, *246*, 47. [CrossRef]

27. Baiocchi, A.; Coletta, A.; Espositi, L.; Lotti, F.; Piscopo, V. Sustainable Groundwater Development in a Naturally Arsenic-Contaminated Aquifer: The Case of the Cimino-Vico Volcanic Area (Central Italy). *Ital. J. Eng. Geol. Environ.* **2013**, *13*, 5–18. [CrossRef]

28. Preziosi, E.; Giuliano, G.; Vivona, R. Natural Background Levels and Threshold Values Derivation for Naturally As, V and F Rich Groundwater Bodies: A Methodological Case Study in Central Italy. *Environ. Earth Sci.* **2010**, *61*, 404. [CrossRef]

29. Vivona, R.; Preziosi, E.; Madé, B.; Giuliano, G. Occurrence of Minor Toxic Elements in Volcanic-Sedimentary Aquifers: A Case Study in Central Italy. *Hydrogeol. J.* **2007**, *15*, 169. [CrossRef]

30. Angelone, M.; Cremisini, C.; Piscopo, V.; Proposito, M.; Spaziani, F. Influence of Hydrostratigraphy and Structural Setting on the Arsenic Occurrence in Groundwater of the Cimino-Vico Volcanic Area (Central Italy). *Hydrogeol. J.* **2009**, *17*, 403. [CrossRef]

31. Armiento, G.; Baiocchi, A.; Cremisini, C.; Crovato, C.; Lotti, F.; Lucentini, L.; Mazzuoli, M.; Nardi, E.; Piscopo, V.; Proposito, M.; et al. An Integrated Approach to Identify Water Resources for Human Consumption in an Area Affected by High Natural Arsenic Content. *Water* **2015**, *7*, 5091. [CrossRef]

32. Cinti, D.; Poncia, P.P.; Brusca, L.; Tassi, F.; Quattrocchi, F.; Vaselli, O. Spatial Distribution of Arsenic, Uranium and Vanadium in the Volcanic-Sedimentary Aquifers of the Vicano-Cimino Volcanic District (Central Italy). *J. Geochem. Explor.* **2015**, *152*, 8. [CrossRef]

33. Casentini, B.; Pettine, M.; Millero, F.J. Release of Arsenic from Volcanic Rocks through Interactions with Inorganic Anions and Organic Ligands. *Aquat. Geochem.* **2010**, *16*, 9090. [CrossRef]

34. Lim, M.S.; Yeo, I.W.; Prabhakar Clement, T.; Roh, Y.; Lee, K.K. Mathematical Model for Predicting Microbial Reduction and Transport of Arsenic in Groundwater Systems. *Water Res.* **2007**, *41*, 17. [CrossRef] [PubMed]

35. Nguyen, K.P.; Itoi, R.; Yamashiro, R. Influence of Redox Potential on Arsenic Release from Soil in the Presence of Iron Oxyhydroxide. *Mem. Fac. Eng. Kyushu Univ.* **2008**, *68*, 754.

36. Razzak, A.; Jinno, K.; Hiroshiro, Y.; Halim, M.A.; Oda, K. Mathematical Modeling of Biologically Mediated Redox Processes of Iron and Arsenic Release in Groundwater. *Environ. Geol.* **2009**, *58*, 1517. [CrossRef]

37. Gao, Y.; Mucci, A. Acid Base Reaction, Phosphate and Arsenate Complexation, and Thier Competitive Adsorption at the Surface of Geothite in 0.7 M Nacl Solution. *Geochim. Cosmochim. Acta* **2001**, *65*, 589. [CrossRef]

38. Frohne, T.; Rinklebe, J.; Diaz-Bone, R.A.; du Laing, G. Controlled Variation of Redox Conditions in a Floodplain Soil: Impact on Metal Mobilization and Biomethylation of Arsenic and Antimony. *Geoderma* **2011**, *160*, 12. [CrossRef]

39. Kumar, M.; Das, N.; Goswami, R.; Sarma, K.P.; Bhattacharya, P.; Ramanathan, A.L. Coupling Fractionation and Batch Desorption to Understand Arsenic and Fluoride Co-Contamination in the Aquifer System. *Chemosphere* **2016**, *164*, 75. [CrossRef]

40. Deuel, L.E.; Swoboda, A.R. Arsenic Solubility in a Reduced Environment. *Soil Sci. Soc. Am. J.* **1972**, *36*, 22. [CrossRef]

41. Hess, R.E.; Blanchar, R.W. Dissolution of Arsenic from Waterlogged and Aerated Soil. *Soil Sci. Soc. Am. J.* **1977**, *41*, 9. [CrossRef]

42. McGeehan, S.L. Arsenic Sorption and Redox Reactions: Relevance to Transport and Remediation. *J. Environ. Sci. Health Part A Toxic Hazard. Subst. Environ. Eng.* **1996**, *31*, 6493. [CrossRef]

43. Tingzong, G.; DeLaune, R.D.; Patrick, W.H. The Influence of Sediment Redox Chemistry on Chemically Active Forms of Arsenic, Cadmium, Chromium, and Zinc in Estuarine Sediment. *Environ. Int.* **1997**, *23*, 33. [CrossRef]

44. Yu, K.; Böhme, F.; Rinklebe, J.; Neue, H.-U.; DeLaune, R.D. Major Biogeochemical Processes in Soils-A Microcosm Incubation from Reducing to Oxidizing Conditions. *Soil Sci. Soc. Am. J.* **2007**, *71*, 155. [CrossRef]

45. Matis, K.A.; Zouboulis, A.I.; Malamas, F.B.; Ramos Afonso, M.D.; Hudson, M.J. Flotation Removal of As (V) onto Goethite. *Environ. Pollut.* **1997**, *97*, 91–93. [CrossRef]

46. Jönsson, J.; Sherman, D.M. Sorption of As (III) and As (V) to Siderite, Green Rust (Fougerite) and Magnetite: Implications for Arsenic Release in Anoxic Groundwaters. *Chem. Geol.* **2008**, *255*, 36. [CrossRef]

47. National Research Council (US) Committee on Medical and Biological Effects of Environmental Pollutants Arsenic. *Medical and Biologic Effects of Environmental Pollutants*; National Academies Press: Washington, DC, USA, 1977.

48. Bombach, G.; Pierra, A.; Klemm, W. Arsenic in Contaminated Soil and River Sediment. *Fresenius' J. Anal. Chem.* **1994**, *350*, 6252. [CrossRef]

49. Voigt, D.E.; Brantley, S.L.; Hennet, R.J.C. Chemical Fixation of Arsenic in Contaminated Soils. *Appl. Geochem.* **1996**, *11*, 9–10. [CrossRef]

50. Kavanagh, P.J.; Farago, M.E.; Thornton, I.; Braman, R.S. Bioavailability of Arsenic in Soil and Mine Wastes of the Tamar Valley, SW England. *Chem. Speciat. Bioavailab.* **1997**, *9*, 3291. [CrossRef]

51. Hudson-Edwards, K.A.; Houghton, S.L.; Osborn, A. Extraction and Analysis of Arsenic in Soils and Sediments. *TrAC Trends Anal. Chem.* **2004**, *23*, 10. [CrossRef]

52. Bacon, J.R.; Davidson, C.M. Is There a Future for Sequential Chemical Extraction? *Analyst* **2008**, *133*, 45. [CrossRef]

53. Tessier, A.; Campbell, P.G.C.; Bisson, M. Sequential Extraction Procedure for the Speciation of Particulate Trace Metals. *Anal. Chem.* **1979**, *51*, 17. [CrossRef]

54. Shuman, L.M. Fractionation Method for Soil Microelements. *Soil Sci.* **1985**, *140*, 3. [CrossRef]

55. Gleyzes, C.; Tellier, S.; Sabrier, R.; Astruc, M. Arsenic Characterisation in Industrial Soils by Chemical Extractions. *Environ. Technol.* **2001**, *22*, 8313. [CrossRef] [PubMed]

56. Keon, N.E.; Swartz, C.H.; Brabander, D.J.; Harvey, C.; Hemond, H.F. Validation of an Arsenic Sequential Extraction Method for Evaluating Mobility in Sediments. *Environ. Sci. Technol.* **2001**, *35*, 1511. [CrossRef] [PubMed]

57. Wenzel, W.W.; Kirchbaumer, N.; Prohaska, T.; Stingeder, G.; Lombi, E.; Adriano, D.C. Arsenic Fractionation in Soils Using an Improved Sequential Extraction Procedure. *Anal. Chim. Acta* **2001**, *436*, 924–926. [CrossRef]

58. Pagnanelli, F.; Moscardini, E.; Giuliano, V.; Toro, L. Sequential Extraction of Heavy Metals in River Sediments of an Abandoned Pyrite Mining Area: Pollution Detection and Affinity Series. *Environ. Pollut.* **2004**, *132*, 2. [CrossRef] [PubMed]

59. Claff, S.R.; Sullivan, L.A.; Burton, E.D.; Bush, R.T. A Sequential Extraction Procedure for Acid Sulfate Soils: Partitioning of Iron. *Geoderma* **2010**, *155*, 2. [CrossRef]

60. Matera, V.; le Hécho, I.; Laboudigue, A.; Thomas, P.; Tellier, S.; Astruc, M. A Methodological Approach for the Identification of Arsenic Bearing Phases in Polluted Soils. *Environ. Pollut.* **2003**, *126*, 146–151. [CrossRef]

61. Costagliola, P.; Bardelli, F.; Benvenuti, M.; di Benedetto, F.; Lattanzi, P.; Romanelli, M.; Paolieri, M.; Rimondi, V.; Vaggelli, G. Arsenic-Bearing Calcite in Natural Travertines: Evidence from Sequential Extraction, MxAS, and MxRF. *Environ. Sci. Technol.* **2013**, *47*, 4953a. [CrossRef]

62. Dold, B. Speciation of the Most Soluble Phases in a Sequential Extraction Procedure Adapted for Geochemical Studies of Copper Sulfide Mine Waste. *J. Geochem. Explor.* **2003**, *80*, 182–183. [CrossRef]

63. Ahn, J.S.; Park, Y.S.; Kim, J.Y.; Kim, K.W. Mineralogical and Geochemical Characterization of Arsenic in an Abandoned Mine Tailings of Korea. *Environ. Geochem. Health* **2005**, *27*, 121–129. [CrossRef]

64. Damris, M.; O'Brien, G.A.; Price, W.E.; Chenhall, B.E. Fractionation of Sedimentary Arsenic from Port Kembla Harbour, NSW, Australia. *J. Environ. Monit.* **2005**, *7*, 393b. [CrossRef]

65. Van Elteren, J.T.; Šlejkovec, Z.; Arčon, I.; Glass, H.J. An Interdisciplinary Physical-Chemical Approach for Characterization of Arsenic in a Calciner Residue Dump in Cornwall (UK). *Environ. Pollut.* **2006**, *139*, 3. [CrossRef] [PubMed]

66. Parrone, D.; Ghergo, S.; Preziosi, E. A Multi-Method Approach for the Assessment of Natural Background Levels in Groundwater. *Sci. Total Environ.* **2019**, *659*, 350. [CrossRef] [PubMed]

67. Conticelli, S.; Francalanci, L.; Manetti, P.; Cioni, R.; Sbrana, A. Petrology and Geochemistry of the Ultrapotassic Rocks from the Sabatini Volcanic District, Central Italy: The Role of Evolutionary Processes in the Genesis of Variably Enriched Alkaline Magmas. *J. Volcanol. Geotherm. Res.* **1997**, *75*, 107–136. [CrossRef]

68. Sottili, G.; Palladino, D.M.; Marra, F.; Jicha, B.; Karner, D.B.; Renne, P. Geochronology of the Most Recent Activity in the Sabatini Volcanic District, Roman Province, Central Italy. *J. Volcanol. Geotherm. Res.* **2010**, *196*, 20–30. [CrossRef]

69. Masotta, M.; Gaeta, M.; Gozzi, F.; Marra, F.; Palladino, D.M.; Sottili, G. H$_2$O- and Temperature-Zoning in Magma Chambers: The Example of the Tufo Giallo Della Via Tiberina Eruptions (Sabatini Volcanic District, Central Italy). *Lithos* **2010**, *118*, 119–130. [CrossRef]

70. Capelli, G.; Mazza, R.; Gazzetti, C. *Strumenti e Strategie per La Tutela e l'Uso Compatibile della Risorsa Idrica nel Lazio. Gli Acquiferi Vulcanici*; Quaderni di Tecniche di Protezione Ambientale; Pitagora Editrice: Bologna, Italy, 2005; Volume 78.

71. Ventriglia, U. *Idrogeologia Della Provincia di Roma, II Volume: Regione Vulcanica Sabatina*; Amministrazione Provinciale di Roma—Assessorato LL.PP.: Rome, Italy, 1989.

72. Nappi, G.; de Casa, G.; Volponi, E. Geologia e Caratteristiche Del Tufo Giallo Della Via Tiberina. *Boll. Soc. Geol. It.* **1979**, *98*, 431–445.

73. De Rita, D.; Di Filippo, M.; Funiciello, R.; Parotto, M.; Sposato, A. Carta Geologica Del Complesso Vulcanico Sabatino. In *Sabatini Volcanic Complex*; Di Filippo, M., Ed.; CNR, Quaderni de La Ricerca Scientifica: Rome, Italy, 1993; Volume 114.

74. Campobasso, C.; Cioni, R.; Salvati, L.; Sbrana, A. Geology and Paleogeographic Evolution of the Peripheral Sector of the Vico and Sabatini Volcanic Complex, between Civita Castellana and Mazzano Romano (Latium Italy). *Mem. Descr. Carta Geol. It.* **1994**, *49*, 277–290.

75. Lombardi, G.; Meucci, C. Il Tufo Giallo Della Via Tiberina (Roma) Utilizzato Nei Monumenti Romani. *Rend. Lincei* **2006**, *17*, 263–287. [CrossRef]

76. Jackson, M.D.; Marra, F.; Hay, R.L.; Cawood, C.; Winkler, E.M. The Judicious Selection and Preservation of Tuff and Travertine Building Stone in Ancient Rome. *Archaeometry* **2005**, *47*, 485–510. [CrossRef]

77. Shapiro, S.S.; Wilk, M.B. An Analysis of Variance Test for Normality (Complete Samples). *Biometrika* **1965**, *52*, 591. [CrossRef]

78. Rosner, B. Percentage Points for a Generalized Esd Many-Outlier Procedure. *Technometrics* **1983**, *25*, 7848. [CrossRef]

79. Huber, P.J.; Ronchetti, E.M. *Robust Statistics*, 2nd ed.; Wiley: West Maitland, FL, USA, 2009.

80. USEPA ProUCL. *Statistical Software for Environmental Applications for Data Sets with and without Nondetect Observations*; Version 5.1.002 Technical Guide, EPA/600/R-07/041; USEPA: Washington, DC, USA, 2015.

81. Hammer, Ø.; Harper, D.A.T.; Ryan, P.D. Past: Paleontological Statistics Software Package for Education and Data Analysis. *Palaeontol. Electron.* **2001**, *4*, 455.

82. Coetzee, P.P.; Gouws, K.; Pluddermann, S.; Yacoby, M.; Howell, S.; den Drijver, L. Evaluation of Sequential Extraction Procedures for Metal Speciation in Model Sediments. *Water SA* **1995**, *21*, 458.

83. Kubová, J.; Streško, V.; Bujdoš, M.; Matúš, P.; Medved, J. Fractionation of Various Elements in CRMs and in Polluted Soils. *Anal. Bioanal. Chem.* **2004**, *379*, 56. [CrossRef] [PubMed]

84. Torres, E.; Auleda, M. A Sequential Extraction Procedure for Sediments Affected by Acid Mine Drainage. *J. Geochem. Explor.* **2013**, *128*, 12. [CrossRef]

85. Dinelli, E.; Lima, A.; Albanese, S.; Birke, M.; Cicchella, D.; Giaccio, L.; Valera, P.; de Vivo, B. Major and Trace Elements in Tap Water from Italy. *J. Geochem. Explor.* **2012**, *112*, 9. [CrossRef]

86. Alexandratos, V.G.; Elzinga, E.J.; Reeder, R.J. Arsenate Uptake by Calcite: Macroscopic and Spectroscopic Characterization of Adsorption and Incorporation Mechanisms. *Geochim. Cosmochim. Acta* **2007**, *71*, 55. [CrossRef]

87. Pfeifer, H.R.; Gueye-Girardet, A.; Reymond, D.; Schlegel, C.; Temgoua, E.; Hesterberg, D.L.; Chou, J.W. Dispersion of Natural Arsenic in the Malcantone Watershed, Southern Switzerland: Field Evidence for Repeated Sorption-Desorption and Oxidation-Reduction Processes. *Geoderma* **2004**, *122*, 765. [CrossRef]

88. Bissen, M.; Frimmel, F.H. Arsenic–A Review. Part I: Occurrence, Toxicity, Speciation, Mobility. *Acta Hydrochim. Hydrobiol.* **2003**, *31*, 25. [CrossRef]

89. Dzombak, D.A.; Morel, F.M.M. *Surface Complexation Modeling: Hydrous Ferric Oxide*; John Wiley & Sons: Toronto, ON, Canada, 1991.

90. Fuller, C.C.; Dadis, J.A.; Waychunas, G.A. Surface Chemistry of Ferrihydrite: Part 2. Kinetics of Arsenate Adsorption and Coprecipitation. *Geochim. Cosmochim. Acta* **1993**, *57*, 568. [CrossRef]

91. Deutsch, W.J. *Groundwater Geochemistry: Fundamentals and Applications to Contamination*; CRC Press: Boca Raton, FL, USA, 2020.

92. Bowles, J.F.W.; Cornell, R.M.; Schwertmann, U. The Iron Oxides: Structure, Properties Reactions Occurrence and Uses. *Mineral. Mag.* **1997**, *61*, 20. [CrossRef]
93. Hinkle, S.R.; Polette, D.J. *Arsenic in Ground Water of the Willamette Basin, Oregon*; Water-Resources Investigations Report 98–4205; USGS: Portland, OR, USA, 1999.
94. Molinari, A.; Guadagnini, L.; Marcaccio, M.; Guadagnini, A. Arsenic Fractioning in Natural Solid Matrices Sampled in a Deep Groundwater Body. *Geoderma* **2015**, *247*, 11. [CrossRef]
95. Dodge, C.J.; Francis, A.J.; Gillow, J.B.; Halada, G.P.; Eng, C.; Clayton, C.R. Association of Uranium with Iron Oxides Typically Formed on Corroding Steel Surfaces. *Environ. Sci. Technol.* **2002**, *36*, 1450. [CrossRef] [PubMed]
96. Parrone, D.; Ghergo, S.; Frollini, E.; Rossi, D.; Preziosi, E. Arsenic-Fluoride Co-Contamination in Groundwater: Background and Anomalies in a Volcanic-Sedimentary Aquifer in Central Italy. *J. Geochem. Explor.* **2020**, *217*, 6590. [CrossRef]
97. Zhu, H.; Xiao, X.; Guo, Z.; Han, X.; Liang, Y.; Zhang, Y.; Zhou, C. Adsorption of Vanadium (V) on Natural Kaolinite and Montmorillonite: Characteristics and Mechanism. *Appl. Clay Sci.* **2018**, *161*, 35. [CrossRef]
98. Gäbler, H.E.; Glüh, K.; Bahr, A.; Utermann, J. Quantification of Vanadium Adsorption by German Soils. *J. Geochem. Explor.* **2009**, *103*, 2. [CrossRef]
99. Luo, X.; Yu, L.; Wang, C.; Yin, X.; Mosa, A.; Lv, J.; Sun, H. Sorption of Vanadium (V) onto Natural Soil Colloids under Various Solution PH and Ionic Strength Conditions. *Chemosphere* **2017**, *169*, 105. [CrossRef]
100. Mikkonen, A.; Tummavuori, J. Retention of Vanadium (V) by Three Finnish Mineral Soils. *Eur. J. Soil Sci.* **1994**, *45*, 520. [CrossRef]
101. Gonzalez-Rodriguez, S.; Fernandez-Marcos, M.L. Sorption and Desorption of Vanadate, Arsenate and Chromate by Two Volcanic Soils of Equatorial Africa. *Soil Syst.* **2021**, *5*, 22. [CrossRef]

Article

Exploring Soil Pollution Patterns Using Self-Organizing Maps

Ilaria Guagliardi [1,*]**, Aleksander Maria Astel** [2] **and Domenico Cicchella** [3]

1   National Research Council of Italy—Institute for Agricultural and Forest Systems in Mediterranean (CNR-ISAFOM), Via Cavour 4/6, 87036 Rende, Italy
2   Environmental Chemistry Research Unit, Institute of Biology and Earth Sciences, Pomeranian University in Słupsk, 22a Arciszewskiego Str., 76-200 Słupsk, Poland; aleksander.astel@apsl.edu.pl
3   Department of Science and Technology, University of Sannio, 82100 Benevento, Italy; domenico.cicchella@unisannio.it
*   Correspondence: ilaria.guagliardi@isafom.cnr.it; Tel.: +39-0984-841427

Abstract: The geochemical composition of bedrock is the key feature determining elemental concentrations in soil, followed by anthropogenic factors that have less impact. Concerning the latter, harmful effects on the trophic chain are increasingly affecting people living in and around urban areas. In the study area of the present survey, the municipalities of Cosenza and Rende (Calabria, southern Italy), topsoil were collected and analysed for 25 elements by inductively coupled plasma mass spectrometry (ICP-MS) in order to discriminate the different possible sources of elemental concentrations and define soil quality status. Statistical and geostatistical methods were applied to monitoring the concentrations of major oxides and minor elements, while the Self-Organizing Maps (SOM) algorithm was used for unsupervised grouping. Results show that seven clusters were identified—(I) Cr, Co, Fe, V, Ti, Al; (II) Ni, Na; (III) Y, Zr, Rb; (IV) Si, Mg, Ba; (V) Nb, Ce, La; (VI) Sr, P, Ca; (VII) As, Zn, Pb—according to soil elemental associations, which are controlled by chemical and mineralogical factors of the study area parent material and by soil-forming processes, but with some exceptions linked to anthropogenic input.

Keywords: soil; potentially harmful elements; contamination; multidimensional spatial analysis; Calabria

Citation: Guagliardi, I.; Astel, A.M.; Cicchella, D. Exploring Soil Pollution Patterns Using Self-Organizing Maps. *Toxics* **2022**, *10*, 416. https://doi.org/10.3390/toxics10080416

Academic Editor: Fayuan Wang

Received: 6 July 2022
Accepted: 22 July 2022
Published: 25 July 2022

Publisher's Note: MDPI stays neutral with regard to jurisdictional claims in published maps and institutional affiliations.

1. Introduction

Soil is a dynamic natural resource which, being the basic constituent of the trophic system, has a variety of vital functions for human and environmental life [1–3]. These functions are the result of the soil's ability to control and maintain the materials and energy cycles between the atmosphere, groundwater and plant cover.

Many factors are responsible for the content, distribution and the behaviour of the chemical elements in soil, the first of which is the mineralogical and geochemical composition of the bedrock [4–6], followed by weathering [7,8] and soil formation processes (physical, chemical and biological), In addition, soils can be affected by the influence of phenomena such as the anthropogenic pollution [9–14] and the ratio and chemical composition of atmospheric depositions [15,16]. These latter sources of pollutants are widely distributed in urban soil, which is a repository of rainfall and wastewater discharge as well as atmospheric pollutants accumulated via deposition, Hence, the soil is an indicator of environmental contamination [17,18].

Differing to natural soils, which have a profile consisting of degrading vertical horizons, urban soils do not have a profile, and present great variability, both vertical and horizontal, because during their formation there are no pedogenetic processes, but instead the layering of debris, landfill, construction, and the remains of excavations of foundations [19,20]. Therefore, soils in the urban environment are the result of anthropogenic activities. Rapid industrialization and urbanization have occurred in most parts of the world during the last decades, and have stressed the soil with a growing pool of pollutants from different sources, posing a significant risk to humans and ecosystem [21,22]. The difference between soil pollution and air and water pollution lies in the fact that, in the first case, the pollutants remain for a long time in direct contact with the soil. Thus, the soil is continuously subject to pollution by toxic materials and dangerous micro-organisms which enter the air, water and the food chain [23,24]. Contact with contaminated soil may be also direct (inadvertent hand to mouth administration by children from using soil of parks and schools) or indirect (by inhaling soil contaminants which have vaporized) [25–28]. Longer contact with pollutants causes their accumulation in bones and organs. During this exposure, organ activities are disturbed, the nervous system is affected, and tumour diseases mature [29–32]. There are different types of environmental pollutants, and their potentially harmful elements (PHEs) are those that are particularly dangerous due to their ubiquity, toxicity, and persistence [33,34].

Therefore, evaluating soil pollution is of great concern. Due to urban soil spatial heterogeneity, a valuable approach to assess its quality is the application of multivariate statistics, since the environment is considered multivariate. According to many available recommendations [35], among the most effective data mining tools, those which enable unsupervised grouping based on mutual relationships between features of the analysed matrix, both of linear and nonlinear nature, are the most desired. One of the most powerful techniques for this purpose is use of the Self-Organizing Maps of Kohonen [36] because they enable display of the pattern present in multidimensional data sets on two-dimensional surface plots, are resistant against missing data and outliers, and their results are easily interpretable by decision-makers. In light of these issues, the aims of this study are (i) to individuate different types of pollution fonts controlling for a structure of monitoring data sets in a southern Italy area; (ii) to visualize geographical distribution of potentially harmful elements, and (iii) to identify high-risk areas that can be targeted for environmental risks and public health. These outcomes could be used by decision-makers working in the field of sustainable development implementation.

2. Materials and Methods

2.1. Study Area

The study area is located in the NW sector of the Calabria region (southern Italy) inside the Crati graben and covers the Cosenza and Rende municipalities territory (Figure 1). Geologically, the study area represents a tectonic depression extending over 92 km^2 bordered by NS, SW-NE and NW-SE-trending faults [37–39] associated with the horst-graben system of the Sila-Coastal Chain [40,41]. A tick succession of Pliocenic sediments made up of light brown and red sands and gravels, blue grey silty clays and silt interlayers, Pleistocene to Holocene alluvial sands and gravels and very small outcrops of Miocene carbonate rocks characterize the study area [42]. Sediments overlap a Palaeozoic intrusive-metamorphic complex formed by paragneiss, biotite schists, grey-phyllitic schists with quartz, chlorite and muscovite which, in some cases, are in a weathering process [43].

The soil map of the Calabria region at 1:250,000 scale [44], for the study area reports the presence of Fluvisols, Luvisols, Cambisols, Vertisols, Calcisols, Arenosols, Leptosols, Umbrisolsand Phaeozems. Properties, dynamics and functions of the studied soils are highly variable. For these, the average values are 17.59% for clay content, 56.50% for sand content, 6.84 for pH, 2.86% for organic matter, 0.25 μScm^{-1} for electrical conductivity, 16.14 meq 100 g^{-1} for CEC and 1.24 gcm^{-3} for bulk density.

Geomorphologically, a flat part including the urban area surrounded by hills, characterizes the study area. Falling inside the Mediterranean Sea, the Calabrian climate is typically Mediterranean, but the orography of the region affects it [45] with African warm air currents from its Ionian side and a western humid air current from the Tyrrhenian side.

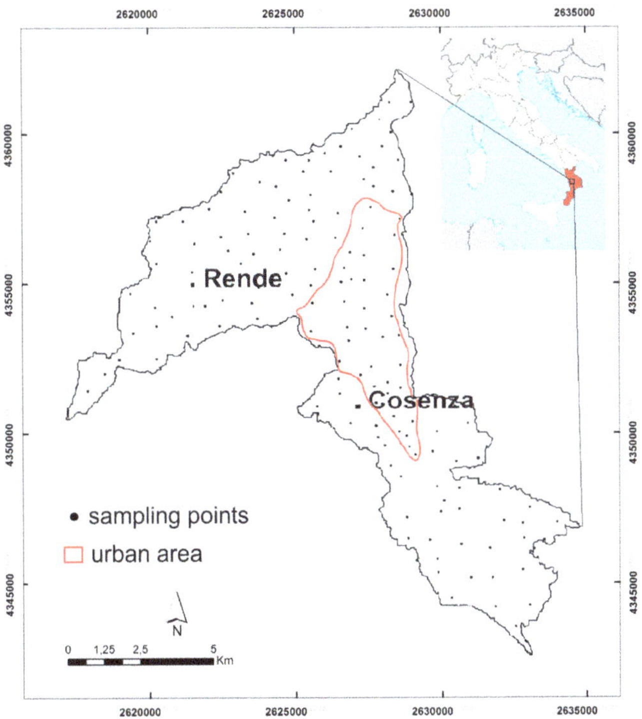

Figure 1. Study area with indication of sampling points and urban areas.

The Cosenza-Rende area has a population of approximately 100,000 inhabitants and typical urban land use, such as housing and intense automobile traffic, with limited presence of industries, commercial activities, parks and gardens. For these characteristics, different potential sources of pollution can be recognized.

2.2. Soil Sampling and Analytical Methods

In this study, 149 soil samples were collected from residual and non-residual topsoil in gardens, parks, flowerbeds and agricultural fields (Figure 1) in the study area. In addition, two duplicate pairs were collected from every 10 sites and split in the laboratory to produce replicates. Before collecting samples, removal of the surface litter at the sampling spot was carried out. At each site, topsoil samples (0–10 cm depth from the surface) were collected from five locations at the corners and at the centre of a 20 × 20 m square with a hand auger and combined to form a bulked sample. Mixing of the samples thoroughly, and removal of foreign materials such as roots, stones, pebbles and gravel, were carried out. The final sample volume was 1–1.5 kg of material, reduced to about half by the following step of quartering. Sample preparation was started in laboratory by drying soil at 40 °C prior to analysis in order to obtain a water-free reference for elemental contents. Prior to further sample processing, the soil was adequately homogenized and then sieved to fine soil of $\leq$2 mm. Successive soil analyses were performed on fine soil, and analyte contents were based on fine soil as common reference for interstudy comparisons.

After appropriate preparation procedures, each soil sample was analysed by X-ray fluorescence spectrometry (XRF) for aluminium (Al), calcium (Ca), iron (Fe), potassium (K), magnesium (Mg), manganese (Mn), sodium (Na), phosphorous (P), silica (Si) and titanium (Ti), and by inductively coupled plasma mass spectrometry (ICP-MS) for arsenic (As), barium (Ba), cerium (Ce), cobalt (Co), chromium (Cr), lanthanum (La), niobium (Nb), nickel (Ni), lead (Pb), rubidium (Rb), strontium (Sr), vanadium (V), yttrium (Y), zinc (Zn) and zirconium (Zr).

Quality of the analysis was monitored by the simultaneous analysis of certified international reference materials AGV-1, BCR-1, BR, DR-N, GA, GSP-1, NIM-G, and analysis duplicates included in analytical procedure in the range of one in twenty in each batch. Errors of the estimate for the measured elements were determined by relative standard deviation (<5%) based on three replicates of one sample randomly chosen.

2.3. Data Processing Methods

Evaluation of the spatial distribution of pollutants is important to assess the anthropogenic burden on the environment. Numerous different chemometric approaches are available for multidimensional data mining; however, methods which can be used for unsupervised exploratory analysis and pattern recognition, as well as able to handle non-linear problems, are the most desired.

Among the different statistical tools applied, an increasing number of studies have used artificial neural networks to probe complex data sets, since the visual output of the SOM analysis provides a rapid and intuitive means to examine covariance between explanatory variables, especially when the relationships among them and phenomena under analysis are unknown, and possibly nonlinear. SOMs, while extensively used in many areas, have only recently been used in ecological applications [46]. Applications can be found in ecological community ordination and gradient analysis [47], and in characterization and prediction of water quality in rivers [48] and coastal areas [49]. Applications of SOMs in oceanography are quite recent, too, and consider mostly feature extractions from univariate data sets [50].

Self-organizing maps (SOMs), in particular, are a kind of unsupervised Artificial Neural Network (ANN) that have been becoming increasingly popular for the analysis of large multivariate data sets, since they provide a topology preserving nonlinear projection of the data set in a regular two-dimensional space, and therefore constitute a methodology for nonlinear ordination analysis.

The SOM technique, known as self-organizing maps of Kohonen, is able to deal with big data sets with the possibility of visually exploring the outcomes of the model in versatile 2D maps in which similar samples are mapped close together on a grid [51]. SOM is often used in association with other algorithms, such as K-means, Principal Component Analysis and Hierarchical Cluster Analysis, for further elaborating its outcomes. However, the majority of those associations are mainly methodological studies aimed at comparing outputs of various data mining strategies. Since the current research is a case-study, it was decided to use only the self-organizing map (SOM) algorithm, considering it one of the most current neural network architectures for exploratory data analysis, clustering, and data visualization.

Among the different statistical tools applied, an increasing number of studies have used artificial neural networks to probe complex data sets, since the visual output of the SOM analysis provides a rapid and intuitive means to examine covariance between variables.

SOM is a kind of artificial neural network performing a non-linear projection of the original data space onto a two-dimensional space of neurons. It consists of two layers: the first represents input nodes (one per variable) connected to the samples, while the second one (an output layer) is a set of neurons organized on an array. A preliminary number of neurons can be determined according to one of the most accepted recommendations where n = (number of samples)$^{\wedge}(-1/2)$ [52], while the final map dimension ratio is usually slightly modified based on analysis of topographic and quantization errors (TE and QE, respectively). In general, a matrix of input vectors representing the variability and relationships of the experimental data is initialized by a series of parameters (i.e., shape of the map, shape of the map units, number of map neurons, map initialization matrix, distance function, neighbourhood function, number of epochs, etc.) retaining the number of variables of the experimental data. This input matrix becomes "the map", usually represented in a two-dimensional plot where the map vectors are called prototypes (or neurons). Then, each vector of the experimental data is presented to the algorithm, and it finds the prototype most similar to the experimental vector and adjusts it together with all surrounding prototypes to be even more similar to the experimental vector. When all the experimental vectors are presented to the algorithm, a single iteration is finished; usually, several iterations are needed to convergence. In the current study, the following initializing parameters were used: rectangular shape of the map, hexagonal shape of the map unit, 66 map units, random initialization, Euclidean distance to find the best prototype and adjust the surrounding neurons, and Gaussian neighbourhood function to establish how the neurons around the best prototype are updated during the training process. Once the SOM has converged, the weight vectors of the elements are fed into a non-hierarchical K-means algorithm to extract the neurons of the best similarity. Separating by K-means requires the user to decide the final number of k clusters the algorithm is converged into. Diverse values of k (predefined number of clusters) were tested and the sum of square for each run was calculated. Lastly, the best classification with the lowest Davies-Bouldin index (D-B) was selected. D-B index is a function of the ratio of the sum of within-cluster scatter and between-cluster separation [53]. The non-parametric Kruskal-Wallis test was performed to evaluate the significance of the cluster pattern.

All calculations in this study were performed by applying Matlab 2020 (Mathworks, Inc., Natick, MA, USA) and TIBCO Statistica 13.0 (TIBCO Software, Palo Alto, CA, USA) running on a Windows 10 platform.

3. Results and Discussion

Table 1 presents the descriptive statistics for the soil data. Except for Na and Si, a positive skewness is observed for all elements (Table 1), and a kurtosis which ranges from slight (0.04) to high (46.74).

Table 1. Basic statistics for soil samples.

	Unit	Min	Max	Mean	Median	Lower Quartile	Upper Quartile	S.D.	Skewness	Kurtosis
Al_2O_3	%	11.19	23.79	15.89	15.38	13.65	17.48	2.80	0.69	0.04
CaO	%	0.66	17.97	4.76	3.84	2.39	6.19	3.39	1.56	3.27
Fe_2O_3	%	3.11	10.58	5.47	5.13	4.42	6.16	1.47	1.03	1.01
K_2O	%	1.32	3.45	2.41	2.40	2.22	2.60	0.34	0.06	1.28
MgO	%	1.45	6.47	2.81	2.74	2.32	3.06	0.81	1.45	3.27
MnO	%	0.05	0.54	0.13	0.10	0.09	0.14	0.08	2.67	8.4
Na_2O	%	0.44	2.08	1.23	1.23	1.02	1.46	0.34	−0.03	0.31
P_2O_5	%	0.10	0.64	0.29	0.26	0.19	0.36	0.12	1.00	0.72
SiO_2	%	33.45	68.98	55.72	56.31	51.98	59.46	5.89	−0.52	0.87
TiO_2	%	0.45	1.18	0.73	0.71	0.61	0.83	0.15	0.47	0.2
As	mg kg^{-1}	3	22	7	7	5	9	3	2	3.47
Ba	mg kg^{-1}	335	2000	603	592	530	643	153	5	46.74
Ce	mg kg^{-1}	34	127	73	70	60	82	19	1	0.57
Co	mg kg^{-1}	6	40	17	16	13	20	6	1	2.02
Cr	mg kg^{-1}	46	309	91	86	73	103	32	3	15.44
La	mg kg^{-1}	13	80	38	37	31	42	11	1	1.91
Nb	mg kg^{-1}	6	35	14	14	11	15	5	2	3.87
Ni	mg kg^{-1}	18	82	35	33	28	40	10	1	2.95
Pb	mg kg^{-1}	8	708	64	31	20	69	85	4	22.56
Rb	mg kg^{-1}	62	154	105	105	92	114	18	0	0.2
Sr	mg kg^{-1}	109	514	234	233	194	271	64	1	1.66
V	mg kg^{-1}	54	239	107	102	87	123	31	1	2.46
Y	mg kg^{-1}	0	55	25	26	19	30	8	0	1.08
Zn	mg kg^{-1}	38	871	167	127	93	189	131	3	8.7
Zr	mg kg^{-1}	121	383	209	209	186	233	41	0	1.74

To analyse the spatial variations of elemental concentrations, the data set, consisting of analytical results from urban and peri-urban soil samples, was arranged in a two-way array of 25 variables, and the SOM algorithm was deployed. Apart from the methodological information presented in the section above, the detailed theoretical background of the SOM approach can be found elsewhere [54–57]; however, it is worth mentioning that here the SOM was successfully applied in assessment of soil pollution with PHEs [58], and heavy metals [59–63] as well as PCDD and PCDFs [64]. In Tao et al. [58] the distribution of PHEs in surface soil was examined. Yotova et al. [59] focused on toxic elements present in soil and their phytoavailability in an industrial area with copper mining factories and a smelter. In Yang et al. [60], soil samples were collected in several sites in a vast Chinese region and analysed for toxic elements presence. Kosiba et al. [61] compared the use of SOM with three other statistical techniques for assessing soil quality in a Polish area and its impact on the diffusion of a pathogen on a specific plant species. Dai et al. [64] evaluated the dioxin content in soil at different depths and in different years in a river floodplain, while in Nadal et al. [62] the use of the SOM allowed identification sites differently impacted by heavy metal pollutants in a petrochemical industrial area. Cheng et al. [63] proposed a SOM model built from a dataset composed of toxic metal content of soil and sediment samples collected at different depths from cascading reservoir catchments of a Chinese river. Having in mind the facts mentioned above, in the present study, exploratory data analysis, clustering and data imagining were approached by the self-organizing map (SOM) algorithm, which represents a powerful neural network architecture for these topics.

According to one of the most accepted recommendations [52], the total number of Kohonen's map neurons was estimated as $n = 5 * (149)^{(-1/2)} \approx 61$. Since there was more than one possible combination of the final dimension which was close to the dimension obtained by Vesanto's formula (i.e., 10×6, 8×7, 9×7, 11×6), quantization (QE) and topographic errors (TE) were calculated in all cases. Finally, the chosen dimensionality of the 11×6 had the lowest values of errors (QE = 0.231, TE = 0.011). Once the SOM's grid has been optimized, the U-matrix and the individual variable planes based on hexagonal lattice were visualized (Figure 2).

A component plane, scaled to represent the range of changeability of a parameter, is associated to each variable while the corresponding hexagon (i.e., top-left one of coordinates row $\times$ column = 1×1) of the consecutive plane represents the changeability of the given parameters for the same set of samples. Based on this, the component planes can be used to visualize possible correlation among the variables, while the U-matrix can be used to identify the possible presence of different clusters of data. By the analysis of planes, high concentration values of Al, Ti, Fe, Y, Rb, Cr, V, La and Ce, which are generally located in the top of the planes, and the highest concentration values of Pb and Zn in the bottom-left part of the planes, were observed.

Since PHE concentration in soils depends both on the nature of bedrock, on abiotic and biotic factors, and human activities, accurately extracting key features and characteristic patterns of variability from an elemental large data set is essential to correctly determining the sources. For this, the relationship between elements in the soil matrix gives information on PHE sources and pathways in the geo-environment. In fact, positive correlations between elements, inspected by comparing component planes, suggest that pairs in the soil samples are from the same source. Conversely, negative correlations suggest different origins between the element's pairs which, therefore, can be considered unrelated to their geochemical dynamics.

Scaling the weights vectors of each plane in the range between 0 (the least positive) and 1 (the most positive), the set of variables could be separated into several groups of similarity representing their mutual directly or inversely proportional correlations.

I.   Cr, Co, Fe, V, Ti, and Al with clear consistent patterns of the highest weights in the top-left part of the planes and the lowest weights in the middle-bottom section of the planes. These variables are all positively correlated and probably not associated with anthropogenic sources, but supposedly related to the predominant rock-forming elements constituting the soil parental materials. Indeed, the higher values of this group's element concentrations were located mostly in the NW and SE sectors of the study area where a very low road network density occurs and where there is the occurrence of ultrabasic rocks, found below the Pliocene deposits, in which these elements are predominant. The igneous-metamorphic complex can be ascribed to the pile of tectonic nappes forming the mountain chain of the northern Calabrian Arc, described in [65], which includes an intermediate structural element made up of ophiolite-bearing units [66] that mostly extend along the Tyrrhenian side of the arc to form a westward convex arc-shaped belt separated from the southern Apennines by the roughly E-W trending left-lateral strike-slip fault zone. This unit is represented by a tectonic mélange constituted by a monotonous sequence of phyllites, quartzites, and calcschists, including metric to kilometric lens-shaped blocks of ophiolitic rocks. These rocks are mainly constituted by serpentinized ultramafics, and by glaucophane-bearing meta-basites, with remnants of their sedimentary cover and rare meta-gabbros [65]. In particular, the geochemical behaviour of V resembles that of Fe which can substitute in Fe-Mg silicates (amphiboles, pyroxenes, micas). This elemental association confirms that the soils are controlled by the same typically lithogenic elements associated with silicate minerals.

II.   Ni and Na with clear opposite patterns of the highest and the lowest weights of Ni and Na occur, respectively, in the top-left triangle of hexagons. By contrast, the lowest and the highest weights of Ni and Na, respectively, cover the bottom-right triangle of hexagons. One important observation that arises from the calculation of the correlations for these elements is that Ni does not have any positive relation with Na. The absence of this correlation could be attributed to the influence of the distribution of these elements by anthropogenic activities.

III.   Y, Zr and Rb with consistently increasing weights occur in the top-half part of the planes and descending weights in the bottom-half. Such a pattern indicates that Y, Zr and Rb are positively correlated and considered to indicate provenance compositions as a consequence of their immobile behaviour [67]. Zr is enriched in silica rich sediments compared to the associated shales, which suggests its propensity to be preferentially concentrated in coarser sediments. Many soil samples can be attributed to the compositional field in which the local content of marbles, sandstones, and gneisses are part, indicating a strong lithological influence on element concentrations. Therefore, Y, Zr and Rb association prove to be of undoubted geogenic origin.

IV.   Si, Mg and Ba have patterns that, in general, are similar to the patterns observed in the case of Ni and Na. The highest weights for Mg and Ba are observed only for single hexagons of coordinates 1×1, 1×2 and 2×1, while for those hexagons, relatively low weights of Si occur. By contrast, in the bottom-right triangle of hexagons, high weights for Si correspond with low weights for Mg and Ba. Generally, such patterns indicate that Si is negatively correlated with Mg and Ba, while Ba is positively correlated with Mg. Ba is a trace element common in alkali feldspars and biotite. The lack of a clear correlation between Al, Rb and Sr and Ba indicates a relationship between Ba and mica components, or that Ba was lost at an early stage in weathering of feldspars.

V.   Nb, Ce and La, with the highest values of weights, occur in only a few hexagons in the top-right triangle of the planes. Nb, Ce and La, belonging to the rare earth elements (REEs), show positive correlations, explaining their similar behaviour in soil samples. Their primary source is accessory minerals in magmatic rocks, e.g., monazite, xenotime and allanite. This could explain their common geogenic sources.

VI.   Sr, P and Ca, with compatibly the highest weights, occur in a thinly vertical belt of hexagons located on the left-hand side of the planes. Such a pattern indicates a strong positive correlation between Sr, P and Ca, which confirms a mineralogical common source of elemental association. This may be due to Sr geochemical affinity with Ca [68]. Sr is a relatively common element that substitutes for Ca in crystal lattices of rock-forming minerals, including feldspars and plagioclase, as in the study area.

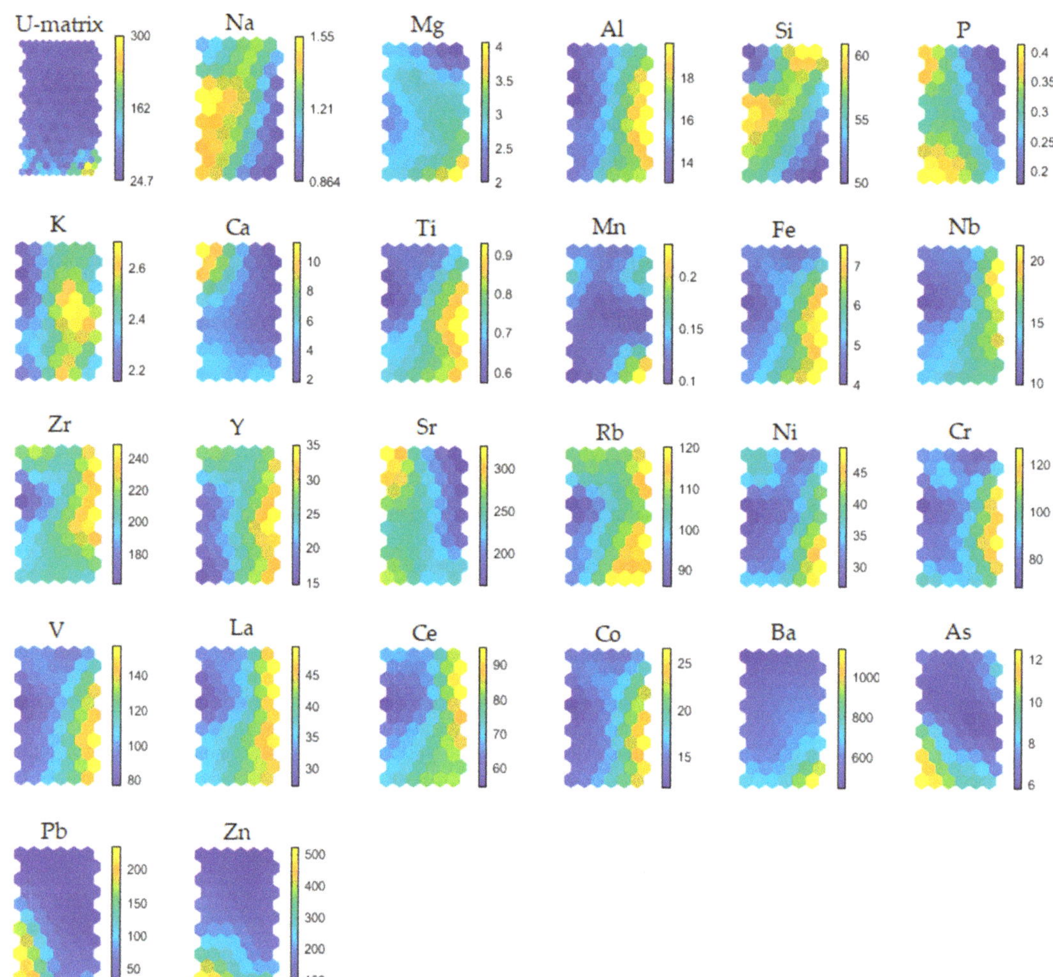

Figure 2. Component planes for all sampling sites and parameters. U-matrix visualizes distances between neighbouring map units and helps to identify the cluster structure of the map. High values of the U-matrix indicate a cluster border, uniform areas of low values indicate clusters themselves; each component plane shows the values of one variable in each map unit. Both grey-tone pattern and grey-tone bar labelled as "d" deliver information regarding compounds/element abundance calculated through the SOM learning process.

VII. As, Zn, Pb, with compatibly the highest weights. occur in only a few hexagons located in the bottom-left triangle of the planes. The consistent colour indicates that As, Zn and Pb have a strong positive correlation, and their concentrations are higher in soil next to roads than in the soils away from them. This indicates that larger concentrations of these elements are related to road traffic. Consequently, their positive correlation allows us to draw conclusions about their common source linked to anthropogenic activities conducted in urban environments. These elements are, indeed, present in vehicle fuel, being used for increasing gasoline antiknock.

The set of component planes, with weights scaled in the range 0–1 grouped according to their correlations, is presented in Figure 3.

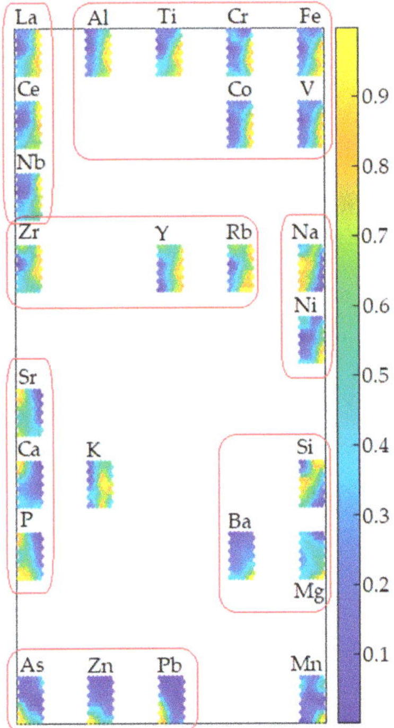

Figure 3. Soil quality parameter similarity pattern obtained by self-organizing mapping. An analysis of the distance between variables on the map connected with an assessment of the colour-tone patterns provides semi-quantitative information about the nature of correlations between them.

The significant information deriving from the SOM theory, that each node of the SOM map could be consecutively referred to one or more samples, leads to the conclusion that the differentiated structure of PHEs abundance (reflected in different colour scales in the planes) revealed the presence of numerous similarity clusters in the set of samples. Consequently, weight vectors of the converged map were clustered based on a K-means clustering mode. Some predefined numbers of clusters were tested, and the sum of squares for each run was calculated. The best partition was gained for a seven-cluster configuration having the lowest Davies-Bouldin index value (Figure 4).

According to SOM theory, the node (map neuron) with a weight vector closest to the input sample vector is identified as the best matching unit, and the number of tagging is summarized. Lastly, the distribution of the sample vectors along a Kohonen map can be analysed by decoding the best matching unit selection events. Clusters I-VII (consecutively named as C_I-C_VII) include numerous numbers of 149 soil samples (C_I-20, C_II-11, C_III-22, C_IV-26, C_V-28, C_VI-19, C_VII-23). Cluster distribution of investigated soil samples in the study area according to the local geological setting is presented in Figure 5. Comparison of initially determined PHEs concentrations in soil samples with the clustering results allowed for the assignment of clustering patterns to factors impacting soil quality. Comparison of analyte concentration values according to clustering pattern is presented in

Figure 6 (concentration at % level) and Figure 7 (concentration at mg kg^{-1} level) together with a statistical assessment from the non-parametric Kruskal-Wallis test.

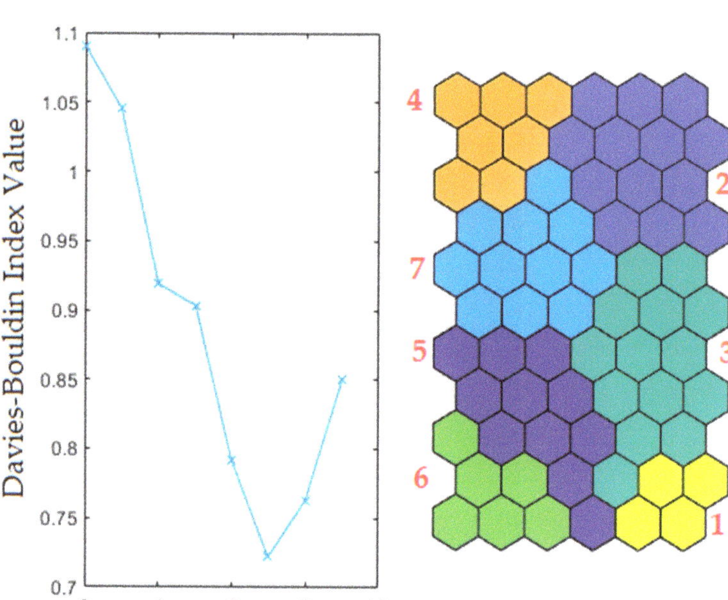

Figure 4. Clustering patterns according to the Davies-Bouldin index minimum value.

Among the seven clusters, C_I includes 20 samples (13.4%) with the highest concentration of Mg, Al, Ti, Ni, Fe, Y, Cr, V, Co and Ba. Most of the samples included in C_I was collected in peri-urban soils and in areas in which Paleozoic paragneiss and biotite schists occur. More precisely, the observed association clustered in C_I can be clarified considering the presence in the study area of ultrabasic rocks in which these elements are principal. Highest baseline concentrations of these elements seem to be highly associated with the igneous-metamorphic complex found below the Pliocene deposits that outcrop mostly in the NW and SE sectors of the territory. This structure represents the pile of tectonic nappes forming the mountain chain of the northern Calabrian Arc and contains an intermediate structural element made up of ophiolite-bearing units.

C_II consists of only 11 (7.4%) samples collected in peri-urban soils. These samples were characterized by the lowest concentration of Mg and Ca, with the highest abundance of REEs such as La, Ce, Nb and Zr. Such a phenomenon indicates that REE content is associated with alkaline igneous rocks and carbonatites, which are igneous rocks derived from carbonate-rich magma rather than silica-rich magma [69].

C_III includes 22 samples (14.8%) with the highest concentration of K and relatively low abundance of Zn, Mn, Ni, Ba, Pb, As. The majority of these samples were collected in soils along the part of the Crati river falling in the study area, and their composition indicates the presence of organic matter, suggesting that this might play a role in increasing K adsorption rate. As can be seen, samples clustered in C_I-C_III as a set, in comparison to the rest of clusters, were characterized by higher concentrations of Al, K, Ti, and Fe, with lower concentration of P and Sr. Moreover, samples from C_I and C_II were characterized by the highest concentration range for REEs. The content of REE in soil, without other inputs, is influenced by the parent material and on geochemical processes such as mineral weathering, which is an important input of elements into the soils [70]. Twenty-six samples clustered in C_IV (17.4%) were, in general, grouped together based on the lowest content of

Al and Si, relative to minimal concentrations among of the other samples, and the highest concentration of P, Ca, Sr. According to their location, this suggests that the underlying rocks are the major source of P. C_V, consisting of 28 (18.8%) samples, represents soils with moderate concentrations of the majority of investigated elements. It seems they are clustered separately due to a relatively large range of determined concentrations for Mg and Mn. C_VI includes 19 (12.7%) soil samples in which their chemical composition is dominated by relatively high concentration of P, Zn, and Sr. These samples were additionally characterized by the highest concentration and range of values for Pb and As, and lowest the abundance of Y. The soil samples characterized by these elemental contents are distributed in the urban area, where road networks and vehicular traffic are intense, and, consequently, higher Pb contents occur. Particularly, soils close to high traffic roads of the study area showed the highest Pb and Zn baseline values. These elements are included in vehicle fuel for increasing gasoline antiknock. The last C_VII includes 23 samples characterized by the lowest concentration of the majority of elements, such as Ti, Mn, Rb, Ni, Fe, Zr, Y, Cr, V, La, Ce and Co. In general samples clustered in C_V-C_VII show consistent chemical composition with the exception of some elements, determining their separation in a single cluster. As can be seen in Figure 6, a monotonic increasing trend of determined concentration values from C_I to C_VII is observed for Si, Na, and Sr, while much more frequently observed was a decreasing trend for Al, Ti, Rb, Ni, Fe, Y, Cr, V and Co.

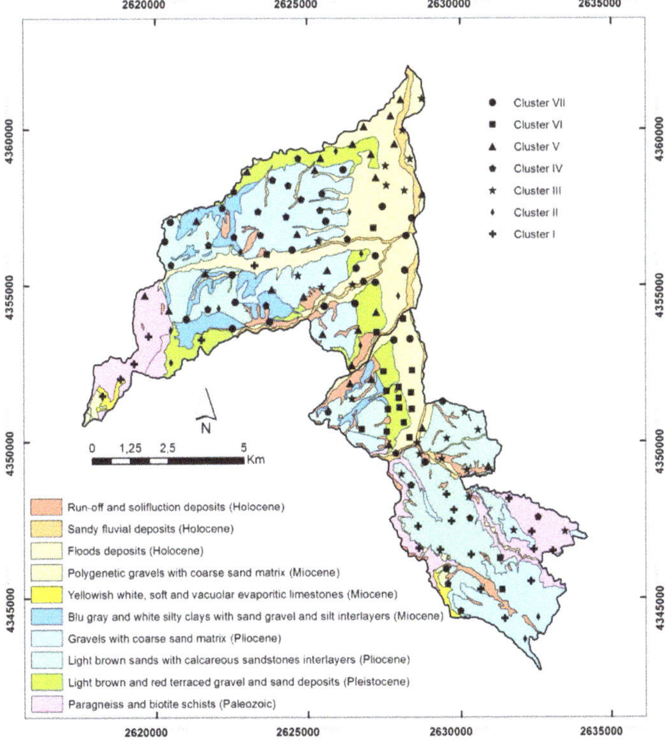

Figure 5. Geological setting of the study area and localization of sampling according to cluster classification. The seven identified clusters—(I) Cr, Co, Fe, V, Ti, Al; (II) Ni, Na; (III) Y, Zr, Rb; (IV) Si, Mg, Ba; (V) Nb, Ce, La; (VI) Sr, P, Ca; (VII) As, Zn, Pb—representing the seven groups in which the elements are associated according to local lithologies, are indicated.

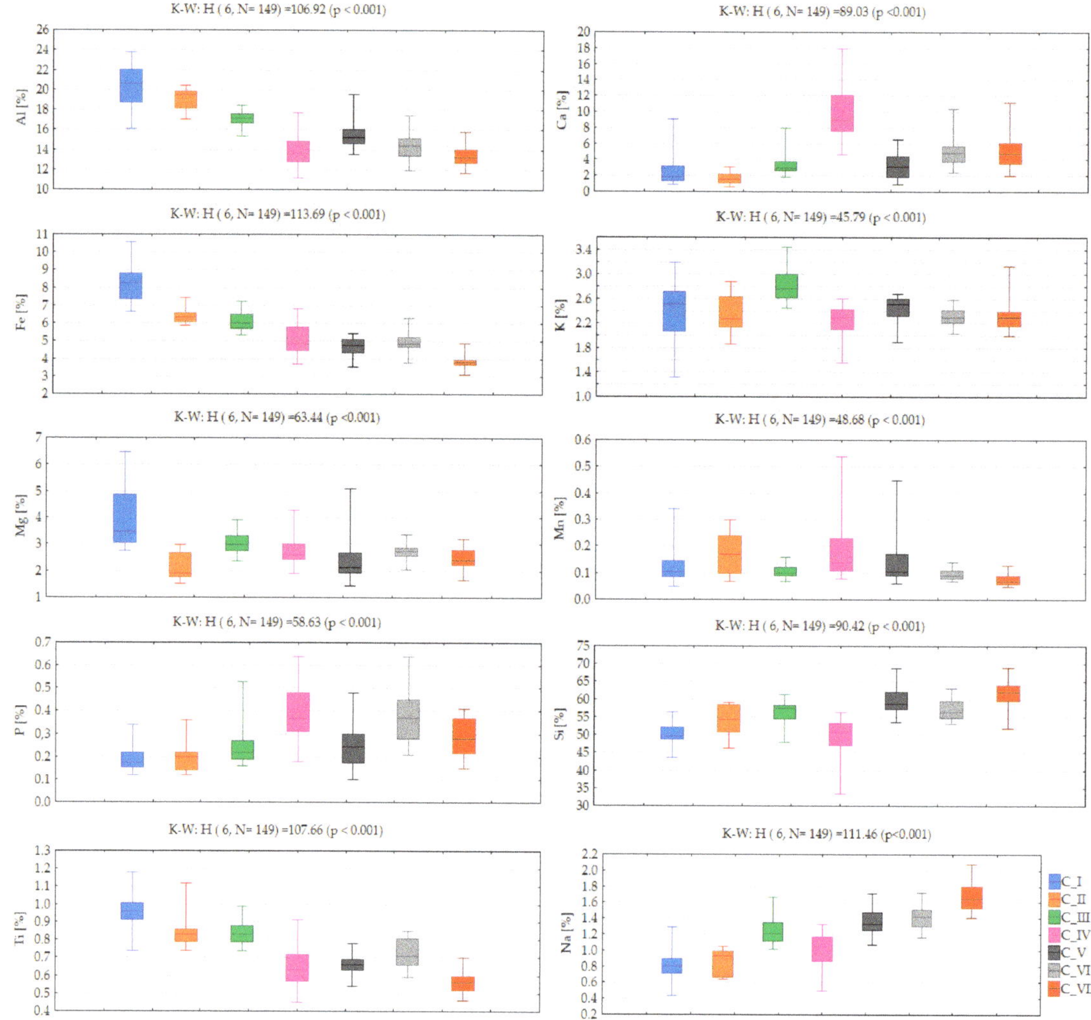

Figure 6. Compound concentration values according to clustering patterns (central line: median, box: 25–75% percentile, whiskers: minimum-maximum) with statistical assessment of differences between clusters based on the Kruskall-Wallis non-parametric test (K-W).

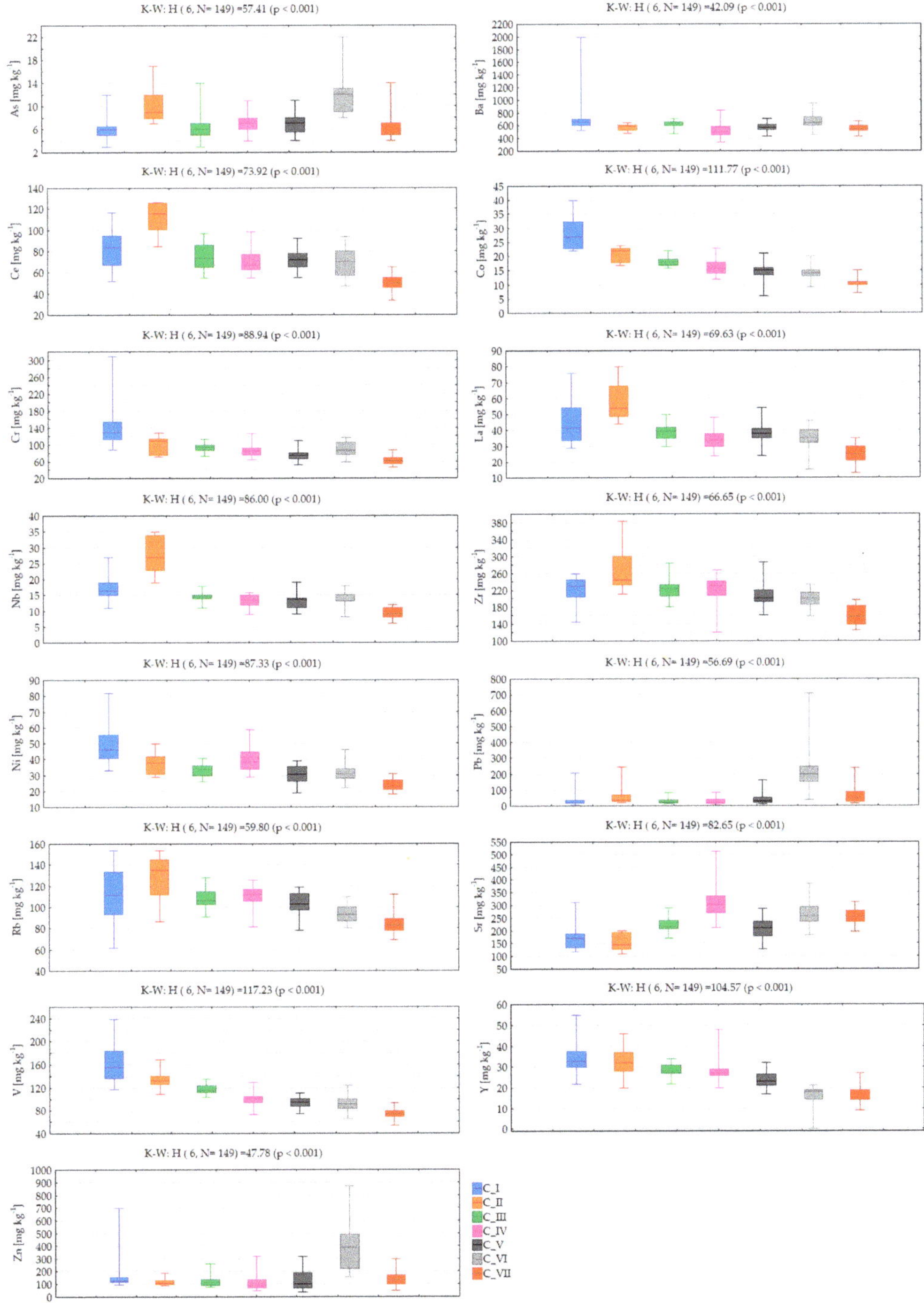

Figure 7. PHE concentration values according to clustering patterns (central line: median, box: 25–75% percentile, whiskers: minimum-maximum) with statistical assessment of differences between clusters based on the Kruskall-Wallis non-parametric test (K-W).

4. Conclusions

Correct monitoring and management of potentially harmful elements are key issues for urban and peri-urban soil knowledge, linking PHE concentrations at sites in which geogenic or anthropogenic input occur. In this study, evaluation of the usefulness of a powerful approach, such the SOM algorithm, for multidimensional geochemical data analysis and modelling problems of environmental pollution, was performed using data sets obtained by comprehensive monitoring of PHE content in the municipalities of the Cosenza-Rende area (Calabria, southern Italy). In the study area, a total of 149 soil samples, collected in residual and non-residual areas, parks, flowerbeds and agricultural fields, were investigated for 25 elements in order to better understand influences on soil geochemistry.

A self-organizing map (SOM) was selected as a powerful approach in soil science application for spatial distribution and geochemical mapping. A combination of the analysis of major metals, minor metals and PHEs, with the statistical treatment of SOMs, showed the geolithological formations and anthropogenic pressure on the territory. The association between the neurons and variables achieved by an unsupervised procedure performed by the SOM technique, allows recognition of high-risk areas which can represent environmental hazards and public health risks. By using the SOM method, the occurrence of anomalies ascribable to anthropogenic input in urban soils, referring to elements such as Pb and Zn, and of some geogenic anomalous high values of As, Cr, and V mainly identified in peri-urban areas, was recognized. The SOM was employed to cluster the data, and results presented a classification in seven clusters—(I) Cr, Co, Fe, V, Ti, Al; (II) Ni, Na; (III) Y, Zr, Rb; (IV) Si, Mg, Ba; (V) Nb, Ce, La; (VI) Sr, P, Ca; (VII) As, Zn, Pb—mainly determined by the chemical and mineralogical factors typical of the geological setting of the study area, and by soil forming and weathering processes. Among them, C_II and C_VII can be linked to anthropogenic input. However, in general, more contamination was identified in urban soils than in peri-urban ones.

In summary, the main outcomes of the study are as follows:

1. SOM was verified as a promising approach for pattern recognition and, in particular, for delineating pollution patterns of soil;
2. the main factors that influence PHE concentration in the Cosenza-Rende area were associated with geological setting and human activities;
3. classification of soil patterns provides a great deal of information enhancing risk status source identification, which can be used for decision making.

The paper contains an important methodological novelty. In fact, it proposes the application of an existing methodology for data analysis to a new class of problems. Its results can have a valuable role in identifying polluted areas and proposing remedial action aimed at reducing health risks to people. Further development of this tool should also help soil scientists to identify novel relationships about already studied phenomena, and act as a hypothesis generator for traditional research, as well as supplying clear and intuitive visualization of the environmental phenomena studied.

Author Contributions: Conceptualization, I.G., A.M.A. and D.C.; methodology, I.G. and D.C.; software, A.M.A.; validation, I.G., A.M.A. and D.C.; formal analysis, A.M.A.; investigation, I.G. A.M.A. and D.C.; resources, I.G. and D.C.; data curation, I.G.; writing—original draft preparation, I.G. and A.M.A.; writing—review and editing, I.G., A.M.A. and D.C.; visualization, I.G. and A.M.A.; supervision, I.G.; project administration, I.G. and D.C.; funding acquisition, I.G. and D.C. All authors have read and agreed to the published version of the manuscript.

Funding: This research received no external funding.

Informed Consent Statement: Not applicable.

Data Availability Statement: Not applicable.

Conflicts of Interest: The authors declare no conflict of interest.

References

1. Astel, A. Soil contamination interpretation by the use of monitoring data analysis. *Water Air Soil. Pollut.* **2011**, *216*, 375–390. [CrossRef] [PubMed]
2. Karim, Z.; Qureshi, B.A.; Mumtaz, M. Geochemical baseline determination and pollution assessment of heavy metals in urban soils of Karachi, Pakistan. *Ecol. Indic.* **2015**, *48*, 358–364. [CrossRef]
3. Kelepertzis, E. Accumulation of heavy metals in agricultural soils of Mediterranean: Insights from Argolida basin, Peloponnese, Greece. *Geoderma* **2014**, *221–222*, 82–90. [CrossRef]
4. Guagliardi, I.; Zuzolo, D.; Albanese, S.; Lima, A.; Cerino, P.; Pizzolante, A.; Thiombane, M.; De Vivo, B.; Cicchella, D. Uranium, thorium and potassium insights on Campania region (Italy) soils: Sources patterns based on compositional data analysis and fractal model. *J. Geochem. Explor.* **2020**, *212*, 106508. [CrossRef]
5. Tarvainen, T.; Ladenberger, A.; Snöälv, J.; Jarva, J.; Andersson, M.; Eklund, M. Urban soil geochemistry of two Nordic towns: Hämeenlinna and Karlstad. *J. Geochem. Explor.* **2018**, *187*, 34–56. [CrossRef]
6. Dinter, T.C.; Gerzabek, M.H.; Puschenreiter, M.; Strobel, B.W.; Couenberg, P.M.; Zehetner, F. Heavy metal contents, mobility and origin in agricultural topsoils of the Galápagos Islands. *Chemosphere* **2021**, *272*, 129821. [CrossRef]
7. Guagliardi, I.; Rovella, N.; Apollaro, C.; Bloise, A.; De Rosa, R.; Scarciglia, F.; Buttafuoco, G. Modelling seasonal variations of natural radioactivity in soils: A case study in southern Italy. *J. Earth Sys. Sci.* **2016**, *125*, 1569–1578. [CrossRef]
8. Mazurek., R.; Kowalska, J.; Gąsiorek, M.; Zadrożny, P.; Józefowska, A.; T. Zaleski, T.; Kępka, W.; MarylaTymczuk, M.; Orłowska, K. Assessment of heavy metals contamination in surface layers of Roztocze National Park forest soils (SE Poland) by indices of pollution. *Chemosphere* **2017**, *168*, 839–850. [CrossRef]
9. Buttafuoco, G.; Guagliardi, I.; Tarvainen, T.; Jarva, J. A multivariate approach to study the geochemistry of urban topsoil in the city of Tampere, Finland. *J. Geochem. Explor.* **2017**, *181*, 191–204. [CrossRef]
10. Cicchella, D.; Zuzolo, D.; Albanese, S.; Fedele, L.; Di Tota, I.; Guagliardi, I.; Thiombane, M.; De Vivo, B.; Lima, A. Urban soil contamination in Salerno (Italy): Concentrations and patterns of major, minor, trace and ultra-trace elements in soils. *J. Geochem. Explor.* **2020**, *213*, 106519. [CrossRef]
11. Ding, F.; He, Z.; Liu, S.; Zhang, S.; Zhao, F.; Li, Q.; Stoffella, P.J. Heavy metals in composts of China: Historical changes, regional variation, and potential impact on soil quality. *Environ. Sci. Pollut. Res.* **2017**, *24*, 3194–3209. [CrossRef]
12. Marrugo-Negrete, J.; Enamorado-Montes, G.; Durango-Hernández, J.; Pinedo-Hernández, J.; Díez, S. Removal of mercury from gold mine effluents using Limnocharis flava in constructed wetlands. *Chemosphere* **2017**, *167*, 188–192. [CrossRef]
13. Zuzolo, D.; Cicchella, D.; Lima, A.; Guagliardi, I.; Cerino, P.; Pizzolante, A.; De Vivo, B.; Albanese, S. Potentially toxic elements in soils of Campania region (Southern Italy): Combining raw and compositional data. *J. Geochem. Explor.* **2020**, *213*, 106524. [CrossRef]
14. Minkina, T.; Sushkova, S.; Yadav, B.; Rajput, V.; Mandzhieva, S.; Nazarenko, O. Accumulation and transformation of benzo[a]pyrene in Haplic Chernozem under artificial contamination. *Environ. Geochem. Health* **2020**, *42*, 2485–2494. [CrossRef]
15. Cozza, V.; Guagliardi, I.; Rubino, M.; Cozza, R.; Martello, A.; Picelli, M.; Zhupa, E. Esopo: Sensors and social pollution measurements. *CEUR Workshop Proc.* **2015**, *1478*, 52–57.
16. Pellicone, G.; Caloiero, T.; Guagliardi, I. The De Martonne aridity index in Calabria (Southern Italy). *J. Maps* **2019**, *15*, 788–796. [CrossRef]
17. Buttafuoco, G.; Caloiero, T.; Guagliardi, I.; Ricca, N. Drought assessment using the reconnaissance drought index (RDI) in a southern Italy region. In Proceedings of the 6th IMEKO TC19 Symposium on Environmental Instrumentation and Measurements, Reggio Calabria, Italy, 24–25 June 2016; pp. 52–55.
18. Wu, H.; Wang, J.; Guo, J.; Hu, X.; Bao, H.; Chen, J. Record of heavy metals in Huguangyan Maar Lake sediments: Response to anthropogenic atmospheric pollution in Southern China. *Sci. Total Environ.* **2022**, *831*, 154829. [CrossRef]
19. Calzolari, C.; Tarocco, P.; Lombardo, N.; Marchi, N.; Ungaro, F. Assessing soil ecosystem services in urban and peri-urban areas: From urban soils survey to providing support tool for urban planning. *Land Use Policy* **2020**, *99*, 105037. [CrossRef]
20. Ricca, N.; Guagliardi, I. Multi-temporal dynamics of land use patterns in a site of community importance in Southern Italy. *Appl. Ecol. Environ. Res.* **2015**, *13*, 677–691.
21. Mehmood, K.; Bao, Y.; Abbas, R.; Saifullah.; Petropoulos, G.P.; Ahmad, H.R.; Abrar, M.M.; Mustafa, A.; Abdalla, A.; Lasaridi, K.; et al. Pollution characteristics and human health risk assessments of toxic metals and particle pollutants via soil and air using geoinformation in urbanized city of Pakistan. *Environ. Sci. Pollut. Res.* **2021**, *28*, 58206–58220. [CrossRef]
22. O'Riordan, R.; Davies, J.; Stevens, C.; Quinton, J.N.; Boyko, C. The ecosystem services of urban soils: A review. *Geoderma* **2021**, *395*, 115076. [CrossRef]
23. Guagliardi, I.; Cicchella, D.; De Rosa, R.; Ricca, N.; Buttafuoco, G. Geochemical sources of vanadium in soils: Evidences in a southern Italy area. *J. Geochem. Explor.* **2018**, *184*, 358–364. [CrossRef]
24. Zuzolo, D.; Cicchella, D.; Catani, V.; Giaccio, L.; Guagliardi, I.; Esposito, L.; De Vivo, B. Assessment of potentially harmful elements pollution in the Calore River basin (Southern Italy). *Environ. Geochem. Health* **2017**, *39*, 531–548. [CrossRef]
25. Kim, N.; Fergusson, J. Concentrations and sources of cadmium, copper, lead and zinc in house dust in Christchurch, New Zealand. *Sci. Total Environ.* **1993**, *138*, 1–21. [CrossRef]
26. Gupta, S.K.; Vollmer, M.K.; Krebs, R. The importance of mobile, mobilisable and pseudo total heavy metal fractions in soil for three-level risk assessment and risk management. *Sci. Total Environ.* **1996**, *178*, 11–20. [CrossRef]

27. Charlesworth, S.; Everett, M.; McCarthy, R.; Ordonez, A.; Miguel, E. A comparative study of heavy metal concentration and distribution in deposited street dusts in a large and small urban area: Birmingham and Coventry, West Midlands, UK. *Environ. Int.* **2003**, *29*, 563–573. [CrossRef]

28. Imperato, M.; Adamo, P.; Naimo, D.; Arienzo, M.; Stanzione, D.; Violante, P. Spatial distribution of heavy metals in urban soils of Naples city (Italy). *Environ. Pollut.* **2003**, *124*, 247–256. [CrossRef]

29. IPCS. *Cadmium. Environmental Health Criteria 134*; World Health Organization: Geneva, Switzerland, 1992.

30. IPCS. *Lead, Environmental Health Criteria 85*; World Health Organization: Geneva, Switzerland, 1995.

31. Komarnicki, G.J.K. Lead and cadmium in indoor air and the urban environment. *Environ. Pollut.* **2005**, *136*, 47–61. [CrossRef]

32. Galušková, I.; Borůvka, L.; Drábek, O. Urban Soil Contamination by Potentially Risk Elements. *Soil Water Res.* **2011**, *6*, 55–60. [CrossRef]

33. Guney, M.; Zagury, G.J.; Dogan, N.; Onay, T.T. Exposure assessment and risk characterization from trace elements following soil ingestion by children exposed to playgrounds, parks and picnic areas. *J. Hazard. Mater.* **2010**, *182*, 656–664. [CrossRef]

34. Burges, A.; Epelde, L.; Blanco, F.; Becerril, J.M.; Garbisu, C. Ecosystem services and plant physiological status during endophyte-assisted phytoremediation of metal contaminated soil. *Sci. Total Environ.* **2017**, *584*, 329–338. [CrossRef] [PubMed]

35. Astel, A.; Tsakovski, S.; Barbieri, P.L.; Simeonov, V. A comparison of SOM classification approach with cluster analysis and PCA for large environmental data sets. *Wat. Res.* **2007**, *41*, 4566–4578. [CrossRef] [PubMed]

36. Kohonen, T. Self-organizing formation of topologically correct feature maps. *Biol. Cybern.* **1982**, *43*, 56–69. [CrossRef]

37. Tansi, C.; Muto, F.; Critelli, S.; Iovine, G. Neogene-Quaternary strike-slip tectonics in the central Calabrian Arc (Southern Italy). *J. Geodyn.* **2007**, *43*, 393–414. [CrossRef]

38. Van Dijk, J.P.; Bello, M.; Brancaleoni, G.P.; Cantarella, G.; Costa, V.; Frixa, A.; Golfetto, F.; Merlini, S.; Riva, M.; Torricelli, S.; et al. A regional structural model for the northern sector of the Calabrian Arc (Southern Italy). *Tectonophysics* **2000**, *324*, 267–320. [CrossRef]

39. Vignaroli, G.; Minelli, L.; Rossetti, F.; Balestrieri, M.L.; Faccenna, C. Miocene thrusting in the eastern Sila Massif: Implication for the evolution of the Calabria-Peloritani orogenic wedge (Southern Italy). *Tectonophysics* **2012**, *538–540*, 105–119. [CrossRef]

40. Gaglioti, S.; Infusino, E.; Caloiero, T.; Callegari, G.; Guagliardi, I. Geochemical characterization of spring waters in the Crati River Basin, Calabria (Southern Italy). *Geofluids* **2019**, *2019*, 3850148. [CrossRef]

41. Iovine, G.; Guagliardi, I.; Bruno, C.; Greco, R.; Tallarico, A.; Falcone, G.; Lucà, F.; Buttafuoco, G. Soil-gas radon anomalies in three study areas of Central-Northern Calabria (Southern Italy). *Nat. Hazards* **2018**, *91*, 193–219. [CrossRef]

42. Fabbricatore, D.; Robustelli, G.; Muto, F. Facies analysis and depositional architecture of shelf-type deltas in the Crati Basin (Calabrian Arc, south Italy). *Ital. J. Geosci.* **2014**, *133*, 131–148. [CrossRef]

43. Le Pera, E.; Critelli, S.; Sorriso-Valvo, M. Weathering of gneiss in Calabria, southern Italy. *Catena* **2001**, *42*, 1–15. [CrossRef]

44. ARSSA (Agenzia Regionale per lo Sviluppo e per i Servizi in Agricoltura). I suoli della Calabria. Carta dei suoli in scala 1:250,000 della Regione Calabria. I suoli della Calabria. In *Monografia Divulgativa: Programma Interregionale Agricoltura-Qualità e Misura 5*; Rubbettino, Ed.; ARSSA, Servizio Agropedologia: Catanzaro, Italy, 2003.

45. Buttafuoco, G.; Caloiero, T.; Ricca, N.; Guagliardi, I. Assessment of drought and its uncertainty in a southern Italy area (Calabria region). *Meas. J. Int. Meas. Confed.* **2018**, *113*, 205–210. [CrossRef]

46. Giraudel, J.L.; Lek, S. A comparison of self-organizing map algorithm and some conventional statistical methods for ecological community ordination. *Ecol. Modell.* **2001**, *146*, 329–339. [CrossRef]

47. Park, Y.-S.; Verdonschot, P.F.M.; Chon, T.-S.; Lek, S. Patterning and predicting aquatic macroinvertebrate diversities using artificial neural network. *Water Res.* **2003**, *37*, 1749–1758. [CrossRef]

48. Brodnjak-Vončina, D.; Dobčnik, D.; Novič, M.; Zupan, J. Chemometrics characterisation of the quality of river water. *Anal. Chim. Acta* **2002**, *462*, 87–100. [CrossRef]

49. Aguilera, P.A.; Garrido Frenich, A.; Torres, J.A.; Castro, H.; Martinez Vidal, J.L.; Canton, M. Application of the Kohonen neural network in coastal water management: Methodological development for the assessment and prediction of water quality. *Water Res.* **2001**, *35*, 4053–4062. [CrossRef]

50. Liu, Y.; Weisberg, R.H. Pattern of ocean current variability on the West Florida Shelf using the Self-Organizing Map. *J. Geophys. Res.* **2005**, *110*, C06003. [CrossRef]

51. Vesanto, J. SOM-based data visualization methods. *Intell. Data. Anal.* **1999**, *3*, 111–126. [CrossRef]

52. Vesanto, J. Neural network tool data mining: SOM Toolbox. In Proceedings of the Symposium on Tool Environments and Development Methods for Intelligent Systems (TOOL-MET2000), Oulu, Finland, 13–14 April 2000; pp. 184–196.

53. Davies, D.L.; Bouldin, D.W. A cluster separation measure. *IEEE Trans. Pattern Anal. Mach. Intell.* **1979**, *1*, 224–227. [CrossRef]

54. Kohonen, T. *Self-Organizing Maps, 3rd ed*; Springer: Berlin/Heidelberg, Germany, 2001.

55. Fort, J.C. SOM's mathematics. *Neural Netw.* **2006**, *19*, 812–816. [CrossRef]

56. Cottrell, M.; Fort, J.C.; Pagès, G. Theoretical aspects of the SOM algorithm. *Neurocomputing* **1998**, *21*, 119–138. [CrossRef]

57. Ultsch, A.; Siemon, H.P. Kohonen's self organizing feature maps for exploratory data analysis. In Proceedings of the International Neural Network Conference (INNC'90), Kluwer, Dordrecht, Paris, France, 9–13 July 1990; pp. 305–308.

58. Tao, S.-Y.; Zhong, B.-Q.; Lin, Y.; Ma, J.; Zhou, Y.; Hou, H.; Zhao, L.; Sun, Z.; Qin, X.; Shi, H. Application of a self-organizing map and positive matrix factorization to investigate the spatial distributions and sources of polycyclic aromatic hydrocarbons in soils from Xiangfen County, northern China. *Ecotoxicol. Environ. Saf.* **2017**, *141*, 98–106. [CrossRef]

59. Yotova, G.; Zlateva, B.; Ganeva, S.; Simeonov, V.; Kudłak, B.; Namieśnik, J.; Tsakovski, S. Phytoavailability of potentially toxic elements from industrially contaminated soils to wild grass. *Ecotoxicol. Environ. Saf.* **2018**, *164*, 317–324. [CrossRef]
60. Yang, C.; Guo, R.; Wu, Z.; Zhou, K.; Yu, Q. Spatial extraction model for soil environmental quality of anomalous areas in a geographic scale. *Environ. Sci. Pollut. Res.* **2014**, *21*, 2697–2705. [CrossRef]
61. Kosiba, P. Self-organizing feature maps and selected conventional numerical methods for assessment of environmental quality. *Acta Soc. Bot. Pol.* **2009**, *78*, 335–343. [CrossRef]
62. Nadal, M.; Schuhmacher, M.; Domingo, J.L. Metal pollution of soils and vegetation in an area with petrochemical industry. *Sci. Total Environ.* **2004**, *321*, 59–69. [CrossRef]
63. Cheng, F.; Liu, S.; Yin, Y.; Zhang, Y.; Zhao, Q.; Dong, S. Identifying trace metal distribution and occurrence in sediments, inundated soils, and non-flooded soils of a reservoir catchment using Self-Organizing Maps, an artificial neural network method. *Environ. Sci. Pollut. Res.* **2017**, *24*, 19992–20004. [CrossRef]
64. Dai, D.; Oyana, T.J. Spatial variations in the incidence of breast cancer and potential risks associated with soil dioxin contamination in Midland, Saginaw, and Bay Counties, Michigan, USA. *Environ. Health Glob. Access Sci. Sour.* **2008**, *7*, 49. [CrossRef]
65. Tortorici, L.; Catalano, S.; Monaco, C. Ophiolite-bearing melanges in southern Italy. *Geol. J.* **2009**, *44*, 153–166. [CrossRef]
66. Infusino, E.; Guagliardi, I.; Gaglioti, S.; Caloiero, T. Vulnerability to Nitrate Occurrence in the Spring Waters of the Sila Massif (Calabria, Southern Italy). *Toxics* **2022**, *10*, 137. [CrossRef]
67. Taylor, S.R.; McLennan, S.M. *The Continental Crust: Its Composition and Evolution*; Scientific Publications: Oxford, UK, 1985; p. 312.
68. Aubert, H.; Pinta, M. *Trace Elements in Soils*; Elsevier: Amsterdam, The Netherlands, 1977; Volume 7, p. 396.
69. Chakhmouradian, A.R.; Zaitsev, A.N. Rare earth mineralization in igneous rocks—Sources and processes. *Elements* **2012**, *8*, 347–353. [CrossRef]
70. Price, D.G. Weathering and weathering processes. *Q. J. Eng. Geol.* **1995**, *28*, 243–252. [CrossRef]

Article

Impact Factors on Migration of Molybdenum(VI) from the Simulated Trade Effluent Using Membrane Chemical Reactor Combined with Carrier in the Mixed Renewal Solutions

Liang Pei [1,2,3] and Liying Sun [1,3,*]

1   Key Laboratory of Water Cycle and Related Land Surface Processes, Institute of Geographic Sciences and Natural Resources Research, Chinese Academy of Sciences, Beijing 100101, China; peiliang@ms.xjb.ac.cn
2   Xinjiang Institute of Ecology and Geography, Chinese Academy of Sciences, Urumqi 830011, China
3   University of Chinese Academy of Sciences, Beijing 100049, China
*   Correspondence: sunliying@igsnrr.ac.cn; Tel.: +86-106-4883-172

Abstract: Molybdenum is harmful and useful. The efficiency of molybdenum trade effluent treatment is low and it is difficult to extract and recycle. To solve this problem, a novel membrane chemical reactor with mixed organic-water solvent(MCR-OW) had been used for the investigation of impact factors on the migration characteristics of Mo(VI) in the simulated trade effluent. The novel MCR-OW contains three parts, such as feeding pool, reacting pool and renewal pool. Flat membrane of polyvinylidene fluoride(PVDF) membrane was used in the reacting pool, the mixed solutions of diesel and NaOH with N, N′-di(1-methyl-pentyl)-acetamide(N-503) as the carrier in the renewal pool and the simulated trade effluent with Mo(VI) as feeding solution. The influencing factors of pH and the ion strength in the feeding solutions, the volume ratio of diesel to NaOH solution and N-503 concentration in the renewal solutions were investigated for the testing of the migration efficiency of Mo(VI). It was found that the migration efficiency of Mo(VI) could reach 94.3% in 225 min, when the concentration of carrier(N-503) was 0.21 mol/L, the volume ratio of diesel to NaOH in the renewal pool was 4:3, pH in the feeding pool was 3.80 and the initial concentration of Mo(VI) was 2.50×10^{-4} mol/L. Moreover, the stability and feasibility of MCR-OW were discussed according to Mo(VI) retention on the membrane and the reuse of the membrane.

Keywords: flat membrane; N,N′-di(1-methyl-pentyl)acetamide; molybdenum; membrane chemical reactor; renewal solution

Citation: Pei, L.; Sun, L. Impact Factors on Migration of Molybdenum(VI) from the Simulated Trade Effluent Using Membrane Chemical Reactor Combined with Carrier in the Mixed Renewal Solutions. *Toxics* **2022**, *10*, 438. https://doi.org/10.3390/toxics10080438

Academic Editor: Ilaria Guagliardi

Received: 13 July 2022
Accepted: 29 July 2022
Published: 31 July 2022

Publisher's Note: MDPI stays neutral with regard to jurisdictional claims in published maps and institutional affiliations.

1. Introduction

China has the largest reserves of molybdenum in the world, concentrated in central and north of China with the maximum percentage of porphyry molybdenum deposits (77.5%), following by the porphyry–skarns, skarns and veins [1–3]. Molybdenum is widely used in aerospace, metallurgy, steel, electronics, military, petrochemical and agricultural fields, especially playing significant roles in iron and steel industry for alloys [4].

China has so many metal mines, many of which are rich in various toxic and harmful metals. At present, for some political reasons, we need to excavate these minerals. However, these metal mines will also produce a large amount of trade effluent that is difficult to treat. There is a molybdenum mine in a province in northern China. Molybdenum trade effluent is large, which has a certain impact on human and natural environment. The discharge of molybdenum trade effluent is not up to standard. We detected a certain amount of molybdenum in the plant roots around the mine. If this problem is not solved, human consumption of these plants will cause damage to human viscera. The removal effect of traditional effluent treatment methods was not good. This study attempted to explore the method of extracting molybdenum and provide theoretical support for the treatment of molybdenum containing trade effluent in the future. The dramatic economic development has resulted in the rapid consumption of raw materials and increased the related environmental hazards greatly in China [3]. With the large consumptions of molybdenum resources in different industries, the molybdenum contaminations have attracted large attentions in recent years [5–7]. Varying degrees of molybdenum pollution was reported in many areas of China and leading to the excessive amounts of molybdenum(VI) in rivers and reservoirs [8–11]. The transformation of molybdenum contaminants in water can cause toxicological risks to aquatic wildlife and threaten human health [6,11,12]. It was reported that the normal human requirement of molybdenum is up to 300 μg per day with the maximum tolerable does as 2000 μg per day from food, whilst the higher intakes are detrimental to human health [13]. Therefore, the effective method is not only to remove the molybdenum ion from the trade effluent, but also to enrich the molybdenum for the reusing, where understanding the migration of molybdenum ions is the basis.

Great efforts have been made in recent years to develop technologies for the removal and separation of molybdenum in trade effluent, mainly included the chemical precipitation [14–16], sub-exchange resin method [17,18], adsorption method [19–21], membrane separation [22–24], constructed wetland [23,25,26], solvent extraction [27–29], biological-methods [30]. However, these technologies are still in the development stage, and there are many challenges to be addressed, such as instability, long running time, low removal rate, high energy consumption, secondary pollution and high cost etc., due to the high complexity of the migration processes of the molybdenum ions under different conditions [26].

In the past decades, the migration, removal and reuse of heavy metals from waste water by liquid membranes(LMs), like supported liquid membrane(SLM), emulsion liquid membrane(ELM) and bulk liquid membrane(BLM), coupled with variations of carriers(N-503, P507 and TBP) were recognized to has broad application prospects, due to its advantages of high speed and short time in operation, high enrichment ratio, less reagent consuming and low cost [31–36].Among these technologies, SLM is widely used, particularly due to its convenience in operation and low costs in surfactant.

However, the unstable membrane with the less absorption capability of carrier in the SLM technologies remains to over come. It was found that the migration efficiency of heavy metal was higher when organic solvent was introduced into the SLM system [36–38]. Still, the stability of the membrane should be addressed. Thus, we designed the renewable membrane chemical reactor with mixed organic-water solvent(MCR-OW) on the basis of SLM for the purposes of both improving the efficiency of molybdenum(VI) ions and the stability of the membrane. N-503 was used as carrier and the diesel was used as the organic solvent. The influencing factors on the migration of Mo(VI) and the stability of the novel MCR-OW were discussed with the expectation of breakthrough industrial applications in future. So there are still some gaps in ore trade effluent treatment technology, which we will just fill the gap with the MCR-OW in the field of ore trade effluent treatment.

To summarize, the purpose of this study is to solve the problems of molybdenum pollution and difficult extraction and recovery in industrial wastewater, and to develop more efficient, more environmental friendly and more stable molybdenum extraction and recovery technology. At the same time, it provides a theoretical and scientific basis for future industrial applications

2. Materials and Methods

2.1. The Design of MCR-OW and Reactionmechanisms

The framework of MCR-OW was shown in Figure 1. Basically, MCR-OW contained three pools, namely feeding pool (400 mL), reacting pool (300 mL) and enrichment pool (200 mL), connected with the feeding pump and renewal pump. The reacting pool was separated into feeding part and renewal part by the flat membrane, with the effective volume of each part of the reacting pool as 150 mL. Mo(VI) and buffer solution (HAC-NaAc) in the feeding pool was poured into the feeding part of the reacting pool by the feeding pump. The certain volume ratio of the mixed organic-water solution, included membrane solution(diesel) with carrier N-503 and stripping solution(NaOH), in the renewal pool was placed into the renewal part of the reacting pool by the renewal pump for the extraction of Mo(VI) and renewal of the membrane simultaneously, as to improve the stability and increase the extraction rate of Mo(VI) at the same time.

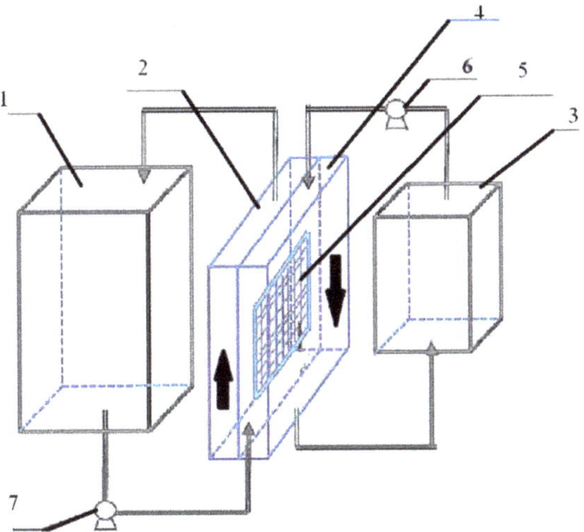

1—feeding pool; 2—feeding part of reacting pool; 3—renewal part of reacting pool;

4—renewal pool; 5—flat membrane; 6- renewal pump; 7- feeding pump

Figure 1. The structure of the membrane chemical reactor with mixed organic-water solvent (MCR-OW).

The co-removing involves various (equilibrium) reactions, as Equations (1)–(3):

$$[N-503]_m + H_f^+ + Cl_f^- = ([N-503]H^+Cl^-)_m \tag{1}$$

$$6([N-503]H^+Cl^-)_m + (Mo_7O_{24}^{6-})_f = [([N-503]H)_6Mo_7O_{24}]_m + 6Cl_f^- \tag{2}$$

$$[([N-503]H)_6Mo_7O_{24}]_m + 6(OH^-)_r = 6[N-503]_m + (Mo_7O_{24}^{6-})_r + 6(H_2O)_r \tag{3}$$

Accordingly, the mechanisms included: (a) the diffusion of Mo(VI) from the feeding part into the interface between membrane and the feeding solution; (b) the extraction of Mo(VI) from the feeding solution with carrier N-503 in diesel, resulted in the metal-complex $[([N-503]H)_6Mo_7O_{24}]$, which expressed as chemical Equations (1) and (2); (c) the diffusion of metal-complex $[([N-503]H)_6Mo_7O_{24}]$ through the membrane from the interface between membrane and feeding solution of the feeding part to the interface between membrane and renewal solution of the renewal part in the reacting pool; (d) the decomplexation of $[([N-503]H)_6Mo_7O_{24}]$ into N-503 and Mo(VI) in the interface between membrane and renewal solution, with the NaOH as stripping agent of Mo(VI) and diesel as the solvent of N-503 in the renewal part of the reacting pool, as Equation (3); (e) the enrichment of Mo(VI) in the renewal pool; (f) the re-input of the N-503 by the renewal pump from the renewal pool into the reacting pool, which could react with the feeding solution to increase the stability of the system, due to the diffusion of the N-503 through membrane.

2.2. Materials and Reagent

Flat membrane of porous polyvinylidene fluoride(PVDF) membrane was used in our new design, with pore size as 0.24 μm., thickness as 75 μm, tortuosity as 1.67, andporosity as 70~80% (Shanghai Yadong nuclear grade resin Co., Ltd. Shanghai, China). N, N′-di(1-methyl-pentyl)-acetamide(N-503) was used as the carrier in this work, with the density as 0.865, purity as 96% (Shanghai laiyashi Chemical Co., Ltd. Shanghai, China). The HCl and Mo(VI) solutions were mixed as feeding solution to simulate the trade effluent containing

Mo(VI). The HAc-NaAc buffer solution was used for the pH adjustment (2.0–6.0) of the feeding solution and the mixed solution of NaCl and KNO$_3$ were used for the regulation of the ion strengthin feeding solution to simulate the waste industrial water. NaOH is selected as the stripping solution and the self-made diesel is used as the organic solvent. The mixed solutions ofdiesel with N-503 and NaOH solution were used as the renewal solution. All the reagents (except diesel) were of analytical grade.

2.3. Test Method

Digital Acidity Ion Meter(pHS-3C, Shanghai Kangyi Instrument Co., Ltd. Shanghai, China) is used to determine the pH of the solution. The concentration of Mo(VI) was determined by spectrophotometry with 4-(2-pyridine-azo) resorcinol(PAR) as chromogenic agent and the absorbance was measured at 530 nm. The migration efficiency(Re) and separation coefficient are calculated as the logarithm value of the ratio of Mo(VI) concentration (C_t) to the initial Mo(VI) concentration(C_0) in the feeding pool, shown in Equation (4) and the migration rate(Rr) is calculated as Equation (5).

$$Me = -\ln\left(\frac{C_t}{C_0}\right) \tag{4}$$

$$Mr = \frac{C_0 - C_t}{C_0} \times 100 \tag{5}$$

2.4. Experimental Procedure

All experiments were accomplished at 20 $\pm$ 5 °C in the MCR-OW. The effective area of the device is 20 cm^2. The flow rates of two pumps are all 11.3 mL/min. Before the experiment, the PVDF membrane was firstly immersed into the diesel solvent with N-503 for an hour and then dried naturally and fixed into the reacting pool. The prepared feeding solution and renewal solution were poured into feeding pool and renewal pool separately. Then the experiments were initiated formally with the starting of both feeding pump and renewal pump. Samples were taken from the feeding pool and renewal pool for the tests of the Mo(VI) concentration, at 30, 60, 100, 165, 225 min, respectively.

3. Results and Discussion

3.1. Effects of pH and Ion Strength in the Feeding Pool

Basically, the concentration difference between feed part and renewal part in the reacting pool is the driving power of mass removing process, which should be impacted by the pH of the feeding solution [39]. To investigate the effects of pH of the feeding solution on the migration efficiency(Me) of Mo(VI) in our work, the initial experimental conditions were that: ratio of diesel toNaOHin the renewal pool ast1:1, the concentration of NaOH solution at 0.30 mol/L, the concentration of carrier at 0.21 mol/L in the mixed solutions of the renewal pool. The initial concentration of Mo(VI) was adjusted to 2.50 $\times$ 10^{-4} mol/L in the feeding pool. As shown in Figure 2, Me increased when pH of the feeding solution increased from 2.2 to 4.4, and increased with the increase of the running time as well.

According to Equations (1) and (2), when H$^+$ increased, it is benefit for both the formations of ([N-503]H$^+$Cl$^-$) and [([N-503]H)$_6$Mo$_7$O$_{24}$]. Also, pH is the driving power of the mass diffusion of the complex [([N-503]H)$_6$Mo$_7$O$_{24}$], which is theoretically increased the migration efficiency of Mo(VI) when pH decreased. However, in our experiment, we found that molybdenum existed in the form of MoO$_2$$^{2-}$when pH decreased, which hinder the reactions of Equation (2) and lower the migration efficiency of Mo(VI). As a result, the lower the pH is, the more inefficiency the Me is. We chose pH of 3.80 as the optimum pH condition of the feed part during the following experiments.

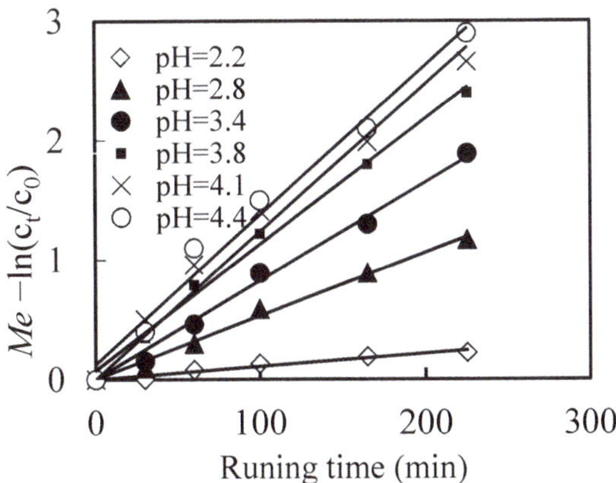

Figure 2. Effects of pH of the feeding solution on the migration efficiency (*Me*) of Mo(VI).

In the trade effluent, there always existed various ions, which may affect the migration of Mo(VI). Thus the NaCl and KNO_3 were used to simulate and regulation ion strength of the simulated trade effluent. The effects of the initial ion strength in the feeding pool on the migration rate (*Mr*) of Mo(VI) was thus examined. As shown in Figure 3, the *Mr* of Mo(VI) increased when the initial ion strength changed from 0.2 mol/L to 1.3 mol/L, kept higher than 80%. Ionic strength has little impact on the MCR-OW system, which is more suitable for industrial application and lays a foundation for future industrial application [40].

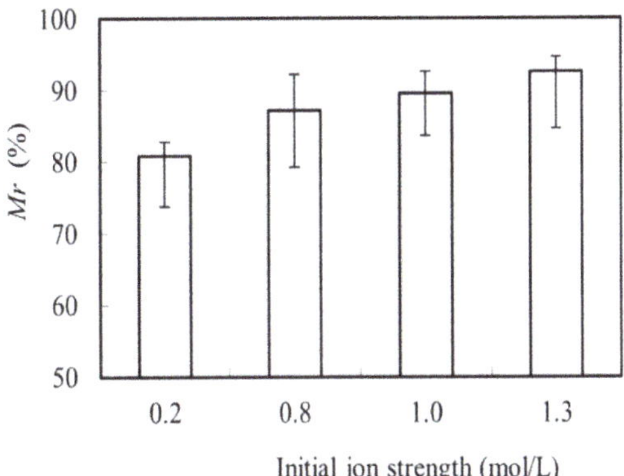

Figure 3. Effects of the initial ion strength in the feeding pool on the migration rate (*Mr*) of Mo(VI).

3.2. Effects of the Volume Ratio and Carrier Concentration in the Renewal Pool

As we described above, the mixed solutions in the renewal pool was a certain volume ratio of diesel with N-503 to NaOH. To investigate the volume ratio effects of the mixed solutions, the ratio of diesel to NaOH was adjusted as 1:5, 1:2, 1:1, 4:3, 5:2, and the results were shown in Figure 4. The migration efficiency(*Me*) of Mo(VI) firstly increased with the volume ratio of diesel to NaOH, when it lower than 4:3. And then *Me* of Mo(VI) decreased sharply when the volume ration of diesel to NaOH reached 5:2.

In our work, the diesel with N-503 was selected as membrane solution, which could recover the stability of the membrane and benefit the formation of the complex compounds, by increasing the chances of the recycling of the N-503. Simultaneously, the NaOH was selected as stripping solution for the migration of Mo(VI), the increase of which not only increased the stripping rate for the decomplexation of $[([N\text{-}503]H)_6Mo_7O_{24}]$, but also increased the concentration differences of the H^+ concentration in the feeding part and renewal part of the reacting pool, to increase the diffusion of both $[([N\text{-}503]H)_6Mo_7O_{24}]$ and the recycled N-503. When ratio of diesel to NaOH increased, the complexation rate increased and the stripping rate and diffusion rate decreased, the balance of these reactions resulted in the maximum migration efficiency of Mo(VI) when volume ratio of diesel to NaOH was 4:3.

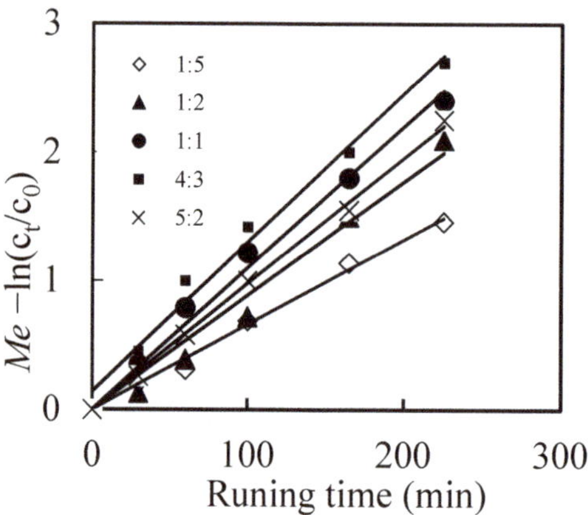

Figure 4. Effects of the volume ratio of diesel to NaOH on the migration efficiency (*Me*) of Mo(VI).

Moreover, N-503 played important roles for the migration efficiency of the Mo(VI). Based on Equations (1) and (2), the higher N-503 is, the higher chances of the formation of complexation $[([N\text{-}503]H)_6Mo_7O_{24}]$ are, which increased the migration efficiency (*Me*) of Mo(VI) (Figure 5). It was noticed that the differences on *Me* of Mo(VI) was much when the N-503 concentration changed from 0.21 mol/L to 0.25 mol/L. Considering the removal efficiency and solvent cost, 0.21 mol/L was selected as the optimum carrier concentration.

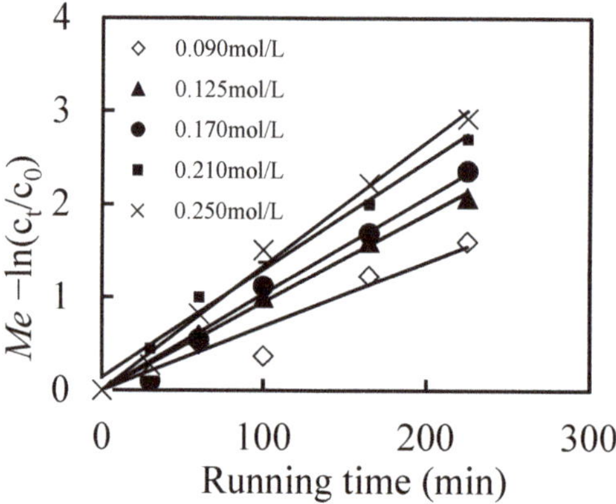

Figure 5. Effects of N-503 concentration on the migration efficiency (*Me*) of Mo(VI).

3.3. Effects of Mo(VI) Retention on the Reuse of the Membrane

In the previous investigations of the removal of heavy metal using SLM, the retention of the heavy metal ion on the membrane was observed and concerned [32,33]. In this work, the Mo(VI) ion was also concerned. According to the concentration of Mo(VI) in both feed pool and renewal pool, the concentration of Mo(VI) on the membrane can be calculated. As shown in Figure 6, the retention of Mo(VI) on the membrane increased with the running time. However, the increase rate decreased with the running time, which resulted in the

the stable concentration of Mo(VI) percentage on the membrane when running time was longer than 150 min. This is mainly due to the decrease of the decomplexation rate of [([N-503]H)$_6$Mo$_7$O$_{24}$] in the interface between membrane and renewal solution with increase of Mo(VI) in the renewal part of the reacting pool when running time increased [41]. With the increasing of the running time, the balance reached and the retention of Mo(VI) no longer increased, at the approximately 18% of the Mo(VI) on the membrane of the reacting pool.

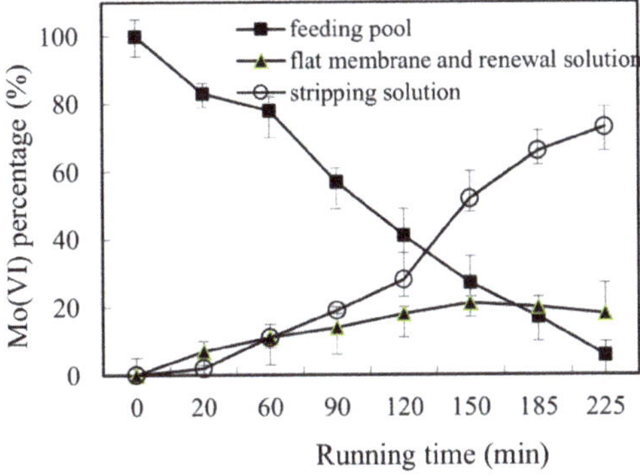

Figure 6. The retention of Mo(VI).

However, the retention of Mo(VI) on the membrane did not have significant negative effects on the renewal of themembrane. As shown in Figure 7, the migration rate(*Mr*) kept higher than 80% when the experiment repeated 7 times, which verified the stability and feasibility of the MCR-OW. The stability of the membrane and the migration rate of Mo(VI) was also enhanced by the separation of the reacting pool with the feeding and renewal pools in the novel design. This is precisely the advantage of our new designed MCR-OW, which avoid the falls off of the carrier that may be caused by the stirrer in the traditional SLM method. This part of the research has great reference significance for our future industrial application of the MCR-OW.

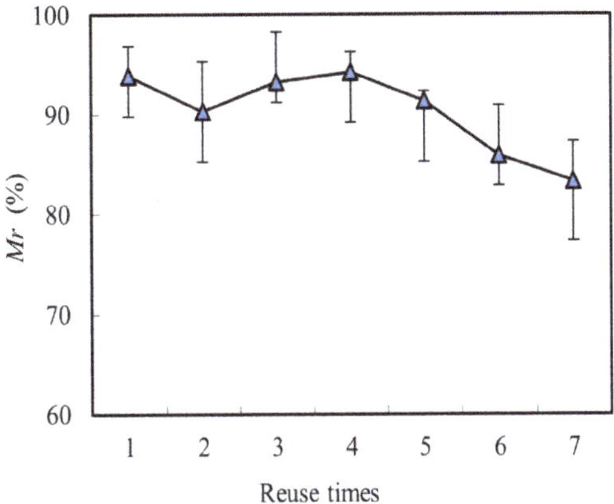

Figure 7. The effects of reuse time on the migration rate (*Mr*) of Mo(VI).

4. Conclusions

The influencing factors on the migration characteristics of Mo(VI) from the simulated trade effluent was examined using our new designed membrane chemical reactor with mixed organic-water solvent(MCR-OW) in this work. The results showed that MCR-OW was able to improving the efficiency of Mo(VI) migration from the simulated trade effluent, using the mixed diesel and NaOH(stripping solution) with the carrier of N-503 as renewal solution. The migration efficiency(*Me*) of Mo(VI) increased with the increase of the pH in the feeding pool, the concentration of the carrier(N-503) in the renewal pool and the running time and the concentration. The migration rate(*Mr*) kept higher than 80% when the initial ion strength of other ions increased from 0.2–1.3 mol/L. *Me* of Mo(VI) firstly increased with the volume ratio of diesel to NaOH ($\leq$4:3) and then decreased sharply. Considering on both the removal efficiency and the solvent costs, the optimum parameters of the MCR-OW were chosen as follows: the concentration of carrier(N-503) at 0.21 mol/L, the volume ratio of diesel to NaOH in the renew alpool at 4:3, pH in the feeding pool at 3.80, the running time at 225 min. Under the optimum conditions, the migration rate(*Mr*) could reach 94.3% when the initial Mo(VI) concentration was 2.50×10^{-4} mol/L. Although there are detainment or retention of Mo(VI) on the membrane during the operation, it did not hinder the reuse of the membrane with the *Mr* of Mo(VI) higher than 80% even though the membrane was used seven times. However, the retention of the membrane should be further tackled in the future for the application of the MCR-OW in larger scale.

Although the research is still in the experimental stage at present, as long as the state pays more attention to the research and development of new technologies for the treatment of trade effluent and environmental protection policies support the extraction and recycling of special substances, this technology will be widely used. In the future, we will do some pilot research with enterprises to explore its practical application effect. In the future, the development of material technology will greatly reduce the price of membrane, and the cost of this technology will be greatly reduced. At that time, there will be no problem to widely promote and apply it.

Author Contributions: All authors contributed to the study conception and design. Material preparation, experiment executive and data collection were performed by L.P. Data analysis was performed by L.S. and the first draft of the manuscript was written by L.S. All authors commented on previous versions of the manuscript. All authors have read and agreed to the published version of the manuscript.

Funding: This work was supported by the Independent research projects of Xinjiang Institute of ecology and geography, Chinese Academy of Sciences (Grant No.: E2500201) and National Natural Science Foundation of China (Grant No.: 41977069).

Institutional Review Board Statement: Not applicable.

Informed Consent Statement: Not applicable.

Data Availability Statement: The datasets used and/or analyzed during the current study are avail-able from the corresponding author on reasonable request.

Acknowledgments: The authors would like to acknowledge the kind help and suggestions of all the anonymous reviewers.

Conflicts of Interest: The authors declare that they have no known competing financial interests or personal relationships that could have appeared to influence the work reported in this paper.

References

1. Zhang, Z.; Yang, X.; Dong, Y.; Zhu, B.; Chen, D. Molybdenum deposits in the eastern Qinling, central China: Constraints on the geodynamics. *Int. Geol. Rev.* **2011**, *53*, 261–290. [CrossRef]
2. Zeng, Q.D.; Liu, J.M.; Qin, K.Z.; Fan, H.; Chu, S.; Wang, Y.; Zhou, L. Types, characteristics, and time-space distribution of molybdenum deposits in China. *Int. Geol. Rev.* **2013**, *55*, 1311–1358. [CrossRef]
3. Han, C.M.; Xiao, W.J.; Su, B.X.; Sakyi, P.A.; Ao, S.; Zhang, J.; Wan, B.; Song, D.; Zhang, Z.; Wang, Z.; et al. Late Paleozoic metallogenesis and evolution of the Chinese Western Tianshan Collage, NW China, Central Asia orogenic belt. *Ore. Geol. Rev.* **2020**, *124*, 103643. [CrossRef]
4. Chen, X.-R.; Zhai, Q.-J.; Dong, H.; Dai, B.-H.; Mohrbacher, H. Molybdenum alloying in cast iron and steel. *Adv. Manuf.* **2020**, *8*, 3–14. [CrossRef]

5. Tsai, K.-S.; Chang, Y.-M.; Kao, J.C.M.; Lin, K.-L. Groundwater Molybdenum from Emerging Industries in Taiwan. *Bull. Environ. Contam. Tox.* **2016**, *96*, 102–106. [CrossRef]
6. Ghazaryan, K.A.; Movsesyan, H.S.; Khachatryan, H.E.; Ghazaryan, N.P. Geochemistry of potentially toxic trace elements in soils of mining area: Acase study from zangezurcopper and molybdenum combine. *Armenia. Bull. Environ. Contam. Tox.* **2018**, *101*, 732–737. [CrossRef]
7. Solongo, T.; Fukushi, K.; Altansukh, O.; Fukushi, K.; Altansukh, O.; Takahashi, Y.; Akehi, A.L.; Baasansuren, G.; Ariuntungalag, Y.; Enkhjin, O.; et al. Distribution and chemical speciation of molybdenum in river and pond sediments affected by mining activity in Erdenet City, Mongolia. *Minerals* **2018**, *8*, 288. [CrossRef]
8. Han, Z.; Wan, D.; Tian, H.; He, W.; Wang, Z.; Liu, Q. Pollution assessment of heavy metals in soils and plants around a Molybdenum mine in Central China. *Pol. J. Environ. Stud.* **2019**, *28*, 123–133. [CrossRef]
9. Song, Z.; Song, G.; Tang, W.; Yan, D.; Zhao, Y.; Zhu, Y.; Wang, J.; Ma, Y. Molybdenum contamination dispersion from mining site to a reservoir. *Ecotoxicol. Environ. Saf.* **2021**, *208*, 111631. [CrossRef]
10. Li, F.P.; Wang, Y.; Mao, L.C.; Tao, H.; Chen, M. Molybdenum background and pollution levels in the TaipuRiver, China. *Environ. Chem. Lett.* **2022**, *20*, 1009–1015. [CrossRef]
11. Zhang, Y.; Zhang, M.; Yu, W.; Li, J.; Kong, D. Ecotoxicological risk ranking of 19 metals in the lower Yangtze River of China based on their threats to aquatic wildlife. *Sci. Total Environ.* **2022**, *812*, 152370. [CrossRef] [PubMed]
12. Davies, T.D.; Pickard, J.; Hall, K.J. Acute molybdenum toxicity to rainbow trout and other fish. *J. Environ. Eng. Sci.* **2005**, *4*, 481–485. [CrossRef]
13. Pichler, T.; Koopmann, S. Should monitoring of molybdenum (Mo) in groundwater, drinking water and well permitting made mandatory? *Environ. Sci. Technol.* **2020**, *54*, 1009–1015. [CrossRef] [PubMed]
14. Liu, J.; Qiu, Z.; Yang, J.; Cao, L.; Zhang, W. Recovery of Mo and Ni from spent acrylonitrile catalysts using an oxidation leaching-chemical precipitation technique. *Hydrometallurgy* **2016**, *164*, 64–70. [CrossRef]
15. Hamza, M.F.; Roux, J.C.; Guibai, E. Metal valorization from the waste produced in the manufacturing of Co/Mo catalysts: Leaching and selective precipitation. *J. Mater. Cycles. Waste Manag.* **2019**, *21*, 525–538. [CrossRef]
16. Zeng, L.; Zhao, Z.; Huo, G.; Wang, X.; Pu, H. Mechanism of selective precipitation of molybdenum from tungstate solution. *JOM* **2020**, *72*, 800–805. [CrossRef]
17. Joo, S.H.; Kim, Y.U.; Kang, J.G.; Kumar, J.R.; Yoon, H.-S.; Parhi, P.L.; Shin, S.M. Recovery of Rhenium and Molybdenum from Molybdenite Roasting Dust Leaching Solution by Ion Exchange Resins. *Mater. Trans.* **2012**, *53*, 2034–2037. [CrossRef]
18. Gao, Z.G.; Zhang, B.; Liu, H.Z.; Wang, W.; Cao, Y. Recovery of molybdenum from alkali leaching solution of low-grade molybdenum concentrate by ion exchange. *Green Process. Synth.* **2015**, *4*, 317–321. [CrossRef]
19. Okutani, T.; Noshiro, K.; Sakuragawa, A. Determination of a trace amount of molybdenum(VI) by graphite-furnace atomic absorption spectrometry after adsorption and elution as a molybdenum-Pyrocatechol Violet complex on activated carbon. *Anal. Sci.* **1998**, *14*, 621–624. [CrossRef]
20. Li, M.M.; Wang, Y.J. Determination of molybdenum in alloy steel by flame atomic absorption spectrometry with mixed surfactant. *Chinese J. Anal. Chem.* **2000**, *28*, 428–431.
21. Mukherjee, B.; Simsek, E. Plasmonics Enhanced Average Broadband Absorption of Monolayer MoS2. *Plasmonics* **2016**, *11*, 285–289. [CrossRef]
22. Yobilishetty, S.M.; Marathe, K.V. Removal of molybdenum (VI) from effluent waste water streams by cross flow micellar enhanced ultrafiltration (MEUF) using anionic, non-ionic and mixed surfactants. *Indian J. Chem. Techn.* **2014**, *21*, 321–327.
23. Lian, J.J.; Xu, S.G.; Zhang, Y.M.; Han, C.W. Molybdenum(VI) removal by using constructed wetlands with different filter media and plants. *Water Sci. Technol.* **2013**, *67*, 1859–1866. [CrossRef] [PubMed]
24. Rouhani, S.H.R.; Davarkhah, R.; Zaheri, P.; Mousavian, S.M.A. Separation of molybdenum from spent HDS catalysts using emulsion liquid membrane system. *Chem. Eng. Process.* **2020**, *153*, 107958. [CrossRef]
25. Kröpfelová, L.; Vymazal, J.; Švehla, J.; Štíchová, J. Removal of trace elements in three horizontal sub-surface flow constructed wetlands in the Czech Republic. *Environ. Pollut.* **2009**, *157*, 1186–1194. [CrossRef]
26. Chen, B.; Zhou, F.J.; Yang, F.; Lian, J.J.; Ye, T.R.; Wu, H.Y.; Wang, L.M.; Song, N.; Liu, Y.Y.; Hui, A.Y. Enhanced sequestration of molybdenum(VI) using composite constructed wetlands and responses of microbial communities. *Water Sci. Technol.* **2022**, *85*, 1065–1078. [CrossRef] [PubMed]
27. Bailey, S.E.; Olin, T.J.; Bricka, R.; Adrian, D. A review of potentially low-cost sorbents for heavy metals. *Water Res.* **1999**, *33*, 2469–2479. [CrossRef]
28. Wu, X.; Zhang, G.; Zeng, L.; Zhou, Q.; Li, Z.; Zhang, D.; Cao, Z.; Guan, W.; Li, Q.; Xiao, L. Study on removal of molybdenum from ammonium tungstate solutions using solvent extraction with quaternary ammonium salt extractant. *Hydrometallugy* **2019**, *186*, 218–225. [CrossRef]
29. Wu, X.; Zhang, G.; Zeng, L.; Guan, W.; Wu, S.; Li, Z.; Zhou, Q.; Zhang, D.; Qing, J.; Long, Y.; et al. Continuous solvent extraction operations for the removal of molybdenum from ammonium tungstate solution with quaternary ammonium salt extractant. *Hydrometallugy* **2020**, *195*, 105401. [CrossRef]
30. Kasra-Kermanshahi, R.; Tajer-Mohammad-Ghazvini, P.; Bahrami-Bavani, M. A biotechnological strategy for molybdenum extraction using acidithiobacillusferrooxidans. *Appl. Biochem. Biotechnol.* **2020**, *193*, 884–895. [CrossRef]

31. Yang, X.J.; Fane, A.G.; MacNaughton, S. Removal and recovery of heavy metals from wastewaters by supported liquid membranes. *Water Sci. Technol.* **2001**, *43*, 341–348. [CrossRef] [PubMed]
32. Pei, L.; Wang, L.M.; Ma, Z.Y. Modelling of Ce(IV) Transport through a Dispersion Combined Liquid Membrane with Carrier p507. *Front. Env. Sci. Eng.* **2014**, *8*, 503–509. [CrossRef]
33. Pei, L.; Yao, B.H.; Zhang, C.J. Transport of Tm(III) through dispersion supported liquid membrane containing PC-88A in kerosene as the carrier. *Sep. Purif. Technol.* **2009**, *65*, 220–227. [CrossRef]
34. Valenzuela, F.; Fonseca, C.; Basualto, C.; Correa, O.; Tapia, C.; Sapag, J. Removal of copper ions from a waste water by a liquid emulsion membrane method. *Miner. Eng.* **2015**, *18*, 33–40. [CrossRef]
35. Albarak, Z. Carrier-mediated liquid membrane systems for lead (II) ion separations. *Chem. Pap.* **2020**, *74*, 77–88. [CrossRef]
36. Ahmad, A.; Tariq, S.; Zaman, J.U.; Perales, A.I.M.; Mubashir, M.; Luque, R. Recent trends and challenges with the synthesis of membranes: Industrial opportunities towards environmental remediation. *Chemosphere* **2022**, *306*, 135634. [CrossRef]
37. Zhang, J.; Li, Z.; Zhan, K.; Sun, R.; Sheng, Z.; Wang, M.; Wang, S.; Hou, X. Two dimensional nanomaterial-based separation membranes. *Electrophoresis* **2019**, *40*, 2029–2040. [CrossRef] [PubMed]
38. Mal'tseva, E.E.; Blokhin, A.A.; Murashkin, Y.V.; Mikhaylenko, M.A. Sorptiveparation of molybdenum(VI)from rhenium-containing solutions. *Russ. J. Appl. Chem.* **2017**, *90*, 528–532. [CrossRef]
39. Bhatluri, K.K.; Manna, M.S.; Ghoshal, A.K.; Saha, P. Separation of cadmium and lead from wastewater using supported liquid membrane integrated with in-situ electrodeposition. *Electrochim. Acta.* **2017**, *229*, 1–7. [CrossRef]
40. Foong, C.Y.; Wirzal, M.D.H.; Bustam, M.A. A review on nanofibers membrane with amino-based ionic liquid for heavy metal removal. *J. Mol. Liq.* **2020**, *297*, 111793. [CrossRef]
41. He, D.S.; Ma, M.; Wang, H.; Zhou, P. Effect of kinetic synergiston transport of Copper(II) through a liquid membrane containing P507 in Kerosene. *Can. J. Chem.* **2001**, *79*, 1213–1219. [CrossRef]

Article

Spatial Distribution, Contamination Levels, and Health Risk Assessment of Potentially Toxic Elements in Household Dust in Cairo City, Egypt

Ahmed Gad [1,*], Ahmed Saleh [2,*], Hassan I. Farhat [3], Yehia H. Dawood [1] and Sahar M. Abd El Bakey [4]

[1] Geology Department, Faculty of Science, Ain Shams University, Cairo 11566, Egypt
[2] National Research Institute of Astronomy and Geophysics (NRIAG), Cairo 11421, Egypt
[3] Geology Department, Faculty of Science, Suez University, El Salam City 43518, Egypt
[4] Department of Biological and Geological Sciences, Faculty of Education, Ain Shams University, Cairo 11341, Egypt
* Correspondence: A.gad@sci.asu.edu.eg (A.G.); ahmed.saleh@nriag.sci.eg (A.S.)

Abstract: Urban areas' pollution, which is owing to rapid urbanization and industrialization, is one of the most critical issues in densely populated cities such as Cairo. The concentrations and the spatial distribution of fourteen potentially toxic elements (PTEs) in household dust were investigated in Cairo City, Egypt. PTE exposure and human health risk were assessed using the USEPA's exposure model and guidelines. The levels of As, Cd, Cr, Cu, Hg, Mo, Ni, Pb, and Zn surpassed the background values. Contamination factor index revealed that contamination levels are in the sequence Cd > Hg > Zn > Pb > Cu > As > Mo > Ni > Cr > Co > V > Mn > Fe > Al. The degree of contamination ranges from considerably to very high pollution. Elevated PTE concentrations in Cairo's household dust may be due to heavy traffic emissions and industrial activities. The calculated noncarcinogenic risk for adults falls within the safe limit, while those for children exceed that limit in some sites. Cairo residents are at cancer risk owing to prolonged exposure to the indoor dust in their homes. A quick and targeted plan must be implemented to mitigate these risks.

Keywords: potentially toxic elements; indoor dust; pollution; exposure; risk assessment; urban; Cairo

Citation: Gad, A.; Saleh, A.; Farhat, H.I.; Dawood, Y.H.; Abd El Bakey, S.M. Spatial Distribution, Contamination Levels, and Health Risk Assessment of Potentially Toxic Elements in Household Dust in Cairo City, Egypt. *Toxics* **2022**, *10*, 466. https://doi.org/10.3390/toxics10080466

Academic Editor: Ilaria Guagliardi

Received: 26 July 2022
Accepted: 9 August 2022
Published: 11 August 2022

Publisher's Note: MDPI stays neutral with regard to jurisdictional claims in published maps and institutional affiliations.

1. Introduction

Over the past few decades, a tremendous amount of hazardous waste materials has been released into various environmental media at increasing levels because of the rapid urbanization and globalization of economic and industrial activity [1–5]. Because the air in common is the primary carrier of fine particles, air pollution has produced a significant environmental impact (e.g., climate change and human health). The concentration of suspended particles in the air, which transports contaminants, especially potentially toxic elements (PTEs), has progressively increased, endangering humans. Because of their genotoxicity, carcinogenicity, chemical persistence, and non-degradability, PTEs attached to suspended particles would enrich in surface environments and have an acute or chronic impact on the health of vulnerable residents once they get into the human body [6–10]. PTEs can go through a human body via respiratory inhalation, ingestion of contaminated media, and dermal contact and accumulate over time [1,6,11,12].

Because indoor air can be significantly more polluted than outdoor air, it has captures remarkable attention from researchers. Imperfect air exchange and specific indoor emission sources combined with outdoor sources seems to be the leading causes of indoor air being a complex and contaminated environment [1,12]. People in megacities typically spend 80–90% of their own time indoors, in private homes, schools, and offices, potentially increasing their exposure to toxic substances being emitted from construction materials, household equipment, and electronic products, in conjunction with anthropogenic sources [13]. In this regard, the indoor ambiance and potential health risks inextricably associated with toxic substances' exposure in the indoor environment must be considered. Many scientific studies over the last decades have sufficiently demonstrated that prolonged exposure to contaminated indoor environments has undeniable fingerprints on serious health problems that result from direct and indirect exposure [6,7,12,13]. This direct impact on public health is extremely significant for children, who are more vulnerable to contaminant exposure due to increased hand-to-mouth interactions [1,4,14]. Furthermore, considerable advancements in analytical techniques used to investigate various biological samples will progressively improve exposure estimates for both healthy and at-risk populations [15,16].

There are numerous sources of indoor contaminants, the most significant of which is settled and suspended dust. Most of these hazardous and toxic pollutants are adsorbed to suspended particulates in indoor air and later deposited as house dust. Because of this process, the concentrations of contaminants in indoor dust are higher than their natural crustal concentrations [6,12]. Indoor dust is a motley mixture of inorganic and organic materials that can adsorb and concentrate PTEs [17,18]. This admixture would settle on the surfaces of residential objects (e.g., floors, carpets, furniture, and others) [19,20]. The main pathway for PTEs from outdoor sources into homes is the entry of contaminated suspended particles into outdoor air [20,21]. Many transporting methods bring street dust and soil materials indoors as a consequence of residents' activities (e.g., attached to shoes, clothes, bags, their pets, etc.) [7,12,22–24]. Moreover, considered external contaminated sources of indoor and household dust are suspended grains generated by industrial activities, road dust, traffic emissions, park soil, and other particles that are produced by outdoor activities [6,12]. Indoor dust PTE contamination has received a lot of attention owing to its significant effects on both residents' health and the environment [7,13,25]. One of the serious issues with indoor dust is that it is not exposed to the same processes that reduce its PTE concentrations as those that affect outdoor dust (e.g., diluting, leaching, or weathering). Consequently, indoor dust could be used as a long-term indicator of indoor environmental status [19].

Egypt has experienced severe soil, water, and air pollution in recent decades as deleterious consequences of rapid economic growth, urbanization, and increased energy demands [26–29]. Different studies were conducted to assess Egypt's air pollution. The vast majority of these studies have concentrated on the gaseous (CO, CO_2, SO_2, $H_2 S$, and NO_2) and particulate matter [30–35]. Studies on indoor dust in Egypt have typically focused on major ions (SO4, NO_3, Cl, NH_4, Ca, Mg, Na, and K) [36], organic pollutants [37–40], and microorganisms [41–43]. There are limited studies on PTE contamination in indoor dust and their health risk assessments [23,24,44,45]. Moreover, no comprehensive geochemical study of Cairo City's indoor dust and the potential health risks for PTE exposure have been conducted. A notable lack of such necessary data might hinder the proper development of short and long-term policy initiatives towards reducing air pollution. More extensive research must be directed to thoroughly comprehend the detrimental health impacts of PTE air pollution. Findings and the conclusion of these surveys will be properly utilized to support the national policies and will contribute to the public health improvements. Therefore, the current study's specific objectives are to (1) detect the PTE levels in household dust and identify their spatial distribution in Cairo City; (2) assess the contamination levels using environmental indices; (3) identify the possible sources of PTEs in household dust using multivariate statistical analysis; and (4) assess the potential health risk for children and adults' exposure to PTEs.

2. Materials and Methods

2.1. The Study Area

Cairo (Al-Qhirah) is located in northern Egypt on the River Nile's right bank. It is Egypt's administrative center and the most sizable city in both Africa and the Middle East, and one of the world's most densely populated cities (9.9 million inhabitants). Many issues plague the city, including traffic congestion, air, soil, and water pollution, and ineffective waste management [46]. Cairo is administratively divided into five chief regions (New Cairo, Eastern, Northern, Western, and Southern) (Figure 1; Table S1).

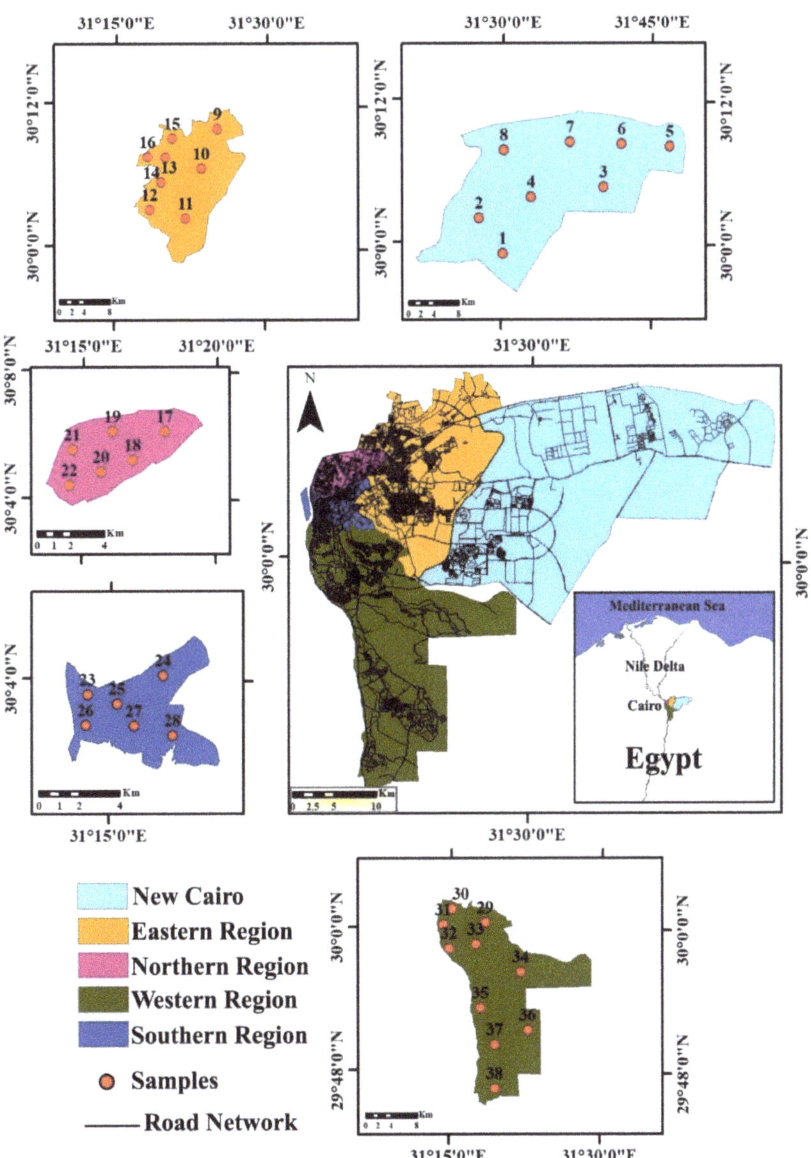

Figure 1. Map displaying Cairo City and its administrative regions and sampling site's locations.

Cairo City has a typical Mediterranean climate, with different temperatures through seasons: winter 14 °C, spring 21 °C, summer 36 °C, and fall 23 °C. Most of the year, wind speeds range from 3 to 8 m/s. The north and northeast were dominant wind directions [47,48]. It is surrounded by agricultural and industrial activities. It contains the main industrial zones that exist in the Northern and Southern regions which host cement manufacturing plants, steel, oil and gas, quarrying, rubber, petrochemicals, metallurgical, textile, and plastic products [26,48].

2.2. Sampling and Samples Preparation

A total of 38 composite household settled dust samples were collected from different regions and districts in Cairo City in 2021 (Figure 1). The sample size was selected based on the major districts in Cairo City, in conjunction with budgetary constraints. To ensure a collection of representative samples at least 10 subsamples were collected from each main district representing a total of 473 private houses (1 sample per house) (Table S1 in Supplementary Materials). The undisturbed surfaces, such as cupboards, fans, bookshelves, and refrigerators, were slowly brushed using precleaned polyethylene brushes and plastic dustpans to collect dust samples, which were then carefully blended and placed into transparent, zip-locked, and labeled plastic bags. The collected dust subsamples were carefully mixed and homogenized into 38 composite samples. The samples were then dried at 50 °C for 24 h in an oven followed by sieve analysis using a standard stainless-steel sieve (63 microns).

2.3. Chemical Analyses

The chemical analyses were performed using the ICP-ES/MS (AQ200) technique in ACME Lab, Vancouver, Canada (ISO 17025 and ISO/IEC 17025). An exact amount of 0.5 g of each household dust sample was leached in modified aqua regia (1: 1: 1 HNO_3: HCl: H_2O) [49]. Detection limits of Al, As, Cd, Co, Cr, Cu, Fe, Hg, Mn, Mo, Ni, Pb, V, and Zn were 0.01%, 0.5 ppm, 0.1 ppm, 0.1 ppm, 1 ppm, 0.1 ppm, 0.01%, 0.01 ppm, 1 ppm, 0.1 ppm, 0.1 ppm, 0.1 ppm, 1 ppm, and 1 ppm, respectively.

2.4. Contamination Levels

2.4.1. Contamination Factor (C_f)

Anthropogenic activities' contribution to PTE contamination has been evaluated using the contamination factor (C_f). C_f is calculated by the Equation (1) [50].

$$C_f^i = \frac{C_s^i}{C_b^i} \tag{1}$$

where C_s^i is the PTE concentration in analyzed samples, and C_b^i is the background value of the investigated PTE. In this investigation the Upper Continental Crust (UCC) element concentrations [51] were considered as the background values. The C_f values are typically categorized in four distinct classes; class 1 ($C_f < 1.0$ = low contamination); class 2 ($1.0 \leq C_f < 3.0$ = moderate contamination); class 3 ($3.0 \leq C_f \leq 6.0$ = considerable contamination); and class 4 ($C_f > 6.0$ = very high contamination) [50].

2.4.2. Contamination Degree (C_{deg})

To detect multielement contamination, C_{deg} was used. It was calculated for each sampling site using Equation (2) [50].

$$C_{deg} = \sum_{i=1}^{n} C_f \tag{2}$$

where C_f is contamination factor, and n is the number of the examined PTEs. The C_{deg} values are typically categorized in four distinct classes; class 1 ($C_{deg} < 6.0$ = low contamination); class 2 ($6.0 \leq C_{deg} < 12.0$ = moderate contamination); class 3 ($12.0 \leq C_{deg} \leq 24.0$ = considerable contamination); and class 4 ($C_{deg} > 24.0$ = very high contamination) [50].

2.5. Health Risk Assessment

PTEs measured in household dust in this investigation are typically known to possess noncarcinogenic effects on human health [52–54]. As, Cd, Cr, Ni, and Pb are believed to possess both noncarcinogenic and carcinogenic effects [52–54]. In the current study, health risks for children and adults in Cairo City were assessed using the noncarcinogenic Hazard Quotient (HQ) of a single element and Hazard Index (HI) of multiple elements via ingestion, inhalation, and dermal routes of exposure. Furthermore, the Cancer Risk (CR) was calculated using the concentrations of As, Cd, Cr, Ni, and Pb in the collected household dust samples. HQ, HI, and CR were calculated using the calculation model of exposure adopted by USEPA [52–55].

The average daily intakes (ADI) of PTEs in the household dust via nondietary inadvertent ingestion (noncarcinogenic) (ADI_{ing}), dust inhalation (noncarcinogenic) (ADI_{inh}), and dermal contact (noncarcinogenic) (ADI_{der}) routes are calculated using Equations (3)–(5) as follows:

$$ADI_{ing} = \frac{C_s \times IngR \times EF \times ED \times CF}{BW \times AT} \tag{3}$$

$$ADI_{inh} = \frac{C_s \times InhR \times EF \times ED}{PEF \times BW \times AT} \tag{4}$$

$$ADI_{der} = \frac{C_s \times SA \times SL \times ABS \times EF \times ED \times CF}{BW \times AT} \tag{5}$$

The noncarcinogenic risk *HQ* and *HI* of PTEs in the household dust is calculated using Equations (6)–(9) as follows:

$$HQ_{ing} = \frac{ADI_{ing}}{RfD_{ing}} \tag{6}$$

$$HQ_{inh} = \frac{ADI_{inh}}{RfD_{inh}} \tag{7}$$

$$HQ_{der} = \frac{ADI_{der}}{RfD_{der}} \tag{8}$$

$$HI = \sum HQ_{ing} + \sum HQ_{inh} + \sum HQ_{der} \tag{9}$$

The lifetime average daily dose (carcinogenic) (LADD) and the carcinogenic risk (CR) of As, Cd, Cr, Ni, and Pb in household dust is calculated using Equations (10)–(13) as follows:

$$LADD_{ing} = \left(\frac{C_s \times EF \times CF}{AT}\right) \times \left(\left(\frac{IngR \times ED}{BW}\right)_{Child} + \left(\frac{IngR \times ED}{BW}\right)_{Adult}\right) \tag{10}$$

$$LADD_{inh} = \left(\frac{C_s \times EF}{AT \times PET}\right) \times \left(\left(\frac{InhR \times ED}{BW}\right)_{Child} + \left(\frac{InhR \times ED}{BW}\right)_{Adult}\right) \tag{11}$$

$$LADD_{der} = \left(\frac{C_s \times SL \times ABS \times EF \times CF}{AT}\right) \times \left(\left(\frac{SA \times ED}{BW}\right)_{Child} + \left(\frac{SA \times ED}{BW}\right)_{Adult}\right) \tag{12}$$

$$R = \left(\sum LADD_{ing} \times SLF_{ing}\right) + \left(\sum LADD_{inh} \times SLF_{inh}\right) + \left(\sum LADD_{der} \times SLF_{der}\right) \tag{13}$$

where all the abbreviations, definitions, and reference values are given and explained in Table 1. If *HI* is less than one, there is no risk of noncarcinogenic effect; if *HI* is greater than one, there is a risk of noncarcinogenic effect. A value of CR less than 1×10^{-6} is regarded as modest, a value of CR between 1×10^{-4} and 1×10^{-6} is regarded within the permissible level, and a value of CR greater than 1×10^{-4} is likely to be harmful to humans [52–55].

Table 1. Definitions and reference values of human health risk model.

Term	Definition	Value	Refs.
C_S	PTE concentration	Site specific	
$IngR$	Dust ingestion rate (mg day^{-1})	200 (Children); 100 (Adults)	
$InhR$	Dust inhalation rate (m^3 day^{-1})	7.6 (Children); 20 (Adults)	
PEF	Particle emission factor (m^3 kg^{-1})	1.36×10^9	
SA	Exposed skin area (cm^2)	2699 (Children); 3950 (Adults)	
SL	Skin adherence factor (mg cm^2 day^{-1})	0.2 (Children); 0.07 (Adults)	[52–57]
ABS	Dermal absorption factor (year)	0.001 except for As (0.03)	
ED	Exposure duration (year)	6 (Children); 24 (Adults)	
EF	Exposure frequency (day year^{-1})	350	
BW	Average body weight (kg)	18.6 (Children); 70 (Adults)	
AT	Average life span for heavy metals (day)	noncarcinogens = ED$\times$365; carcinogens = 70 $\times$ 365	
CF	Transformation factor	1×10^{-6}	
RfD_{ing}	Ingestion reference dose (mg kg day^{-1})	Al (1.00), As (3.00×10^{-4}, Cd (1.00×10^{-3}), Co (2.00×10^{-2}), Cr (3.00×10^{-3}), Cu (4.00×10^{-2}), Hg (3.00×10^{-4}), Mn (4.60×10^{-2}), Mo (5.00×10^{-3}), Ni (2.00×10^{-2}), Pb (3.50×10^{-3}), V (7.00×10^{-3}), Zn (3.00×10^{-1})	[57–59]
RfD_{inh}	Inhalation reference dose (mg m^3 $^{-1}$)	Al (1.43×10^{-3}), As (3.00×10^{-4}), Cd (1.00×10^{-3}), Co (5.71×10^{-6}), Cr (2.86×10^{-5}), Cu (4.02×10^{-2}), Hg (8.75×10^{-5}), Mn (1.43×10^{-5}), Ni (2.06×10^{-2}), Pb (3.25×10^{-3}), V (7.00×10^{-3}), Zn (3.00×10^{-1})	[53–57,59,60]
RfD_{der}	Dermal reference dose (mg kg day^{-1})	Al (1.00×10^{-1}), As (1.23×10^{-4}), Cd (1.00×10^{-5}), Co (1.60 2), Cr (6.00×10^{-5}), Cu (1.20×10^{-2}), Hg (2.10×10^{-5}), Mn (1.84×10^{-3}), Mo (1.90×10^{-3}), Ni (5.40×10^{-3}), Pb (5.25×10^{-4}), V (7.00×10^{-5}), Zn (6.00×10^{-2})	[53–57,59,60]
SLF_{ing}	Ingestion cancer slope factor (mg kg day^{-1})	As (1.5), Cd (0.38), Cr (0.5), Ni (1.7), Pb (0.0085)	[4,11,58]
SLF_{inh}	Inhalation cancer slope factor (mg m^3 $^{-1}$)	As (15.1), Cd (6.3), Cr (0.42), Ni (0.84), Pb (0.042)	[4,57,60]
SLF_{der}	Dermal contact cancer slope factor (mg kg day^{-1})	As (3.66), Cr (2)	[4,12]

2.6. Data Treatment

Arc GIS (version 10.8.1; 2020) with a raster interpolation technique (Spline-Tension) was used to display the measured PTEs' location and spatial distribution maps in Cairo City. OriginLab (version OriginPro 2021) was used to present descriptive statistics, boxplot figures, and multivariate statistical analyses. Excel (version Microsoft Office 365 16.0.15028.20160) was used to calculate contamination levels and health risk assessment.

3. Results and Discussion

3.1. PTE Distribution

This is the first investigation to present a multielement profile of Cairo City household dust. Depending on the study's aims and to guarantee representative sampling, 38 major districts in Cairo City are represented with at least 10 subsamples from each. Table 2 summarizes the descriptive statistical parameters (minimum, maximum, mean, and standard deviation) of the dry weight PTE concentrations in the analyzed indoor household dust samples. Generally, the mean concentrations of these PTEs were ranked in the declining sequence Fe (20,818 ppm) > Al (9092 ppm) > Mn (425 ppm) > Zn (419 ppm) > Cu (116.6 ppm) > Pb (99.3 ppm) > Cr (48.6 ppm) > V (45.7 ppm) > Ni (30.1 ppm) > Co (9.0 ppm) > As (4.0 ppm) > Mo (2.5 ppm) > Cd (1.0 ppm) > Hg (0.30 ppm).

Because there are no PTE guidelines for indoor dusts, our results were compared with the UCC element concentrations [51]. The mean concentrations of As, Cd, Cr, Cu, Hg, Mo, Ni, Pb, and Zn were higher than those of UCC [51], indicating that their sources were affected by anthropogenic activities. CV % indicates the relative variability of element levels in environmental samples. CV of 20% indicates low variability, CV of 20:50% indicates moderate variability, and CV of 50:100% indicates high variability [61,62]. The CV(%) values of the measured PTEs ranged from 23.9% to 148.1% (Table 2). An interesting point in Table 2 is that Hg exhibited the highest CV value (148.1%), indicating extremely high variability through sampling locations. Cd, Cu, and Pb exhibited relatively higher CV values (51.7, 50.3 and 52.2%, respectively) indicating possible pollution. On the other hand, Al, As, Co, Cr, Fe, Mn, Mo, Ni, V, and Zn showed moderate variability.

Table 2. Descriptive statistics of PTEs (ppm) in household dust in Cairo City.

Region		Al	As	Cd	Co	Cr	Cu	Fe	Hg	Mn	Mo	Ni	Pb	V	Zn
New Cairo (*n* = 8)	Min.	6300	2.2	0.3	4.6	28.0	34.4	12,700	0.03	262	1.1	14.7	41.8	28.0	171
	Max.	8200	4.6	0.9	7.4	39.0	96.8	19,400	1.85	347	2.5	21.6	64.1	54.0	266
	Mean	7250	3.2	0.5	6.1	33.4	60.9	16,200	0.37	313	1.6	17.7	53.0	37.4	223
	St.D.	644	0.8	0.2	1.2	4.17	21.4	2389	0.63	34	0.5	2.5	7.9	9.5	40
Eastern (*n* = 8)	Min.	5400	2.9	0.6	6.1	40.0	54.5	16,200	0.15	326	1.2	20.6	56.6	38.0	234
	Max.	15,700	6.2	2.8	12.0	81.0	249.5	24,200	0.92	507	4.0	58.9	267.8	66.0	1084
	Mean	8413	3.9	1.4	8.5	60.1	155.7	20,175	0.41	402	2.7	34.2	127.1	44.8	486
	St.D.	3110	1.1	0.8	2.1	14.3	75.1	2796	0.31	64	1.1	12.7	62.6	9.0	279
Northern (*n* = 6)	Min.	7000	2.8	0.9	6.7	41.0	77.3	18,600	0.04	332	1.6	21.3	63.8	42.0	244
	Max.	7900	3.5	1.4	9.8	81.0	188.5	24,700	0.78	468	5.7	44.6	219.2	47.0	509
	Mean	7583	3.2	1.2	7.9	60.5	146.4	21,317	0.27	400	3.7	34.5	147.6	44.2	383
	St.D.	354	0.4	0.2	1.0	13.5	37.4	2187	0.27	46	1.4	8.1	54.0	1.7	93
Western (*n* = 6)	Min.	8100	3.1	0.7	7.2	37.0	86.9	15,100	0.08	319	1.4	24.6	97.3	38.0	485
	Max.	16,700	5.8	2.0	11.3	81.0	212.3	23,100	1.94	541	3.2	44.6	193.9	61.0	833
	Mean	11,683	4.2	1.2	9.2	55.2	149.4	20,233	0.45	418	2.4	35.9	121.9	49.3	605
	St.D.	3653	1.0	0.4	1.7	16.8	53.7	3020	0.73	79	0.7	6.7	36.3	9.8	161
Southern (*n* = 10)	Min.	5800	2.8	0.7	4.3	29.0	56.4	14,000	0.05	277	1.3	23.8	50.9	24.0	300
	Max.	17,700	7.4	1.1	19.8	61.0	164.3	35,200	0.14	866	3.0	34.7	94.9	67.0	515
	Mean	10,460	5.2	0.9	12.2	40.3	92.3	25,080	0.08	554	2.5	30.4	71.6	52.0	430
	St.D.	3434	1.5	0.2	4.8	9.5	30.4	6594	0.03	176	0.6	3.79	14.1	14.6	75
All Samples (*n* = 38)	Min.	5400	2.2	0.3	4.3	28.0	34.4	12,700	0.03	262	1.1	14.7	41.8	24.0	171
	Max.	17,700	7.4	2.8	19.8	81.0	249.5	35,200	1.94	866	5.7	58.9	267.8	67.0	1084
	Mean	9092	4.0	1.0	9.0	48.6	116.6	20,818	0.30	425	2.5	30.1	99.3	45.7	419
	St.D.	3065	1.3	0.5	3.4	15.9	58.6	4972	0.44	131	1.1	9.8	51.7	11.2	190
	CV (%)	33.7	32.3	51.7	38.3	32.8	50.3	23.9	148.1	31	41.6	32.5	52.1	24.6	45.4
UCC [51]		80,400	1.5	0.09	10	35	25	35,000	0.05	600	1.5	20	20	60	71

Figure 2 depicts the results of plotting PTE concentrations on spatial distribution maps. Elevated levels (hot spots) of Cd, Cr, Cu, Mo, Ni, Pb, and Zn concentrations were found mostly around eastern, northern, and western regions, which are characterized by higher traffic density, population density, and older buildings. On the other hand, high levels (hot spots) of As, Co, Fe, Mn, and V were mostly concentrated in the southern region, which is characterized by intense industrial activity. A high concentration of Hg was distributed in new Cairo and the northern region.

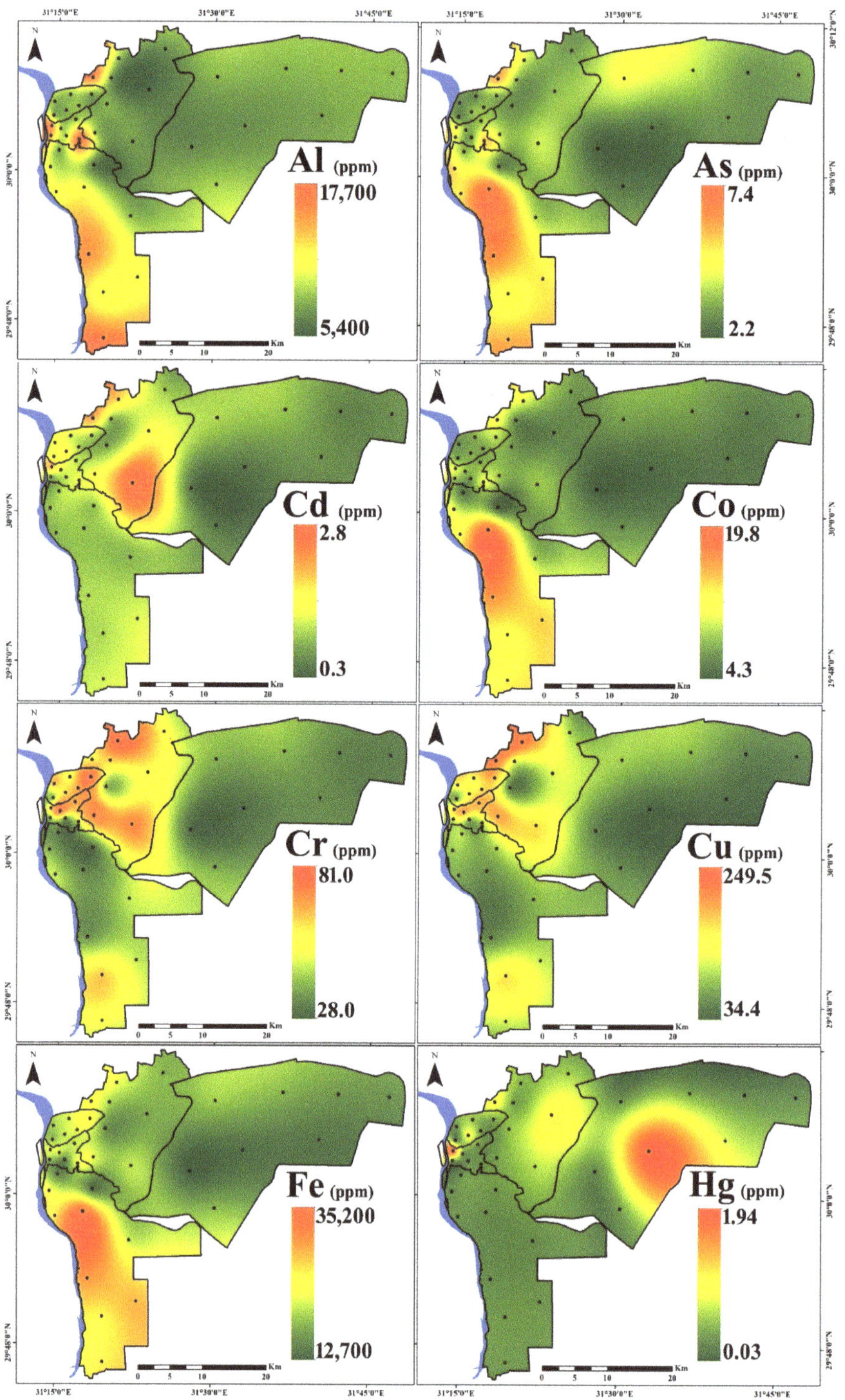

Figure 2. *Cont.*

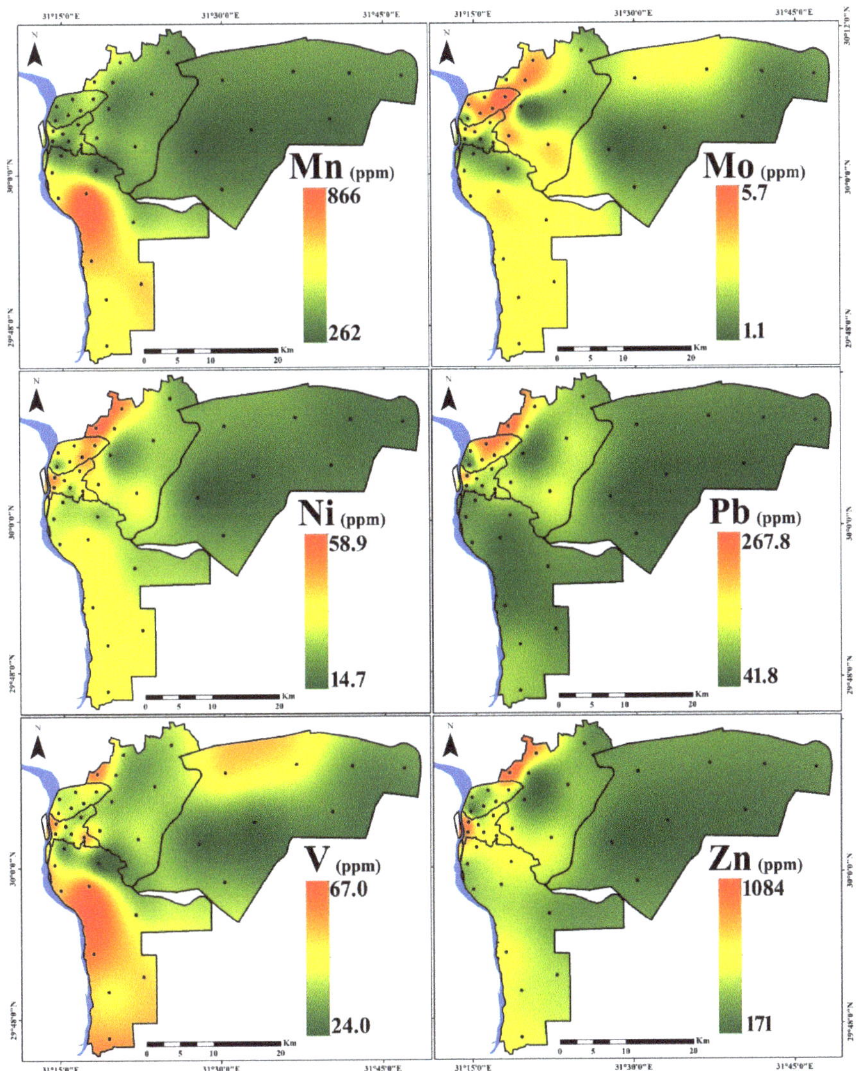

Figure 2. Spatial distribution of PTEs in household dust in Cairo City.

To put the levels of PTEs in Cairo's indoor dust into perspective, they were compared with the levels of the same elements in indoor dust worldwide (Table 3). Table 3 shows that the mean PTE concentrations in our indoor dust samples were both higher and lower than those worldwide. For instance, Al mean concentration was greater than those reported in Slovenia (Maribor) [63], Greece (Athens) [64], and USA (Texas) [7]. As was higher than those reported in Nigeria (Lagos) [22], Nepal [65], and USA (Texas) [7]. Cd was higher than those reported in Alexandria and Kafr El-Sheikh [45], Saudi Arabia (Riyadh) [1], Qatar (Doha) [66], Nigeria (Lagos) [22], Turkey (Istanbul) [67], Iran (Ahvaz) [60], and Greece (Athens) [64]. Cu was higher than those reported in Kafr El-Sheikh [45], Saudi Arabia (Riyadh) [1], Iraq (Al-Fallujah) [68], Nigeria (Lagos) [22], Iran (Ahvaz) [60], and USA (Texas) [7]. Fe was lower than Qatar (Doha) [66], Nigeria (Lagos) [22], and Canada (Alberta) [69]. Mn was far lower than China (Huize) [12] and Nepal [65]. Ni was higher than those reported in Alexandria and Kafr El-Sheikh [45], Saudi Arabia (Riyadh) [1], Nigeria (Lagos) [22], Iran (Ahvaz) [60], Greece (Athens) [64], USA (Texas) [7], and Australia (Sydney) [70]. Pb was lower than Egypt (Alexandria) [45], Saudi Arabia (Jeddah) [14], Kuwait [71], Portugal (Estarreja) [25], China (Huize) [12], Nepal [65], Canada (Alberta) [69], and Australia (Sydney) [70]. Zn was higher than those reported in Egypt (Kafr El-Sheikh) [45], Saudi Arabia (Jeddah and Riyadh) [1,14], Iraq (Al-Fallujah) [68], Nigeria (Lagos) [22], Greece (Athens) [64], and USA (Texas) [7].

Table 3. Comparison between PTE concentrations in the household dust in Cairo City with those for indoor dust in other cities worldwide.

Location	n	Al	As	Cd	Co	Cr	Cu	Fe	Hg	Mn	Mo	Ni	Pb	V	Zn	Ref.
Egypt (Cairo)	$n = 38$	9092	4.0	1.0	9.0	48.6	116.6	20,818	0.3	425	2.5	30.1	99.3	45.7	419	This study
Egypt (Alexandria)	$n = 5$	NA	NA	0.8	3.2	29.2	141.0	NA	NA	237	NA	25.1	260.0	NA	771	[45]
Egypt (Kafr El-Sheikh)	$n = 4$	NA	NA	0.3	8.6	33.4	46.1	NA	NA	438	NA	23.2	24.8	NA	257	[45]
Saudi Arabia (Jeddah)	$n = 10$	NA	8.0	2.1	87.9	40.2	NA	8752	NA	392	NA	35.7	121.2	NA	343	[14]
Saudi Arabia (Riyadh)	$n = 18$	NA	NA	0.1	3.5	NA	59.2	6520	NA	434	NA	15.2	5.0	NA	94	[1]
Kuwait	$n = 50$	12,697	13.0	NA	12.5	90.0	209.0	14,453	NA	441	NA	56.0	158.0	NA	784	[71]
Qatar (Doha)	$n = 12$	19,812	7.2	0.7	12.3	91.8	192.9	20,504	NA	370	15.1	68.7	65.3	52.1	824	[66]
Iraq (Al-Fallujah)	$n = 50$	NA	NA	14.8	NA	289.5	65.0	NA	NA	NA	NA	105.7	75.6	NA	293	[68]
Nigeria (Lagos)	$n = 40$	32,000	3.3	0.5	NA	130.0	28.1	24,500	NA	368	NA	20.9	47.4	52.4	208	[22]
Turkey (Istanbul)	$n = 31$	NA	NA	0.8	5.0	55.0	156.0	NA	NA	136	NA	236.0	28.0	NA	832	[67]
Iran (Ahvaz)	$n = 108$	NA	NA	0.5	8.5	18.0	106.0	NA	NA	100	NA	12.0	74.0	NA	554	[60]
Japan	$n = 100$	15,700	NA	1.0	4.7	67.8	304.0	10,000	NA	226	2.1	59.6	57.9	24.7	920	[72]
Slovenia (Maribor) *	$n = 27$	7400	4.1	1.1	6.2	65.0	140.0	12,700	0.3	306	2.9	38.0	69.0	17.0	716	[63]
Portugal (Estarreja)	$n = 19$	10,500	11.1	1.0	5.5	70.6	261.0	11,900	0.4	178	3.2	67.0	174.0	15.0	1349	[25]
Greece (Athens)	$n = 20$	4217	4.0	0.5	NA	65.2	339.0	4913	0.4	128	NA	29.9	46.1	9.0	401	[64]
China (Huize)	$n = 50$	NA	88.5	25.2	NA	124.0	174.0	NA	1.9	1010	NA	NA	926.8	NA	3029	[12]
Nepal *	$n = 24$	NA	3.0	1.8	28.1	231.0	275.0	838	NA	1650	NA	122.0	233.0	NA	1260	[65]
USA (Texas)	$n = 31$	3738	3.6	1.9	NA	23.0	53.0	2939	NA	48	NA	12.0	38.0	NA	368	[7]
Canada (Windsor)	$n = 60$	11,453	8.1	3.0	NA	65.8	139.0	10,826	NA	171	2.7	50.5	65.0	14.9	677	[20]
Canada (Alberta)	$n = 125$	16,000	13.0	11.0	5.4	92.0	1900.0	26,000	NA	250	8.5	60.0	4500.0	15.0	14,000	[69]
Australia (Sydney)	$n = 82$	NA	NA	4.4	NA	83.6	147.0	5850	NA	76	NA	27.2	389.0	NA	657	[70]

n = Number of Samples; NA = Not Available; * Median.

3.2. Contamination Levels

The UCC element concentrations were used as the background values, and the C_f and integrative C_{deg} indices were applied to objectively analyze the contamination levels in the five administrative regions in Cairo City. The calculated C_f values are presented in Table S2 and Figure 3. Altogether, the five regions were polluted to varying degrees by the measured PTEs. The lowest degrees of pollution were recorded for Al, Co, Fe, Mn, and V, while the highest degrees were recorded for Cd, Cu, Hg, Pb, and Zn, reaching considerably to very high pollution. Hg shows a wide range of C_f values from low to very high pollution.

The calculated C_f-based C_{deg} values in the investigated five regions (Figure 4) indicate that New Cairo recorded the slightest degree of contamination, ranging from considerably to very high pollution. On the other hand, eastern, northern, western, and southern regions' household dust were very highly polluted.

3.3. Correlations between PTEs

The multivariate statistical analysis including Pearson's Correlation Coefficient matrix (PCC), Hierarchical Cluster Analysis (HCA) in Q mode, and Principal Component Analysis (PCA) were utilized to reveal and emphasize the correlation intensity and linkage between the analyzed PTEs.

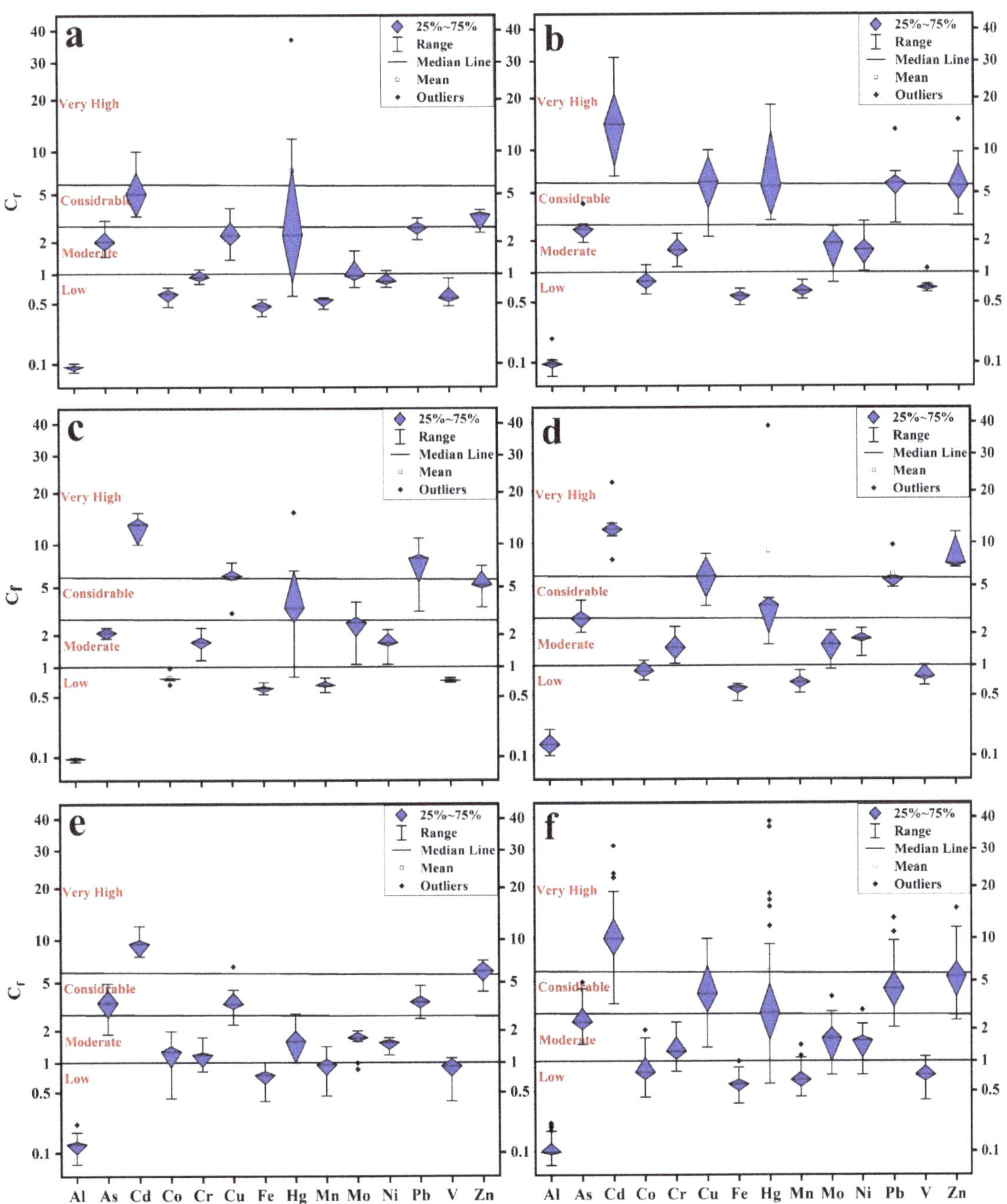

Figure 3. (**a**) Boxplots of C_f values: New Cairo; (**b**) eastern region; (**c**) northern region; (**d**) western region; (**e**) southern region; (**f**) all samples.

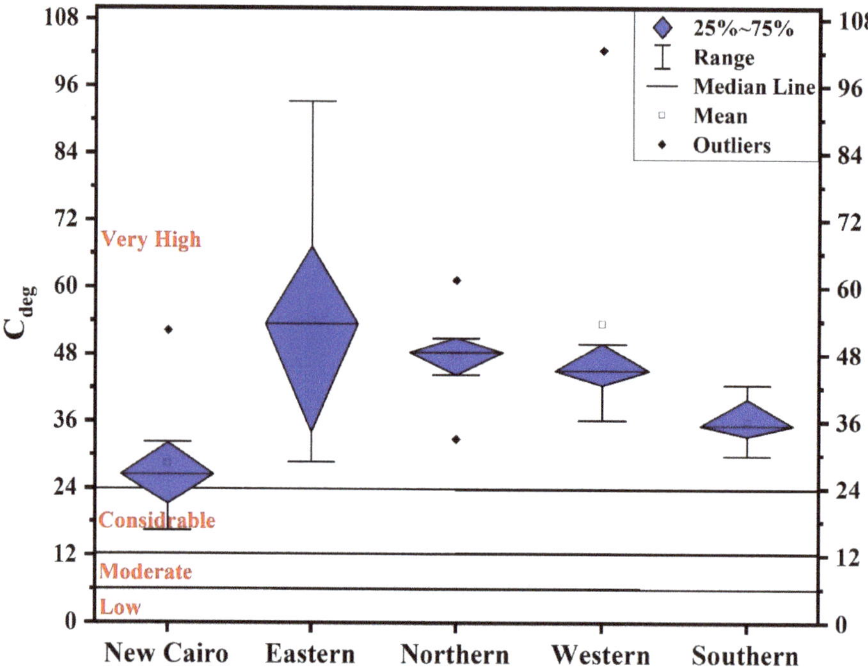

Figure 4. Boxplots of C_{deg} values.

Correlations values 0.00–0.19, 0.20–0.39, 0.40–0.59, 0.60–0.79, and 0.80–1.00 can be considered as very weak, weak, moderate, strong, and very strong correlations, respectively [73]. As shown in Table 4, very strong positive correlations were observed between Al–V (Pearson's R = 0.81), As–Co (R = 0.91), As–Fe (R = 0.84), As–Mn (R = 0.86), As–V (R = 0.90), Co–Fe (R = 0.95), Co–Mn (R = 0.97), Co–V (R = 0.87), Cr–Cu (R = 0.88), Cu–Ni (R = 0.81), Fe–Mn (R = 0.97), Fe–V (R = 0.83), Mn–V (R = 0.81), and Ni–Zn (R = 0.87). Strong positive correlations were observed between Al–As (R = 0.78), Al–Co (R = 0.67), Al–Zn (R = 0.63), Cd–Cu (R = 0.69), Cd–Ni (R = 0.62), Cd–Pb (R = 0.65), Cd–Zn (R = 0.61), Cr–Mo (R = 0.79), Cr–Ni (R = 0.67), Cr–Pb (R = 0.64), Cu–Mo (R = 0.66), Cu–Pb (R = 0.79), Cu–Zn (R = 0.68), Mo–Ni (R = 0.67), Ni–Pb (R = 0.79), and Pb–Zn (R = 0.67). The most significant finding that can be deduced from these positive linear relations is the role played by Al, Fe, and Mn as scavenging elements in the distribution of PTEs, especially As, Co, V, and Zn [48,74]. The strong to very strong positive correlation between the measured PTEs indicates their close distribution and association and may suggest a shared source. It appears to imply that household dust with more elevated levels of one toxic element additionally contain higher levels of other PTEs.

HCA (Figure 5) reduced data into two main clusters. Cluster (1) includes: (a) Al, As, and V and (b) Co, Mn, and Fe. Cluster (2) was subdivided into (c) Cd, Ni, Zn, and Pb; (d) Cr, Cu, and Mo; and (E) Hg. Figure 6 presents the PCA component. Three components, PC1 (49.60%; eigenvalue 6.44), PC2 (24.87%; eigenvalue 3.48), and PC3 (410.40%; eigenvalue 1.46), were extracted from PCA. The 3D plotting of the extracted three components positively confirms the association between Al, As, Co, Fe, Mn, and V (Figure 6a). The 2D plotting of PC1 and PC2 combined with sampling sites (Figure 6b) indicates that Al, As, Co, Fe, Mn, and V are more associated together in the southern region samples. It can be concluded that these elements originated from natural sources; this is in agreement with [22,75]. As enriched from intensive industrial activity in the southern region and adsorbed on Fe–Mn oxides surface [76].

Table 4. PCC matrix for PTEs in the investigated household dust ($n = 38$).

	Al	As	Cd	Co	Cr	Cu	Fe	Hg	Mn	Mo	Ni	Pb	V	Zn
Al	1.00	0.78	0.26	0.67	−0.03	0.10	0.56	0.06	0.57	0.03	0.51	0.27	0.81	0.63
As		1.00	0.28	0.91	−0.01	0.10	0.84	−0.09	0.86	0.22	0.50	0.11	0.90	0.49
Cd			1.00	0.24	0.59	0.69	0.27	0.30	0.22	0.42	0.62	0.65	0.35	0.61
Co				1.00	0.09	0.11	0.95	−0.13	0.97	0.33	0.49	0.09	0.87	0.40
Cr					1.00	0.88	0.27	0.07	0.14	0.79	0.67	0.64	0.11	0.39
Cu						1.00	0.25	0.13	0.14	0.66	0.81	0.79	0.16	0.68
Fe							1.00	−0.16	0.97	0.54	0.58	0.21	0.83	0.35
Hg								1.00	−0.10	−0.01	0.14	0.33	−0.03	0.22
Mn									1.00	0.42	0.51	0.12	0.81	0.35
Mo										1.00	0.67	0.56	0.32	0.30
Ni											1.00	0.79	0.55	0.87
Pb												1.00	0.27	0.69
V													1.00	0.48
Zn														1.00

	Very Weak		Weak		Moderate		Strong		Very Strong

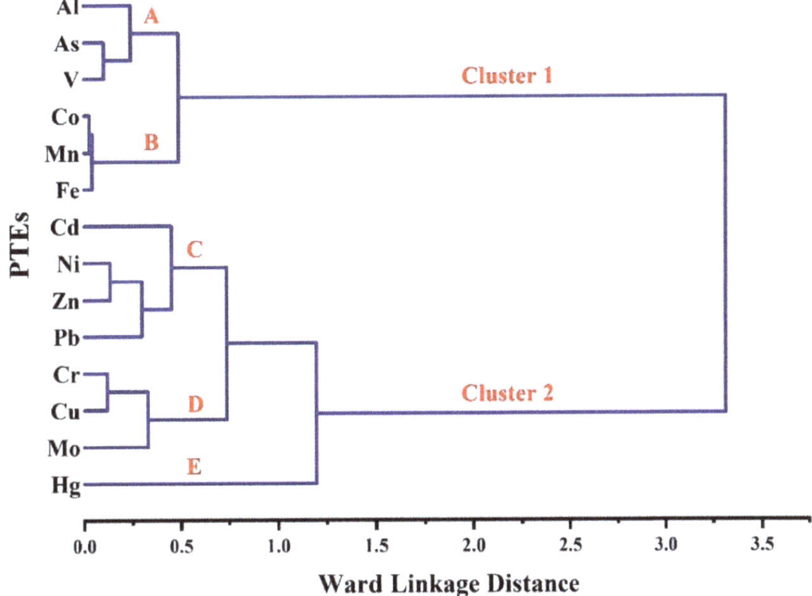

Figure 5. HCA dendrogram.

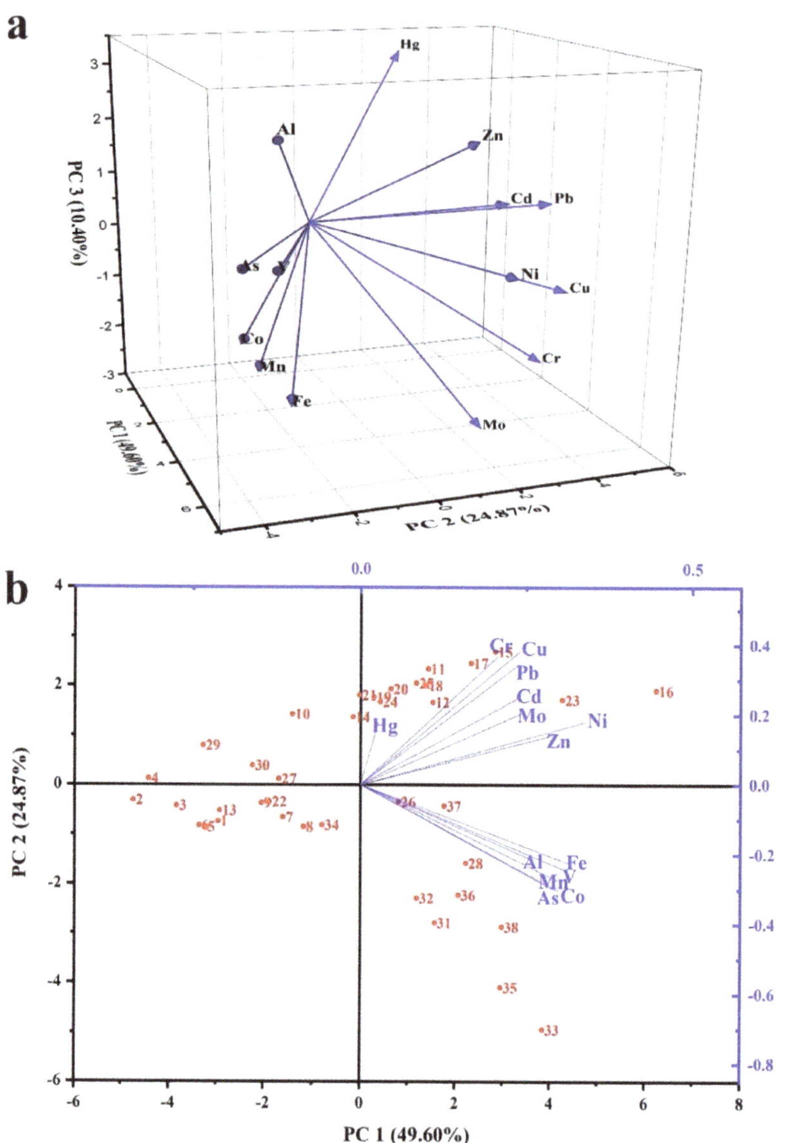

Figure 6. (**a**) PCA variable loading: 3D loading between the extracted 3 components; (**b**) 2D loading between PC1 and PC2 combined with sampling sites.

PTEs in household dust can be attributed to indoor activities such as cooking, smoking, carpet, paper, clothing, cosmetic and personal care products, electric instruments, and cleaning products [65,71,77,78]. A substantial portion of the PTEs emitted by various outdoor activities can travel considerable distances via atmospheric particulate matter and enter the indoor environment in a variety of ways [71]. Al is geochemically stable, while Fe and Mn are geochemically related elements that are abundant in the earth's crust and considered as major elements in soil minerals. The weathering of pre-existing rocks, sediments, and soils primarily releases these major elements [22,71,79] because the levels of Al, Fe, and Mn in the investigated household dust samples are not polluted and relatively deficient. These elements are probably of predominantly geogenic origin and were not enriched in the dust samples by anthropogenic activities. Some exceptions for Mn were recognized in some sites moderately polluted with Mn. Mn can be enriched by many anthropogenic sources such as Mn fungicides [80], Mn–Ni batteries [81], and pigment and paints [82]. Similarly, Co and V concentrations in the majority of the studied samples are below background levels and show a low degree of pollution, indicating that they originated from natural sources before being transported and settling in household dust.

Anthropogenic sources of As, Cd, Cr, Cu, Ni, Pb, and Zn include traffic emissions, braking engine wear, corrosion of vehicle parts, lubricating oils, coal, and fossil fuel combustion, building and construction materials, rubbers, pesticides, and industrial emissions [19,22,63,71,75,77,81,83]. Cr and Zn can be sourced from wood preservative furniture [12,65]. Chemical and pharmaceutical industries, coal combustion, municipal solid waste incineration, and cement manufacture are all anthropogenic sources of Hg. Building materials (interior decorations, paints, and fluorescent lamps), household appliances and electronic devices, LCD displays, monitors, batteries, clothes dryers, irons, washing machines, fluorescent bulbs, neon lights, and thermometers are other potential indoor sources [84].

3.4. Health Risk Assessment

Results of human health risk assessment show that the calculated HQ_{ing}, HQ_{der}, and HQ_{inh} values for individual element (Table S3) and combined PTEs (Table 5; Figure 7a,b) in the household dust were less than one for children and adults. In addition, HI values for adults of the combined PTEs in the household dust were less than one, suggesting no potential noncancer risks (Table 5; Figure 7b). On the other hand, HI values for children were greater than those for adults; one site (site 16; Eastern region) recorded HI values higher than one, suggesting potential noncancer risks for children (Figure 7a). $\sum HQ_{ing}$ was most likely to pose a noncancer risk of more than $\sum HQ_{inh}$ and $\sum HQ_{der}$; this is consistent with several research findings [1,4,12,59,64,85]. The calculated individual element contribution (%) for children and adults noncarcinogenic risk revealed no differences in their contributions in the two age groups. As a result, we discuss them all together. Individual element contribution (%) for noncarcinogenic risk $\sum HQ_{ing}$, $\sum HQ_{der}$, and $\sum HQ_{inh}$ is presented in Figure 7c. In's ingestion route elements' contribution is as follows: Pb (30.53%) > Cr (17.41%) > As (14.38%) > Mn (11.44%) > Al (9.78%) > V (7.03%) > Cu (3.14%) > Ni (1.61%) > Zn (1.50%) > Cd (1.07%) > Hg (1.07%) > Mo (0.54%) > Co (0.48) (Figure 8). For dermal contact route, elements' contribution is as follows: As (31.66%) > Cr (26.19%) > V (21.15%) > Mn (7.48%) > Pb (7.12%) > Cd (3.22%) > Al (2.94%) > Hg (0.46%) > Cu (0.31%) > Zn (0.26%) > Ni (0.18%) > Mo (0.04%) > Co (0.02). For respiratory inhalation route, the contribution of Mn in $\sum HQ_{inh}$ values was the largest, reaching 75.43%, followed by Al (16.12%), Cr (4.30%), and Co (3.99%).

Table 5. Integrated noncancer and cancer risks values.

	Noncancer Risk							
	Children				Adults			
	$\sum HQ_{ing}$	$\sum HQ_{der}$	$\sum HQ_{inh}$	*HI*	$\sum HQ_{ing}$	$\sum HQ_{der}$	$\sum HQ_{inh}$	*HI*
Min	2.62×10^{-1}	4.85×10^{-2}	7.07×10^{-3}	3.18×10^{-1}	1.39×10^{-1}	6.60×10^{-3}	4.94×10^{-3}	1.51×10^{-1}
Max	8.87×10^{-1}	1.28×10^{-1}	2.09×10^{-2}	1.03	4.72×10^{-1}	1.75×10^{-2}	1.46×10^{-2}	4.99×10^{-1}
Mean	4.79×10^{-1}	8.60×10^{-2}	1.14×10^{-2}	5.77×10^{-1}	2.55×10^{-1}	1.17×10^{-2}	7.95×10^{-3}	2.74×10^{-1}
	Cancer Risk							
	$\sum LADD_{ing}$	$\sum LADD_{der}$	$\sum LADD_{inh}$	**CR**				
Min	5.80×10^{-5}	6.35×10^{-7}	5.72×10^{-9}	5.86×10^{-5}				
Max	1.89×10^{-4}	1.87×10^{-6}	1.78×10^{-8}	1.91×10^{-4}				
Mean	1.12×10^{-4}	1.15×10^{-6}	1.09×10^{-8}	1.13×10^{-4}				

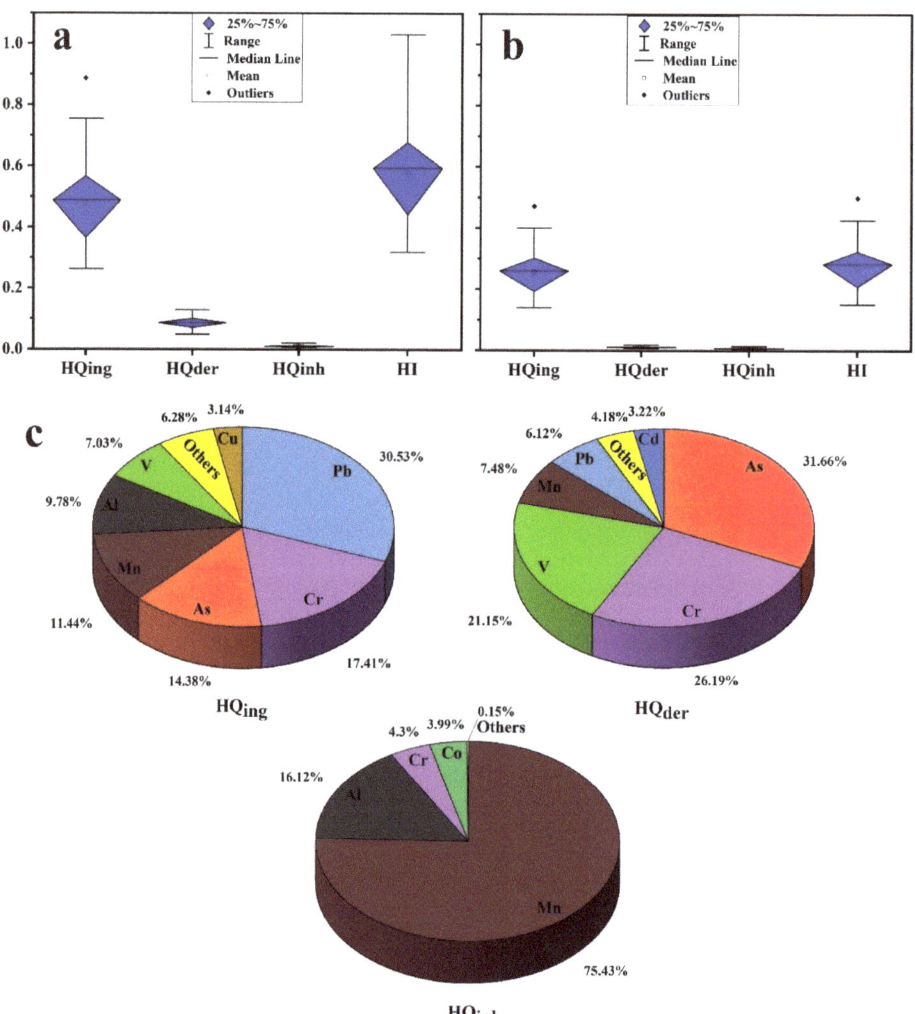

Figure 7. (**a**) Boxplots of noncancer risk for children; (**b**) for adults; (**c**) pie chart showing individual element contribution (%) for noncarcinogenic risk.

In terms of carcinogenic risk, $LADD_{ing}$, $LADD_{der}$, and $LADD_{inh}$ values for individual elements were in the safe limit (Table S4). $\Sigma LADD_{ing}$ values for As, Cd, Cr, Ni, and Pd were higher than 1×10^{-4} in the majority of the investigated sites, indicating a probability of cancer risk. On the other hand, $\Sigma LADD_{der}$ values for As and Cr were higher than 1×10^{-6}, and $\Sigma LADD_{inh}$ values for As, Cd, Cr, Ni, and Pd were lower than 1×10^{-6} (Figure 8a; Table 5). Alarmingly, CR values through the three routs of exposure were higher than 1×10^{-4} in the majority of the investigated sites, indicating a possible cancer risk to inhabitants in Cairo City. The CR risks via various exposure pathways were as follows: hand-to-mouth ingestion > dermal contact > respiratory inhalation. Individual element contribution (%) for carcinogenic risk $\Sigma LADD_{ing}$, $\Sigma LADD_{der}$, and $\Sigma LADD_{inh}$ is presented in Figure 8b. The elements' contributions are Ni (61.84%) > Cr (29.39%) > As (7.28%) > Pb (1.02%) > Cd (0.46%), As (81.93%) > Cr (18.07%), and As (52.14%) > Ni (21.73%) > Cr (17.14%) > Cd (5.40%) > Pb (3.59%) in $\Sigma LADD_{ing}$, $\Sigma LADD_{der}$, and $\Sigma LADD_{inh}$, respectively.

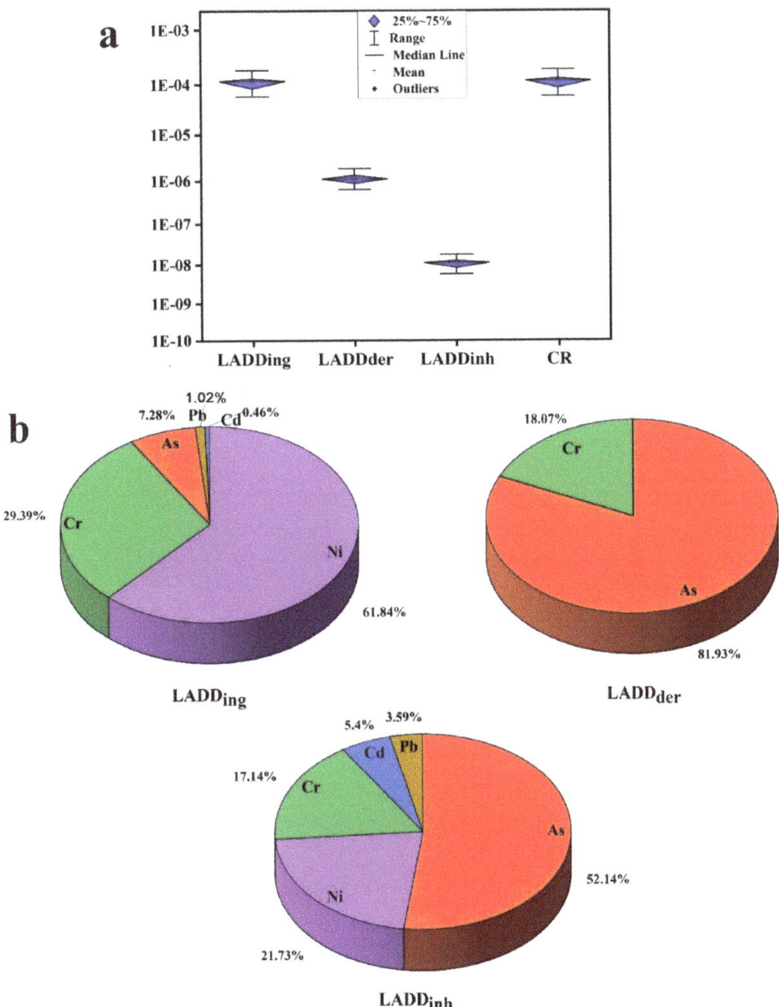

Figure 8. (**a**) Boxplots of cancer risk; (**b**) pie chart showing individual element contribution (%) for cancer risk.

The spatial distribution maps of the calculated HI (children), HI (adults), and CR risks are presented (Figure 9) to inform decision makers about the riskiest districts so that mitigation measures could be implemented. The presented maps show the same distribution for noncancer and cancer risk, with hot spots concentrated in the eastern, northern, and western regions due to condensed road networks in these regions with permanent traffic congestion (Figure 1). In addition, the southern region showed considerable risk distribution due to the intensive industrial activity in this region. One of the most significant limitations of this investigation is the analysis of few composed samples and the undetermined indoor microenvironments. Additional investigation in highly polluted regions should include specific indoor microenvironments such as entrances, kitchens, living rooms, children's rooms, and bedrooms to provide a more comprehensive analysis of household dust geochemistry in various microenvironments and to differentiate between PTE outdoor and indoor sources.

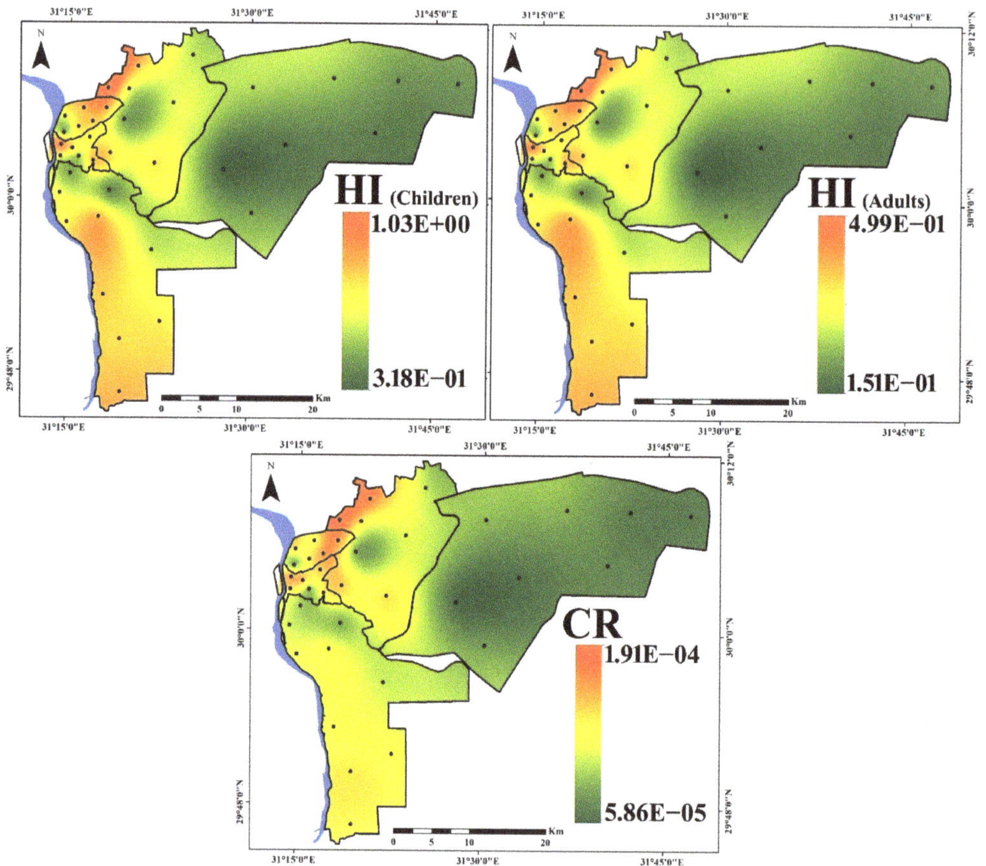

Figure 9. Spatial distribution of HI for children and adults and CR risks for the household dust exposure in Cairo City.

4. Conclusions

This study is the first one to comprehensively measure the chemical composition of household dust in Cairo City, Egypt. In general, the following important conclusions can be gained:

(1) The levels of As, Cd, Cr, Cu, Hg, Mo, Ni, Pb, and Zn surpassed the background values of UCC, indicating anthropogenic influences. The lowest degrees of pollution were recorded for Al, Co, Fe, Mn, and V, while the highest degrees were recorded for Cd, Cu, Hg, Pb, and Zn, reaching considerably to very high pollution.

(2) New Cairo recorded the slightest degree of contamination, ranging from considerably to very high pollution, while in other Cairo regions household dust is very high polluted. Elevated PTE concentrations in Cairo's household dust may be due to industrial activities and heavy traffic emissions.

(3) The health risk assessment model revealed that the vital route of potential PTE exposure that leads to both noncarcinogenic and carcinogenic risks is ingestion, followed by dermal and inhalation pathways. The noncarcinogenic risk was generally in the safe range for adults' exposure. Children are at risk in some sites, where HI values for the measured PTEs in household dust are higher than the recommended safe limit. Prolonged exposure to household dust in Cairo City would produce cancer risk to inhabitants.

(4) The critical contributors to noncancer risk are Pb, As, Cr, Mn, V, and Al. The main causes of cancer risk are Ni, As, and Cr.

(5) The study's findings call for regular detection and assessment of the PTE concentrations and health risk in indoor dust in Cairo City, as well as initiation and facilitation of public health policy development, prevention of anthropogenic source pollutants, and implementation of specific control measures.

Supplementary Materials: The following supporting information can be downloaded at: https://www.mdpi.com/article/10.3390/toxics10080466/s1, Table S1: Samples distribution in administrative regions and districts in Cairo City.; Table S2: Calculated C_f and C_{deg} values; Table S3: Calculated noncancer HQ_{ing}, HQ_{der}, and HQ_{inh} values. Table S4: Calculated cancer $LADD_{ing}$, $LADD_{der}$, and $LADD_{inh}$ values.

Author Contributions: Conceptualization, A.G., H.I.F. and S.M.A.E.B.; methodology, A.G., A.S., H.I.F. and S.M.A.E.B.; software, A.G., H.I.F. and S.M.A.E.B.; validation, A.G., H.I.F. and S.M.A.E.B.; formal analysis, A.G. and A.S.; investigation, A.G., H.I.F. and S.M.A.E.B.; resources, A.G. and A.S.; data curation, A.G., H.I.F. and S.M.A.E.B.; writing—original draft preparation, A.G., H.I.F. and S.M.A.E.B.; writing—review and editing, A.G., A.S. and Y.H.D.; supervision, A.G., A.S. and Y.H.D.; project administration, A.S.; funding acquisition, A.S. All authors have read and agreed to the published version of the manuscript.

Funding: This paper is passed upon work supported by the Science, Technology & Innovation Funding Authority (STDF), Egypt, Under Grant Number: 41628.

Institutional Review Board Statement: Not applicable.

Informed Consent Statement: Not applicable.

Data Availability Statement: Not applicable.

Acknowledgments: This paper is passed upon work supported by Science, Technology & Innovation Funding Authority (STDF), Egypt, under Grant Number: 41628. Therefore, the authors gratefully acknowledge technical and financial support from the Science, Technology & Innovation Funding Authority (STDF), Egypt. Thanks also to A R Abu Zeid (Assiut University, New Valley Branch) for providing Landsat data and helping in the application of the GIS technique.

Conflicts of Interest: The authors declare no conflict of interest. The funders had no role in the design of the study; in the collection, analyses, or interpretation of data; in the writing of the manuscript, or in the decision to publish the results.

References

1. Alotaibi, M.O.; Albedair, L.A.; Alotaibi, N.M.; Elobeid, M.M.; Al-Swadi, H.A.; Alasmary, Z.; Ahmad, M. Pollution Indexing and Health Risk Assessment of Heavy-Metals-Laden Indoor and Outdoor Dust in Elementary School Environments in Riyadh, Saudi Arabia. *Atmosphere* **2022**, *13*, 464. [CrossRef]

2. Chen, R.; Han, L.; Liu, Z.; Zhao, Y.; Li, R.; Xia, L.; Fan, Y. Assessment of Soil-Heavy Metal Pollution and the Health Risks in a Mining Area from Southern Shaanxi Province, China. *Toxics* **2022**, *10*, 385. [CrossRef] [PubMed]

3. Yehia, M.; Baghdady, A.; Howari, F.M.; Awad, S.; Gad, A. Natural radioactivity and groundwater quality assessment in the northern area of the Western Desert of Egypt. *J. Hydrol. Reg. Stud.* **2017**, *12*, 331–344. [CrossRef]

4. Mugudamani, I.; Oke, S.A.; Gumede, T.P. Influence of Urban Informal Settlements on Trace Element Accumulation in Road Dust and Their Possible Health Implications in Ekurhuleni Metropolitan Municipality, South Africa. *Toxics* **2022**, *10*, 253. [CrossRef]

5. Farhat, H.I. Impact of Drain Effluent on Surficial Sediments in the Mediterranean Coastal Wetland: Sedimentological Characteristics and Metal Pollution Status at Lake Manzala, Egypt. *J. Ocean Univ. China* **2019**, *18*, 834–848. [CrossRef]

6. Zhou, L.; Liu, G.; Shen, M.; Liu, Y. Potential Ecological and Health Risks of Heavy Metals for Indoor and Corresponding Outdoor Dust in Hefei, Central China. *Chemosphere* **2022**, *302*, 134864. [CrossRef] [PubMed]

7. Davis, F.R.; Ali, H.H.; Rosenzweig, J.A.; Vrinceanu, D.; Maruthi Sridhar, B.B. Characterization of Chemical and Bacterial Concentrations in Floor Dust Samples in Southeast Texas Households. *Int. J. Environ. Res. Public Health* **2021**, *18*, 12399. [CrossRef] [PubMed]

8. Baghdady, A.; Awad, S.; Gad, A. Assessment of metal contamination and natural radiation hazards in different soil types near iron ore mines, Bahariya Oasis, Egypt. *Arab. J. Geosci.* **2018**, *11*, 506. [CrossRef]

9. Aguilera, A.; Bautista-Hernández, D.; Bautista, F.; Goguitchaichvili, A.; Cejudo, R. Is the Urban Form a Driver of Heavy Metal Pollution in Road Dust? Evidence from Mexico City. *Atmosphere* **2021**, *12*, 266. [CrossRef]

10. Famuyiwa, A.O.; Davidson, C.M.; Ande, S.; Oyeyiola, A.O. Potentially Toxic Elements in Urban Soils from Public-Access Areas in the Rapidly Growing Megacity of Lagos, Nigeria. *Toxics* **2022**, *10*, 154. [CrossRef]
11. Gad, A.; Abd El Bakey, S.M.; Sakr, S. Concentrations of heavy metals and associated human health risk in unrefined salts of inland hypersaline lakes, Egypt. *Int. J. Environ. Anal. Chem.* **2020**, *102*, 1278–1291. [CrossRef]
12. Cao, S.; Wen, D.; Chen, X.; Duan, X.; Zhang, L.; Wang, B.; Qin, N.; Wei, F. Source Identification of Pollution and Health Risks to Metals in Household Indoor and Outdoor Dust: A Cross-Sectional Study in a Typical Mining Town, China. *Environ. Pollut.* **2022**, *293*, 118551. [CrossRef]
13. Tashakor, M.; Behrooz, R.D.; Asvad, S.R.; Kaskaoutis, D.G. Tracing of Heavy Metals Embedded in Indoor Dust Particles from the Industrial City of Asaluyeh, South of Iran. *Int. J. Environ. Res. Public Health* **2022**, *19*, 7905. [CrossRef]
14. Alghamdi, M.A.; Hassan, S.K.; Alzahrani, N.A.; Almehmadi, F.M.; Khoder, M.I. Risk Assessment and Implications of Schoolchildren Exposure to Classroom Heavy Metals Particles in Jeddah, Saudi Arabia. *Int. J. Environ. Res. Public Health* **2019**, *16*, 5017. [CrossRef]
15. Pelfrêne, A.; Cave, M.R.; Wragg, J.; Douay, F. In Vitro Investigations of Human Bioaccessibility from Reference Materials Using Simulated Lung Fluids. *Int. J. Environ. Res. Public Health* **2017**, *14*, 112. [CrossRef] [PubMed]
16. Jurowski, K.; Buszewski, B.; Piekoszewski, W. The analytical calibration in (bio)imaging/mapping of the metallic elements in biological samples–definitions, nomenclature and strategies: State of the art. *Talanta* **2015**, *131*, 273–285. [CrossRef] [PubMed]
17. Cui, X.; Wang, X.; Liu, B. The Characteristics of Heavy Metal Pollution in Surface Dust in Tangshan, a Heavily Industrialized City in North China, and an Assessment of Associated Health Risks. *J. Geochem. Explor.* **2020**, *210*, 106432. [CrossRef]
18. Wang, Y.; Fang, F.; Lin, Y.; Cai, J.; Zhang, C.; Ge, Y. Pollution and Influencing Factors of Heavy Metals from Rural Kitchen Dust in Anhui Province, China. *Atmos. Pollut. Res.* **2020**, *11*, 1211–1216. [CrossRef]
19. Gohain, M.; Deka, P. Trace Metals in Indoor Dust from a University Campus in Northeast India: Implication for Health Risk. *Environ. Monit. Assess.* **2020**, *192*, 741. [CrossRef]
20. Rasmussen, P.E.; Levesque, C.; Chénier, M.; Gardner, H.D. Contribution of metals in resuspended dust to indoor and personal inhalation exposures: Relationships between PM_{10} and settled dust. *Build. Environ.* **2018**, *143*, 513–522. [CrossRef]
21. Cheng, Z.; Chen, L.J.; Li, H.H.; Lin, J.Q.; Yang, Z.B.; Yang, Y.X.; Xu, X.X.; Xian, J.R.; Shao, J.R.; Zhu, X.M. Characteristics and Health Risk Assessment of Heavy Metals Exposure via Household Dust from Urban Area in Chengdu, China. *Sci. Total Environ.* **2018**, *619–620*, 621–629. [CrossRef] [PubMed]
22. Famuyiwa, A.O.; Entwistle, J.A. Characterising and communicating the potential hazard posed by potentially toxic elements in indoor dusts from schools across Lagos, Nigeria. *Environ. Sci. Process. Impacts* **2021**, *23*, 867. [CrossRef]
23. Hassan, S.K.M. Metal concentrations and distribution in the household, stairs and entryway dust of some Egyptian homes. *Atmos. Environ.* **2012**, *54*, 207–215. [CrossRef]
24. Khoder, M.I.; Hassan, S.K.; El-Abssawy, A.A. An Evaluation of Loading Rate of Dust, Pb, Cd, and Ni and Metals Mass Concentration in the Settled Surface Dust in Domestic Houses and Factors Affecting Them. *Indoor Built Environ.* **2010**, *19*, 391–399. [CrossRef]
25. Marinho-Reis, A.P.; Costa, C.; Rocha, F.; Cave, M.; Wragg, J.; Valente, T.; Sequeira-Braga, A.; Noack, Y. Biogeochemistry of Household Dust Samples Collected from Private Homes of a Portuguese Industrial City. *Geosciences* **2020**, *10*, 392. [CrossRef]
26. Osman, R.; Melegy, A.; Dawood, Y.; Gad, A. Distribution of some potentially toxic heavy metals in the soil of Shoubra El Kheima, Egypt. *Egypt. J. Chem.* **2021**, *64*, 1965–1980. [CrossRef]
27. Ibrahim, E.A.; Selim, E.M.M. Pollution and health risk assessment of trace metal in vegetable field soils in the Eastern Nile Delta, Egypt. *Environ. Monit. Assess.* **2022**, *194*, 540. [CrossRef]
28. Osman, R.; Dawood, Y.H.; Melegy, A.; El-Bady, M.S.; Saleh, A.; Gad, A. Distributions and Risk Assessment of the Natural Radionuclides in the Soil of Shoubra El Kheima, South Nile Delta, Egypt. *Atmosphere* **2022**, *13*, 98. [CrossRef]
29. Ramadan, R.S.; Dawood, Y.H.; Yehia, M.M.; Gad, A. Environmental and health impact of current uranium mining activities in southwestern Sinai, Egypt. *Environ. Earth Sci.* **2022**, *81*, 213. [CrossRef]
30. Nasralla, M.M. Studies on indoor air quality in Egypt. *Environ. Int.* **1980**, *4*, 469–473. [CrossRef]
31. Safar, Z.S.; Labib, M.W. Assessment of particulate matter and lead levels in the Greater Cairo area for the period 1998–2007. *J. Advan. Res.* **2010**, *1*, 53–63. [CrossRef]
32. Lowenthal, D.H.; Gertler, A.W.; Labib, M.W. Particulate matter source apportionment in Cairo: Recent measurements and comparison with previous studies. *Int. J. Environ. Sci. Technol.* **2014**, *11*, 657–670. [CrossRef]
33. Hindy, K.T.; Baghdady, A.R.; Howari, F.M.; Abdelmaksoud, A.S. A Qualitative Study of Airborne Minerals and Associated Organic Compounds in Southeast of Cairo, Egypt. *Int. J. Environ. Res. Public Health* **2018**, *15*, 568. [CrossRef] [PubMed]
34. Shaltout, A.A.; Hassan, S.K.; Alomairy, S.E.; Manousakas, M.; Karydas, A.G.; Eleftheriadis, E. Correlation between inorganic pollutants in the suspended particulate matter (SPM) and fine particulate matter ($PM_{2.5}$) collected from industrial and residential areas in Greater Cairo, Egypt. *Air Qual. Atmos. Health* **2019**, *12*, 241–250. [CrossRef]
35. Abdel-Salam, M.M.M. Relationship between residential indoor air quality and socioeconomic factors in two urban areas in Alexandria, Egypt. *Build. Environ.* **2022**, *207*, 108425. [CrossRef]
36. Hassan, S.K.; El-Abssawy, A.A.; Khoder, M.I. Chemical Composition, Characterization and Factors Affecting Household Dust (<20 μm) in Greater Cairo, Egypt. *Open J. Air Pollut.* **2015**, *4*, 184–197. [CrossRef]

37. Hassan, Y.; Shoeib, T. Levels of polybrominated diphenyl ethers and novel flame retardants in microenvironment dust from Egypt: An assessment of human exposure. *Sci. Total Environ.* **2015**, *505*, 47–55. [CrossRef]

38. Shoeib, T.; Hassan, Y.; Rauert, C.; Harner, T. Poly- and perfluoroalkyl substances (PFASs) in indoor dust and food packaging materials in Egypt: Trends in developed and developing countries. *Chemosphere* **2016**, *144*, 1573–1581. [CrossRef]

39. Khairy, M.A.; Lohmann, R. Organophosphate flame retardants in the indoor and outdoor dust and gas-phase of Alexandria, Egypt. *Chemosphere* **2019**, *220*, 275–285. [CrossRef]

40. Hassan, S.K.; El-Mekawy, A.; Alghamdi, M.A.; Khoder, M.I. Insights into the house dust-bound polycyclic aromatic hydrocarbons and their potential human health risk in Greater Cairo, Egypt. *Indoor Built Environ.* **2022**. [CrossRef]

41. Heikal, H.M. Studies on the occurrence, identification and control of house dust mites at rural houses of Shebin El-Kom locality, Egypt. *Pak. J. Biol. Sci.* **2015**, *18*, 179–184. [CrossRef] [PubMed]

42. EL-Morsy, E.S.M. Preliminary survey of indoor and outdoor airborne microfungi at coastal buildings in Egypt. *Aerobiologia* **2006**, *22*, 197–210. [CrossRef]

43. Radwan, S.M.A.; Abdel-Aziz, R.A. Evaluation of microbial content of indoor air in hot arid climate. *Int. J. Environ. Sci. Technol.* **2019**, *16*, 5429–5438. [CrossRef]

44. Hassanien, M.A.; Rieuwerts, J.; Shakour, A.A.; Bitto, A. Seasonal and annual variations in air concentrations of Pb, Cd and PAHs in Cairo, Egypt. *Int. J. Environ. Health Res.* **2001**, *11*, 13–27. [CrossRef]

45. Jadoon, W.A.; Abdel-Dayem, S.M.M.A.; Saqib, Z.; Takeda, K.; Sakugawa, H.; Hussain, M.; Shah, G.M.; Rehman, W.; Syed, J.H. Heavy metals in urban dusts from Alexandria and Kafr El-Sheikh, Egypt: Implications for human health. *Environ. Sci. Pollut. Res.* **2021**, *28*, 2007–2018. [CrossRef] [PubMed]

46. Aly, D.; Dimitrijevic, B. Public green space quantity and distribution in Cairo, Egypt. *J. Eng. Appl. Sci.* **2022**, *69*, 15. [CrossRef]

47. Higazy, M.; Essa, K.S.M.; Mubarak, F.; El-Sayed, E.M.; Sallam, A.M.; Talaat, M.S. Analytical Study of Fuel Switching from Heavy Fuel Oil to Natural Gas in clay brick factories at Arab Abu Saed, Greater Cairo. *Sci. Rep.* **2019**, *9*, 10081. [CrossRef]

48. Saleh, A.; Dawood, Y.H.; Gad, A. Assessment of Potentially Toxic Elements' Contamination in the Soil of Greater Cairo, Egypt Using Geochemical and Magnetic Attributes. *Land* **2022**, *11*, 319. [CrossRef]

49. *ISO 11466*; Soil Quality—Extraction of Trace Elements Soluble in Aqua Regia. ISO: Geneva, Switzerland, 1995.

50. Hakanson, L. An ecological risk index for aquatic pollution control: A sedimentological approach. *Water Res.* **1980**, *14*, 975–1001. [CrossRef]

51. Taylor, S.R.; McLennan, S.M. The geochemical evolution of the continental crust. *Rev. Geophys.* **1995**, *33*, 241–265. [CrossRef]

52. USEPA (United States Environmental Protection Agency). *Risk Assessment Guidance for Superfund. Human Health Evaluation Manual (Part F, Supplemental Guidance for Inhalation Risk Assessment) EPA/540/R/070/002, Volume I*; Office of Superfund Remediation and Technology Innovation: Washington, DC, USA, 2009.

53. USEPA (United States Environmental Protection Agency). *Exposure Factors Handbook 2011 Edition*; EPA/600/R-09/052F; National Center for Environmental Assessment, Office of Research and Development U.S. Environmental Protection Agency: Washington, DC, USA, 2011.

54. USEPA (United States Environmental Protection Agency). *Integrated Risk Information System National Center for Environmental Assessment Office of Research and Development U.S.*; Environmental Protection Agency: Washington, DC, USA, 2017.

55. USEPA (United States Environmental Protection Agency). *Child Specific Exposure Factors Handbook*; EPA-600-P-00–002B; National Center for Environmental Assessment: Washington, DC, USA, 2002.

56. USEPA (United States Environmental Protection Agency). *Soil Screening Guidance: Technical Background Document*; Office of Soild Waste and Emergency Response: Washington, DC, USA, 1996.

57. Ferreira-Baptista, L.; De Miguel, E. Geochemistry and risk assessment of street dust in Luanda, Angola: A tropical urban environment. *Atmos. Environ.* **2005**, *39*, 4501–4512. [CrossRef]

58. USDOE (United States Department of Energy). *The Risk Assessment Information System (RAIS)*; Department of Energy's Oak Ridge Operations Office (ORO): Oak Ridge, TN, USA, 2011.

59. Lin, Y.; Fang, F.; Wang, F.; Xu, M. Pollution distribution and health risk assessment of heavy metals in indoor dust in Anhui rural, China. *Environ. Monit. Assess.* **2015**, *187*, 565. [CrossRef] [PubMed]

60. Neisi, A.; Goudarzi, G.; Akbar Babaei, A.; Vosoughi, M.; Hashemzadeh, H.; Naimabadi, A.; Mohammadi, M.; Hashemzadeh, B. Study of heavy metal levels in indoor dust and their health risk assessment in children of Ahvaz city, Iran. *Toxin Rev.* **2016**, *35*, 16–23. [CrossRef]

61. Nezhad, M.T.K.; Tabatabaii, S.M.; Gholami, A. Geochemical assessment of steel smelter-impacted urban soils, Ahvaz, Iran. *J. Geochem. Explor.* **2015**, *152*, 91–109. [CrossRef]

62. Vosoughi, M.; Shahi Zavieh, F.; Mokhtari, S.A.; Sadeghi, H. Health risk assessment of heavy metals in dust particles precipitated on the screen of computer monitors. *Environ. Sci. Pollut. Res.* **2021**, *28*, 40771–40781. [CrossRef] [PubMed]

63. Gaberšek, M.; Gosar, M. Towards a holistic approach to the geochemistry of solid inorganic particles in the urban environment. *Sci. Total Environ.* **2021**, *763*, 144214. [CrossRef]

64. Stamatelopoulou, A.; Dasopoulou, M.; Bairachtari, K.; Karavoltsos, S.; Sakellari, A.; Maggos, T. Contamination and Potential Risk Assessment of Polycyclic Aromatic Hydrocarbons (PAHs) and Heavy Metals in House Settled Dust Collected from Residences of Young Children. *Appl. Sci.* **2021**, *11*, 1479. [CrossRef]

65. Yadav, I.S.; Devi, N.L.; Singh, V.K.; Li, J.; Zhang, G. Spatial distribution, source analysis, and health risk assessment of heavy metals contamination in house dust and surface soil from four major cities of Nepal. *Chemosphere* **2019**, *218*, 1100–1113. [CrossRef]
66. Mahfouz, M.; Yigiterhan, O.; Elnaiem, A.; Hassan, H.; Alfoldy, B. Elemental compositions of particulate matter retained on air condition unit's filters at Greater Doha, Qatar. *Environ. Geochem. Health* **2019**, *41*, 2533–2548. [CrossRef]
67. Kurt-Karakus, P.B. Determination of heavy metals in indoor dust from Istanbul, Turkey: Estimation of the health risk. *Environ. Int.* **2012**, *50*, 47–55. [CrossRef]
68. Al-Dulaimi, E.M.; Shartooh, S.M.; Al-Heety, E.A. Concentration, Distribution, and Potential Sources of Heavy Metals in Households Dust in Al-Fallujah, Iraq. *Iraqi Geol. J.* **2021**, *54*, 120–130. [CrossRef]
69. Dingle, J.H.; Kohl, L.; Khan, N.; Meng, M.; Shi, Y.A.; Pedroza-Brambila, M.; Chow, C.; Chan, A.W.H. Sources and composition of metals in indoor house dust in a mid-size Canadian city. *Environ. Pollut.* **2021**, *289*, 117867. [CrossRef] [PubMed]
70. Chattopadhyay, G.; Lin, K.C.P.; Feitz, A.J. Household dust metal levels in the Sydney metropolitan area. *Environ. Res.* **2003**, *93*, 301–307. [CrossRef]
71. Al-Harbi, M.; Alhajri, I.; Whalen, J.K. Characteristics and health risk assessment of heavy metal contamination from dust collected on household HVAC air filters. *Chemosphere* **2021**, *277*, 130276. [CrossRef] [PubMed]
72. Yoshinaga, J.; Yamasaki, K.; Yonemura, A.; Ishibashi, Y.; Kaido, T.; Mizuno, K.; Takagi, M.; Tanaka, A. Lead and other elements in house dust of Japanese residences–Source of lead and health risks due to metal exposure. *Environ. Pollut.* **2014**, *189*, 223–228. [CrossRef]
73. Evans, J.D. *Straight Forward Statistics for the Behavioral Sciences*; Brook/Cole Publishing: Pacific Grove, CA, USA, 1996.
74. Khalifa, M.; Gad, A. Assessment of heavy metals contamination in agricultural soil of southwestern Nile Delta, Egypt. *Soil. Sediment. Contam.* **2018**, *27*, 619–642. [CrossRef]
75. Marinho Reis, A.P.; Cave, M.; Sousa, A.J.; Wragg, J.; Rangel, M.J.; Oliveira, A.R.; Patinha, C.; Rocha, F.; Orsiere, T.; Noack, Y. Lead and Zinc Concentrations in Household Dust and Toenails of the Residents (Estarreja, Portugal): A Source-Pathway-Fate Model. *Environ. Sci. Process. Impacts* **2018**, *20*, 1210–1224. [CrossRef]
76. Saleh, A. Evaluation of heavy metals pollution in the soil of Helwan, Cairo, Egypt using magnetic and chemical analysis techniques. In Proceedings of the SUPTM 2022, Online, 17–19 January 2022. [CrossRef]
77. Ferguson, J.E.; Kim, N.D. Trace elements in street and house dusts: Source and speciation. *Sci. Total Environ.* **1991**, *100*, 125–150. [CrossRef]
78. Turner, A. Oral bioaccessibility of trace metals in household dust: A review. *Environ. Geochem. Health* **2011**, *33*, 331–341. [CrossRef]
79. Doležalová Weissmannová, H.; Mihočová, S.; Chovanec, P.; Pavlovský, J. Potential Ecological Risk and Human Health Risk Assessment of Heavy Metal Pollution in Industrial Affected Soils by Coal Mining and Metallurgy in Ostrava, Czech Republic. *Int. J. Environ. Res. Public Health* **2019**, *16*, 4495. [CrossRef]
80. Gunier, R.B.; Jerrett, M.; Smith, D.R.; Jursa, T.; Yousefi, P.; Camacho, J.; Hubbard, A.; Eskenazi, B.; Bradman, A. Determinants of manganese levels in house dust samples from the CHAMACOS cohort. *Sci. Total Environ.* **2014**, *497*, 360–368. [CrossRef] [PubMed]
81. Iwegbue, C.; Obi, G.; Emoyan, O.; Odali, E.; Egobueze, F.; Tesi, G.; Nwajei, G.; Martincigh, B. Characterization of metals in indoor dusts from electronic workshops, cybercafes and offices in southern Nigeria: Implications for on-site human exposure. *Ecotoxicol. Environ. Saf.* **2018**, *159*, 342–353. [CrossRef] [PubMed]
82. Harb, M.; Ebqa'ai, M.; Al-rashidi, A.; Alaziqi, B.H.; Al Rashdi, M.S.; Ibrahim, B. Investigation of selected heavy metals in street and house dust from Al-Qunfudah, Kingdom of Saudi Arabia. *Environ. Earth Sci.* **2015**, *74*, 1755–1763. [CrossRef]
83. Meza-Figueroa, D.; De La O-Villanueva, M.; De la Parra, M.L. Heavy metal distribution in dust from elementary schools in Hermosillo, Sonora, México. *Atmos. Environ.* **2007**, *41*, 276–288. [CrossRef]
84. Dahmardeh Behrooz, R.; Tashakor, M.; Asvad, R.; Esmaili-Sari, A.; Kaskaoutis, D.G. Characteristics and Health Risk Assessment of Mercury Exposure via Indoor and Outdoor Household Dust in Three Iranian Cities. *Atmosphere* **2022**, *13*, 583. [CrossRef]
85. Zhou, L.; Liu, G.; Shen, M.; Hu, R.; Sun, M.; Liu, Y. Characteristics and Health Risk Assessment of Heavy Metals in Indoor Dust from Different Functional Areas in Hefei, China. *Environ. Pollut.* **2019**, *251*, 839–849. [CrossRef]

Article

Composition, Source Apportionment, and Health Risk of PM$_{2.5}$-Bound Metals during Winter Haze in Yuci College Town, Shanxi, China

Lihong Li, Hongxue Qi * and Xiaodong Li

Department of Chemistry and Chemical Engineering, Jinzhong University, Jinzhong 030619, China
* Correspondence: hongxueqi@163.com; Tel.: +86-0351-3985772

Abstract: The composition, source, and health risks of PM$_{2.5}$-bound metals were investigated during winter haze in Yuci College Town, Shanxi, China. The 24-h PM$_{2.5}$ levels of 34 samples ranged from 17 to 174 µg·m^{-3}, with a mean of 81 ± 35 µg·m^{-3}. PM$_{2.5}$-bound metals ranked in the following order: Zn > Cu > Pb > As > Ni > Cr (VI) > Cd > Co. The concentrations of 18% As and 100% Cr (VI) exceeded the corresponding standards of the Ambient Air Quality Standards set by China and the WHO. Subsequently, positive matrix factorization analyses revealed that the three major sources of metals were combustion (37.91%), traffic emissions (32.19%), and industry sources (29.9%). Finally, the non-carcinogenic risks for eight metals indicated that only 2.9% of the samples exceeded a threshold value of one, and As accounted for 45.31%. The total carcinogenic risk values for six metals (As, Cd, Co, Cr (VI), Ni, and Pb) were in the range from 10^{-6} to 10^{-4}, with Cr (VI) and As accounting for 80.92% and 15.52%, respectively. In conclusion, winter haze in Yuci College Town was characterized by higher metal levels and health risks; among the metals, As and Cr (VI) were probably the main contributors.

Keywords: heavy metals; atmosphere; PMF

Citation: Li, L.; Qi, H.; Li, X. Composition, Source Apportionment, and Health Risk of PM$_{2.5}$-Bound Metals during Winter Haze in Yuci College Town, Shanxi, China. *Toxics* **2022**, *10*, 467. https://doi.org/10.3390/toxics10080467

Academic Editor: Ilaria Guagliardi

Received: 26 June 2022
Accepted: 8 August 2022
Published: 11 August 2022

Publisher's Note: MDPI stays neutral with regard to jurisdictional claims in published maps and institutional affiliations.

1. Introduction

Atmospheric haze has attracted considerable attention, especially during winter. Atmospheric fine particulate matter (PM$_{2.5}$) plays an important role in hazy episodes and is the fifth leading cause of death globally after high blood pressure, smoking, diabetes, and hypercholesterolemia [1]. It is significantly associated with the incidence and mortality of bronchitis, asthma, and lung cancer [2]. PM$_{2.5}$ can be attached to a wide variety of chemical contaminants, such as polycyclic aromatic hydrocarbons (PAHs), heavy metals, nitrogen oxides, and emerging pollutants [3]. Some toxicants can be absorbed by the human respiratory system and can affect human health.

Studies on PM$_{2.5}$-bound metals have been performed throughout the world; in some areas, the concentrations of PM$_{2.5}$-bound metals have exceeded the threshold range of the WHO global air quality guidelines [4], such as in Isfahan of Iran [5], Saudi Arabia [6], Kitakyushu of Japan [7], as well as Beijing–Tianjin–Hebei [8], Xi'an [9], Guangzhou [10], and Taiyuan in China [11]. The source apportionment of heavy metals has helped to establish targeted pollution control strategies; for instance, coal burning, industrial pollution, and traffic often have been identified as the major contributors of metals [12–14]. In addition, PM$_{2.5}$-bound metals pose health risks to humans [15]. Some metals have been identified as toxic and hazardous air pollutants in China [16] and some metals (e.g., As, Cd, Co, Cr, Ni, and Pb) have been defined by the WHO as carcinogenic to humans [17]. Therefore, a better understanding of the composition, source apportionment, and carcinogenic risk of metals in PM$_{2.5}$ is crucial to protect human health, especially during haze periods.

Yuci College Town is located in the Yuci District, Jinzhong City, Shanxi Province, China, adjacent to the provincial capital of Taiyuan City. It covers an area of approximate 12 km^2.

It is a campus of provincial colleges and universities built by Shanxi Province. Currently, there are nearly ten colleges and universities, such as Taiyuan University of Technology, Shanxi Medical University, Taiyuan Normal University, and Jinzhong University. The total number of teachers and students in Yuci College Town is about 150,000. The campus has experienced frequent heavy air pollution in the past, particularly in winter. For instance, from January 2016 to December 2018, heavy air pollution events occurred 23 times for a total of 88 days in Jinzhong, Shanxi, China [18]. The predominant source of pollution was $PM_{2.5}$, which accounted for 78.3% of the total pollution.

Shanxi Province is well-known for its coal resources in China, and coal mining is considered to be one of the most significant sources of heavy metal contamination. As previously reported, heavy metal pollution from As and its carcinogenic risk to humans have been reported in Shanxi' mines [19]. Atmospheric $PM_{2.5}$ pollution levels are very important to the health of teachers and students. However, investigations of $PM_{2.5}$-bound heavy metal pollution are lacking in Yuci College Town, Shanxi, China (YCT of China), especially during winter haze periods.

The objectives of the present study were (1) to measure the concentrations of heavy metals (including As, Cd, Co, Cr, Cu, Ni, Pb, and Zn) in $PM_{2.5}$ in YCT of China; (2) to analyze the source apportionment; and (3) to assess the health risks (non-carcinogenic and carcinogenic) of exposure to eight heavy metals, via inhalation exposure, during winter haze periods.

2. Experiments and Methods

2.1. PM$_{2.5}$ Sample Collection

Figure 1 describes the location of the sampling site in Jinzhong University (Yuci district, Jinzhong City, Shanxi Province, China). This site is located at the southwest of Yuci College Town as well as multiple campuses and residential areas, representing an urban area. From 3 November 2020 to 9 December 2020, a total of 34 daily $PM_{2.5}$ samples were collected using a medium flow particle sampler and quartz fiber filters at a gas flow rate of 100 L·min^{-1}. The sampler was placed 15 m from the ground and surrounded by pollution-free emission sources. The daily 24-h mean of $PM_{2.5}$ was calculated using the gravimetric method [20]. The filters were all baked at 450 °C for 4 h before sampling to remove organic contaminants. Then, the filters were placed in boxes with constant humidity (50% ± 5%) and temperature (24 ± 1 °C) for use. The filters were weighed after sampling, and the sensitivity of the balance was 0.01 mg. Finally, the filters were stored at −20 °C, and pollutants were extracted within 2 months.

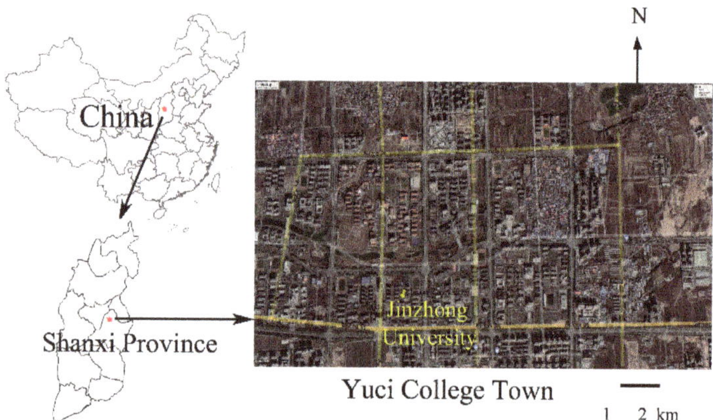

Figure 1. Map of the sampling area in Yuci College Town, Shanxi, China.

2.2. Metals Analysis

The PM$_{2.5}$ filters were digested on mixed acid, dissolved using a microwave digestion system, and the concentrations of eight metals (As, Cd, Co, Cr, Cu, Ni, Pb, and Zn) were analyzed by internal calibration using an inductively coupled plasma–mass spectrometer (ICP-MS, Agilent 7700), according to the method of HJ 657-2013 in China [21]. The concentration of Cr (VI) was calculated to be 1/7 of the measured Cr concentration according to the US EPA regional screening levels [22]. Detailed information on the digestion, internal standards (Table S1), and detection limit for metal analysis are shown in the Supplementary Information (SI).

2.3. PMF Model

The positive matrix factorization (PMF) model is widely used for source apportionment of PM$_{2.5}$. It can simply transform the input data (multiple pollutant data arranged in matrix form) into a factor profile matrix and a factor contribution matrix. Detailed information was obtained according to the US EPA PMF 5.0 Fundamentals and User Guide [23]. Uncertainties (*Unc*) were calculated using the following equation:

If the concentration was higher than the method detection limit (MDL):

$$U_{nc} = \sqrt{(\text{Error fraction} \times \text{concentration})^2 + (0.5 \times \text{MDL})^2} \tag{1}$$

If the concentration was less than or equal to the MDL:

$$U_{nc} = \frac{5}{6} \times \text{MDL} \tag{2}$$

In addition, the rate of $Q_{\text{robust}}/Q_{\text{true}}$ was calculated to determine the optimal number of factors, and bootstrap (BS) and displacement (DISP) analyses were performed to estimate the uncertainties of the PMF model.

2.4. Human Health Risk Assessment

Health risks of heavy metals in PM$_{2.5}$ were estimated via the inhalation pathway, as proposed by the US EPA [11]. The exposure concentration (EC) was calculated to assess the carcinogenic risk using Equation (3):

$$\text{EC} = (\text{CA} \times \text{ET} \times \text{EF} \times \text{ED})/\text{AT} \tag{3}$$

where EC is the exposure concentration ($\mu g \cdot m^{-3}$), CA is the contaminant concentration in air ($\mu g \cdot m^{-3}$), ET is the exposure time (24 h·d^{-1}), EF is the exposure frequency (180 d·y^{-1}), ED is the exposure duration (24 y for adults), and AT is the averaging lifetime: for non-carcinogens (24 × 365 × 24 h) and for carcinogens (70 × 365 × 24 h).

The hazard index (HI) is traditionally used to assess the overall non-carcinogenic risk posed by multiple chemicals, and it was hypothesized that all metal risks were additive effects despite existing synergistic effects, the equation for HI is defined as follow:

$$\text{HI} = \sum \text{HQ}_i = \sum \text{EC}_i/\text{RfC}_i \tag{4}$$

where HQ is the hazard quotient (unitless) and RfC$_i$ is the reference concentration of *i*th heavy metal ($\mu g \cdot m^{-3}$) for inhalation (Table 1). According to the US EPA [24], if the HI value is less than one, the exposed population is unlikely to experience obvious adverse health effects. In contrast, if the HI value exceeds one, an adverse effect may occur for a specific population.

Table 1. Toxicological parameters of the heavy metals used for health risk assessment via the inhalation route.

	Contaminants	Class [a]	RfC_i [b]	IUR [b]
		WHO	$\mu g \cdot m^{-3}$	$(\mu g \cdot m^{-3})^{-1}$
As	Arsenic	1	0.015	0.0043
Cd	Cadmium	1	0.01	0.0018
Co	Cobalt	2B	0.006	0.009
Cr (VI)	Chromium	1	0.1	0.084
Cu	Copper	- [c]	1000 [d]	-
Ni	Nickel	2B	0.014	0.00026
Pb	Lead	2B	0.15 [e]	0.000012
Zn	Zinc	-	300	-

[a] Class: Agents classified by IARC the monographs adapted with permission from Ref. [17]. 2021, International Agency for Research on Cancer. [b] RfC_i: reference concentration of ith heavy metal; IUR: inhalation unit risk, with their values from the US EPA: Regional screening levels (RSLs)—Generic Tables [22]. [c] Not reported. [d] Data from the literature [25]. [e] Data from the literature [26].

Carcinogenic risk (CR) is defined as the probability of an individual developing any type of cancer throughout their lifetime owing to exposure to carcinogenic hazards. CR was summarized by inhalation for an individual over a lifetime according to the following equation:

$$CR = \sum CR_i = \sum IUR \times EC_i \tag{5}$$

where IUR is the inhalation unit risk $(\mu g \cdot m^{-3})^{-1}$ (Table 1). According to the US EPA [24], a CR lower than 10^{-6} indicates an acceptable level, a CR range of 10^{-6} to 10^{-4} is generally considered a potential risk level, and a CR above 10^{-4} is likely to be harmful to the human body.

2.5. Air Mass Backward Trajectory

Backward trajectory was performed using the hybrid single-particle Lagrangian integrated trajectory (HYSPLIT) online model from the National Oceanic and Atmospheric Administration [27]. The start altitude was chosen at three different heights (50, 500, and 1000 m), representing the low, middle, and upper atmosphere, respectively. The global data assimilation system (GDAS) data of 72-h backward trajectories on 25 November 2020 are presented in Figure S1.

2.6. Network Data Collection

The consumption of end-use energy, the number of days that reached the air quality standards, and the annual mean concentrations of $PM_{2.5}$ in Shanxi, China, were collected from the Shanxi Statistics Yearbook [28].

3. Results and Discussion

3.1. Mass Levels of $PM_{2.5}$

The levels of $PM_{2.5}$ are presented in Table S2, and their descriptive statistics are presented in Table 2. The daily $PM_{2.5}$ concentrations of 34 samples ranged from 17 to 174 $\mu g \cdot m^{-3}$, with a median of 74 $\mu g \cdot m^{-3}$ and an average concentration of 81 $\pm$ 35 $\mu g \cdot m^{-3}$; 94% of the samples exceeded Grade I (35 $\mu g \cdot m^{-3}$) of the Chinese ambient air quality standards [29]; none of the samples reached the WHO's AQG level (15 $\mu g \cdot m^{-3}$) in 2021 [30]; however, 53% of the samples reached the Grade II standard (75 $\mu g \cdot m^{-3}$) [29] and Interim Target-1 (75 $\mu g \cdot m^{-3}$) of the WHO global air quality guidelines [30]. During the sampling period, there were two heavy haze episodes, including 11 to 16 November and 24 to 26 November. During haze episodes, the $PM_{2.5}$ concentrations averaged over 100 $\mu g \cdot m^{-3}$ and peaked at 127 and 174 $\mu g \cdot m^{-3}$, respectively. These results indicate that air pollution during winter was severe in this area.

Table 2. Statistical description of the daily $PM_{2.5}$ mass ($\mu g \cdot m^{-3}$) and the concentrations of its metals ($ng \cdot m^{-3}$) during winter in 2020 at Yuci College Town, Shanxi, China ($n = 34$).

	$PM_{2.5}$	As	Cd	Co	Cr (VI)	Cu	Ni	Pb	Zn	Sum
Min	17	0.43	0.04	0.01	0.39	0.69	0.22	1.04	0.71	27.92
Median	74	4.22	0.68	0.15	0.99	12.82	1.48	15.19	164.02	196.46
Max	174	11.36	4.88	1.61	2.91	67.15	6.82	40.52	823.39	905.26
Mean	81	4.71	0.89	0.29	1.31	20.04	1.82	14.95	191.87	235.87
SD [a]	35	2.70	0.86	0.38	0.77	17.35	1.48	9.09	145.92	161.88
WHO guideline value [b]	25	6.6	5	-	0.25	-	25	500	-	
Grade II threshold [c]	75	6	5	- [d]	0.025	-	-	500	-	

[a] SD: standard deviation. [b] WHO global air quality guidelines adapted with permission from Ref. [4]. 2005, World Health Organization. [c] Annual average concentrations of the Chinese ambient air quality standards [29]. [d] Not reported.

3.2. Concentrations of Heavy Metals in $PM_{2.5}$

The levels of eight $PM_{2.5}$-bound heavy metals are presented in Table S2 and their descriptive statistics are presented in Table 2. The total concentration of the eight metals was 235.87 ± 161.88 ng·m^{-3}. The daily mean levels of the heavy metals were ranked in the order Zn > Cu > Pb > As > Ni > Cr (VI) > Cd > Co. Zn was the most abundant metal with a mean of 191.87 ± 145.92 ng·m^{-3}, followed by Cu (20.04 ± 17.35), Pb (14.95 ± 9.09), As (4.71 ± 2.70), Ni (1.82 ± 1.48), Cr (VI) (1.31 ± 0.77), Cd (0.89 ± 0.86), and Co (0.29 ± 0.38) ng·m^{-3}. Moreover, the concentrations of Cd, Ni, and Pb in all samples were lower than the annual values recommended by the WHO [4] and China [29] (Table 2). However, the concentrations of 18% As and 100% Cr (VI) exceeded the Grade II standard of the Chinese Ambient Air Quality Standards [29]. There were no corresponding standards for Co, Cu, and Zn. These results are similar to those found in other areas; for instance, the levels of As and Cr (VI) greatly exceeded their corresponding threshold values in 60 cities in China [31].

In recent years, the mean concentrations of the metals in $PM_{2.5}$ have been obtained from China and abroad, and the results of other studies are summarized in Table 3.

For As, the average concentration during the 2020 winter haze periods in YCT of China was lower than that in Changzhi, Shanxi, China [32]. For other provinces in China, the average concentration of As, during the 2020 winter haze episodes in YCT of China, were higher than those in Beijing [33], Chengdu [34], Guangzhou [10], and Lanzhou [35]. Worldwide, it was higher than those in Tˇrinec-Kosmos of Czech Republic [36] and Kitakyushu of Japan [7]. In contrast, the average concentration of As in the 2020 winter haze episodes in YCT of China was lower than those in Yuci and Taiyuan [11] in 2017 within Shanxi Province. For other provinces in China, it was lower than those in Chongqing [37], Handan [38], Xi'an [9], and Xuanwu [10]. For worldwide, it was lower than those in Iasi of Romania [39], Isfahan of Iran [5], Karaj of Iran [40], and Saudi Arabia [6].

For Cr (VI), since total chromium concentrations were only available from the literature in some regions, the concentrations of Cr (VI) were also calculated to be 1/7 of the Cr concentration according to the US EPA regional screening levels [22]. For comparison, the mean concentration of Cr (VI) during the 2020 winter haze periods in YCT of China was higher than those in Tˇrinec-Kosmos of Czech Republic [36], Iasi of Romania [39], Koldata of India [41], Kitakyushu of Japan [7], Saudi Arabia [6], Los Angeles of USA [12], Beijing [33], and Chongqing [37] of China. Conversely, the concentrations of Cr (VI) in the remaining regions were higher than that of YCT in China. It has been suggested that Cr (VI) in $PM_{2.5}$ is very common and is a common air pollution problem worldwide. More stringent measures are required to control Cr pollution in the future.

Overall, the $PM_{2.5}$-bound heavy metal contents were generally high in the YCT of China, with severe As and Cr (VI) contamination in $PM_{2.5}$ during winter haze periods.

Table 3. A comparison of the results from other studies regarding the mean concentrations (ng·m^{-3}) of PM$_{2.5}$-bound heavy metals in urban regions.

Country	Areas	Year	PM$_{2.5}$	As	Cd	Co	Cr (VI)	Cu	Ni	Pb	Zn	References
Czech Republic	Třinec-Kosmos	2020	28	1.3	0.26	0.05	1.1/7 [d]	4.6	0.81	11	34	[36]
Romania	Iasi	2016	20	**5.70**	0.33	0.09	1.78/7	7.70	24.3	1.99	33.9	[39]
India	Kolkata	2019	111.7	-	1.2	0.3	6.9/7	15	8.1	36	370	[41]
Iran	Isfahan	2015	- [a]	**32.47** [d]	5.77	-	**57.4/7**	13.76	7.43	46.72	-	[5]
Iran	Karaj	2019	67	**32.1**	84	-	**49.5/7**	203	60.8	133	242	[40]
Japan	Kitakyushu	2019	21.3	1.4	-	-	3.0/7	3.6	3.3	10.5	29.5	[7]
Saudi Arabia		2020	-	**83**	17	-	8/7	9	10	119	31	[6]
USA	Los Angeles	2018	13.8	-	6	1	3/7	10	3	5	10	[12]
China	Beijing	2019	-	4.02	-	-	1.79/7	7.37	0.77	21.13	78.99	[33]
	Chengdu	2018	113.2	4.5	-	-	-	7.5	7.7	21.9	60.8	[34]
	Chongqing	2019	97.1	**7.56**	-	-	4.29/7	15.83	1.39	37.93	94.22	[37]
	Guangzhou	2017	55	4.39	0.74	0.53	**10.1/7**	16.37	5.72	25.52	127.31	[10]
	Guilin (haze)	2017	144	-	19.0	-	**11.5/7**	17.4	-	78.8	300.7	[42]
	Handan	2017	-	**11.94**	2.74	-	**11.1/7**	23.17	2.11	104.3	286.9	[43]
	Hefei	2017	81	-	-	-	**10/7**	11.29	-	12.64	273.5	[44]
	Lanzhou	2018	73	3	1	1.3	-	29	-	407	-	[35]
	Xi'an	2016	50.1	**117.2**	16.3	-	**343/7**	-	11.3	35.0	267.1	[9]
	Xuanwu	2016	61.1	6.44	1.88	0.29	**77.5/7**	20.99	3.73	54.72	212.76	[10]
Shanxi, China	Changzhi	2018	56.1	4.9	0.7	0.2	**14.3/7**	7.8	4.2	30.8	82.3	[32]
	Taiyuan	2017	-	**8.15**	1.07	1.20	**29.9/7**	29.56	12.69	94.36	230.57	[11]
	Yuci	2017	-	**9.45**	1.12	0.70	**11.7/7**	14.66	3.56	91.29	263.26	[11]
	Yuci	2020	80.65	4.71	0.89	0.29	1.31	20.04	1.82	14.95	191.87	This study
WHO guideline value [b]				25	6.6	5	-	0.25	-	25	500	-
Grade II threshold [c]				75	6	5	-	0.025	-	-	500	-

[a] Not reported. [b] WHO global air quality guidelines adapted with permission from Ref. [4]. 2005, World Health Organization. [c] Annual average concentrations of the Chinese ambient air quality standards [29]. [d] /7: The concentration of Cr (VI) was calculated to be 1/7 of the Cr concentration according to the US EPA regional screening levels [22]. Values higher than those in Yuci College Town are shown in bold.

3.3. Source Apportionment of PM$_{2.5}$-Bound Elements

The source apportionment of eight heavy metals in PM$_{2.5}$ was conducted using the PMF model [23]. Two to six factors were used for each dataset to determine the optimal solutions. Finally, the optimum result with a three-factor solution (Q_{robust} = 1141.5, Q_{true} = 1255.4, Q_{robust}/Q_{true} = 0.91, 92–100% of BS runs, and no swaps of DISP runs) was selected based on the interpretability of the source profiles and the results of the modeling diagnostics. The correlation coefficient (R^2) values between the predicted data and the input data ranged from 0.69 to 0.92, meaning a good fit of the model.

As shown in the factor profile in Figure 2, the PMF resolved three factor profiles, namely, industry, traffic, and combustion, as the three anthropogenic sources contributing to the total PM2.5-bound metals. Factor 1 was associated with high loadings of Cd (39.9%), Cr (VI) (39.2%), and Zn (80.3%). A previous study reported that Cr (VI) originated from the glassmaking industry, while Ni originated from steelworks [45]. Therefore, Factor 1 was identified as an industrial source. Factor 2 accounted for 48.8% and 77.6% of the Cu and Pb concentrations, respectively. Cu and Zn have been documented as surrogates for brake wear [46]. Pb emissions may result from the use of leaded gasoline. Thus, Factor 2 was labeled as a traffic source. Factor 3 demonstrated high loadings of the metals As (82.7%), Co (62.6%), and Ni (50.8%). These elements are all related to coal combustion [47], and Ni is also typical of oil combustion [14]. Therefore, Factor 3 was regarded as "combustion sources", which included emissions from oil and coal combustion.

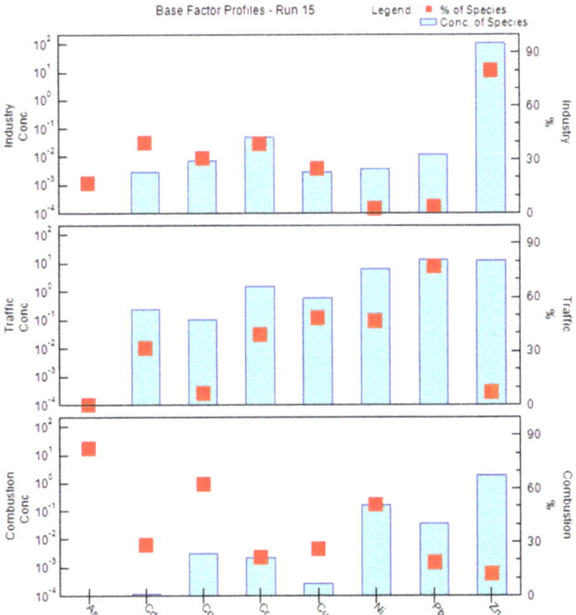

Figure 2. Source profiles obtained by the PMF analyses for Yuci College Town, Shanxi, China. Bars represent mass concentrations and red squares represent contribution percentages from each source factor.

Overall, three major sources of $PM_{2.5}$-bound heavy metals were characterized based on the PMF analyses. As shown in Figure 3, combustion was still the largest contributor to $PM_{2.5}$-bound heavy metals (37.91%) in Jinzhong, China. The contribution of traffic emissions (32.19%) was ranked second, just higher than the industry source (29.9%).

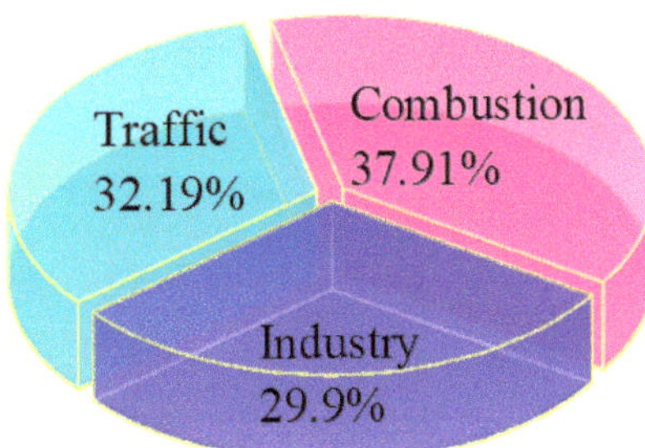

Figure 3. Contributions of identified sources to $PM_{2.5}$-bound metal based on the PMF model in Yuci College Town, Shanxi, China.

Coal combustion was a main contributor that cannot be ignored. Shanxi is a famous coal province in China [48], and coal mining is considered one of the most important sources of heavy metal pollution [49]. More attention should be paid to the burning of coal for heating in winter, and the use of clean energy should be further increased.

The second contribution of traffic sources resulted from the increase in car ownership year by year. For example, the number of cars in Shanxi Province increased from 3.76 million in 2013 to 7.61 million in 2020 [28]. However, in the following years, this situation is expected to improve because of the continuous investment of new energy vehicles [50].

In fact, the formation of $PM_{2.5}$ is very complex, and they involve adverse meteorological conditions, local emission accumulation, and regional transport, etc. [51]. Industrial sources of heavy metals in $PM_{2.5}$ were likely to be regional transport because there were almost no industrial pollution sources around the sampling sites. Evidence of regional transport resulted from the analysis of the reverse trajectory. As shown, 72-h air mass backward trajectories occurred in haze episodes on 25 November 2020 (Figure 4); the results of 50 and 100 m were basically the same, and the air masses passed through Xinzhou-Taiyuan and reached Yuci after a roundabout in Yangquan. At 1000 m, the air mass passed through Linfen–Lvliang–Taiyuan, made a detour through Yangquan before reaching Yuci. These results indicated that most of the long-distance air masses were influenced by northwest winds and reached YTC of China through several industrial cities. This also indirectly suggested that the industrial pollution in YTC might come from regional transport rather than local sources. Due to the implementation of air pollution control policies and the shift of industrial production, industrial pollution is no longer the main source of pollution in YTC of China.

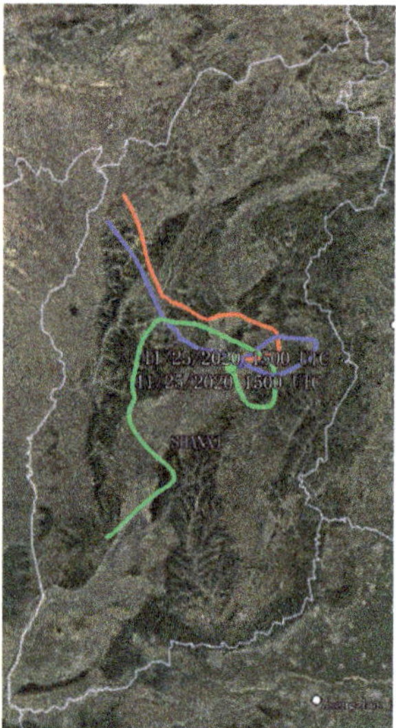

Figure 4. The 72-h back trajectories on 25 November 2020 in Yuci College Town, Shanxi, China. The red color line represents the height of 50 m; blue line: 500 m; green line: 1000 m.

3.4. Human Health Risk Assessment

3.4.1. Non-Carcinogenic Risk Assessment

The non-carcinogenic risks of the eight $PM_{2.5}$-bound metals were calculated via the inhalation route in YCT of China (Figure 5). HI values ranged from 0.052 to 1.12 in 34 samples with a median of 0.30 and a mean of 0.34 ± 0.21. In this study, 2.9% of the samples exceeded an acceptable threshold of one. Among the eight metals, As accounted for 45.31% of the entire HI value, which demonstrated that the metal As was probably the main contributor to non-carcinogenic risk, while the contributions of the remaining metals were negligible. It can be concluded that heavy metals in $PM_{2.5}$ have low non-carcinogenic risks to the public in YCT of China.

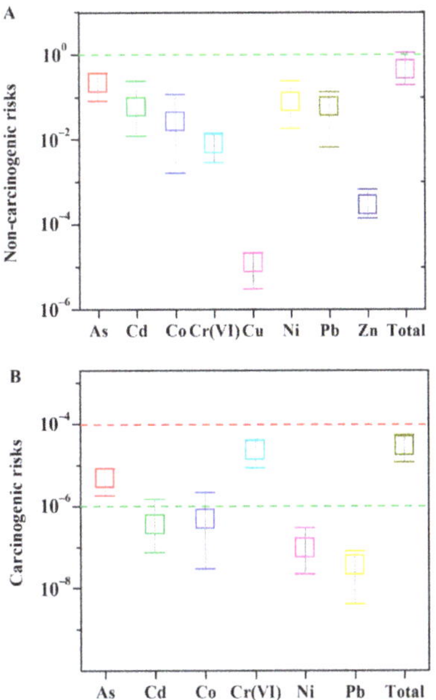

Figure 5. Non-carcinogenic (**A**) and carcinogenic (**B**) risks of exposure to PM$_{2.5}$-bound metals estimated via the inhalation route in Yuci College Town, Shanxi, China (*n* = 34).

3.4.2. Cancer Risk Assessment

The carcinogenic risk of exposure to six metals (As, Cd, Co, Cr, Ni and Pb) was estimated via the inhalation route and are presented in Figure 4. They ranged from 7.56×10^{-6} to 5.36×10^{-5}, with an average value of $(2.29 \pm 1.30) \times 10^{-5}$. The carcinogenic risk values all ranged from 10^{-6} to 10^{-4}. Among the six metals, Cr (VI) was the main contributor to carcinogenic risk, accounting for $80.92 \pm 6.20\%$ of the total CR, and As accounted for $15.52 \pm 5.42\%$ of the total CR. These results demonstrated that the metals As and Cr (VI) were probably the main contributors to carcinogenic risk, while the contribution of the remaining metals was negligible. These values are lower than the total CR of metals in PM$_{2.5}$ reported in Taiwan, China [52], but are higher than the total CR reported in Changzhi, China (10.31×10^{-6}) [32]. These results are consistent with previous studies that indicated the CR also mainly resulted from the contribution of Cr in Shenzhen [53], Taiyuan and Yuci [11], and Changzhi [32] in China. In addition, Cr (VI) also contributed to the highest potential years of life lost in most cities, with a proportion of 72.7% across 60 cities in China [31].

In summary, the non-carcinogenic risks of heavy metals were negligible in PM$_{2.5}$ in YCT of China. However, more attention should be paid to the carcinogenic risks of heavy metals, especially As and Cr (VI).

3.5. Policy Implication

A series of strict control measures have been implemented to prevent and control air pollution in China, and the PM$_{2.5}$ levels have declined since 2013. For instance, China's "Action Plan for the Prevention and Control of Air Pollution" was proclaimed in 2013; the "Blue Sky Protection" campaign was enacted in 2018; the 14th "Five-Year Plan for Modern Energy System" was issued in 2022; and the "Opinions on Further Strengthening the Prevention and Control of Heavy Metal Pollution" was issued in March 2022, which set two goals for 2025 and 2035. By 2025, the emissions of key heavy metal pollutants from key industries will be reduced by 5 percent as compared with those in 2020. By 2035, a heavy metal pollution prevention and control system and long-term mechanism will be established to comprehensively improve the ability to monitor environmental pollution, to control heavy metal pollution, and to prevent environmental risks. In addition, Shanxi has implemented many policies, such as promoting coal energy transformation, developing emerging industries, limiting the traffic volume in winter by odd or even days, and operating new energy vehicles.

The implementation of coal banning was very effective in controlling air pollution during winter. For instance, the proportion of coal consumption in Shanxi Province decreased annually, while the proportion of electricity and natural gas increased. As shown in Figure 6 (original data were presented in Table S3), from 2013 to 2020, the proportion of coal consumption decreased from 27.92% to 16.39%, while electricity increased from 32.90% to 39.33% and natural gas increased from 13.89% to 22.72%.

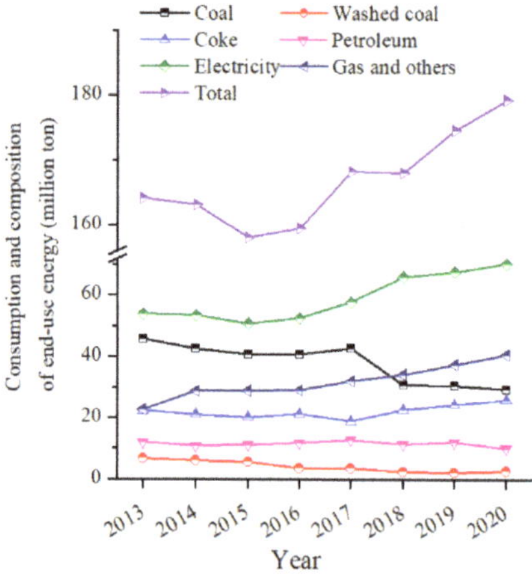

Figure 6. The composition of energy consumption in recent years in Shanxi Province, China.

Air quality has improved significantly, owing to a series of strict control policies in China. The number of days that reached the air quality standards gradually increased, while the average annual PM$_{2.5}$ level gradually decreased. As shown in Figure 7 (the original data are listed in Table S4), from 2014 to 2020, in Jinzhong, Shanxi Province, China, the number of days that reached the air quality standards increased from 241 to 267 days, while the annual mean concentrations of PM$_{2.5}$ decreased from 64 to 42 μg·m^{-3}. These results indicate that Shanxi's energy transformation has achieved initial effects and will continue to control coal consumption and further increase the use of clean energy to win the battle for a blue sky in the future.

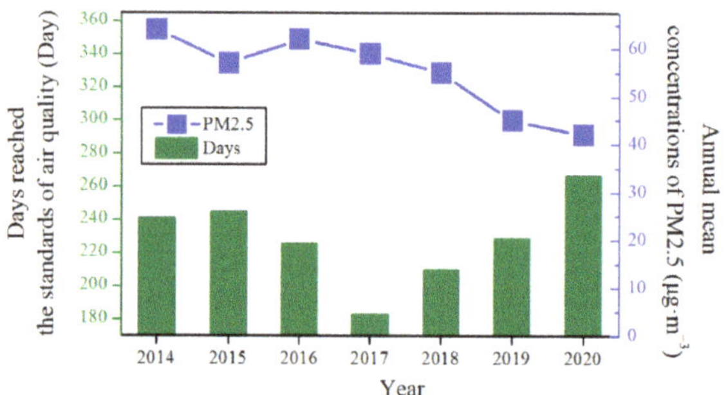

Figure 7. The number of days that reached the air quality standards in recent years and the annual mean concentrations of PM$_{2.5}$ in Jinzhong, Shanxi Province, China.

3.6. Limitations

First, there are two main types of health risk assessments of $PM_{2.5}$-bound heavy metals. One type of health risk assessment is based solely on one exposure route, i.e., the inhalation route, while the other type of health risk assessments considers three exposure routes: inhalation, dermal, and oral intake. The former type was applied for health risk assessments in the industrial cities of Iran [40], Kolkata of India [41], Beijing of China [13], and Tianjin of China [38]. The latter type was used in Saudi Arabia [6], Hebei of China [54], and Shenzhen of China [53]. However, which is more representative? The models and parameters of health risk assessment should be optimized for specific populations and regions. Second, the traditional approach to heavy metals inherently hypothesizes that the carcinogenic risks of all metals are additive effects in a mixture. However, the synergistic or antagonistic effects may also occur during metabolism. For instance, three joint effects (synergistic, antagonistic, or additive effects) of the toxicity were all observed in the different component metal mixtures (Cd, Cr, Cu, Hg, Mn, Ni, Pb, and Zn) [55]. The mixtures of Cd + Pb [56], Cd + Ni [57], and Zn + Al [46] have synergistic effects [58], while Cd + Cu [59] and Cd + Zn [60] mixtures have antagonistic effects. In addition, heavy metals are only a small part of the $PM_{2.5}$, which is also composed of other substances, such as organic matter, nitrate, sulfate, ammonium, elemental carbon, and chloride [61]. In future research, the above insufficiency should be overcome to assess the risk to human health more accurately.

4. Conclusions

In this study, the levels of eight metals in $PM_{2.5}$ were detected during the 2020 winter haze periods in YCT of China. The 24-h $PM_{2.5}$ levels of 34 samples ranged from 17 to 174 $\mu g \cdot m^{-3}$, with a mean of 81 $\pm$ 35 $\mu g \cdot m^{-3}$. The $PM_{2.5}$-bound heavy metals ranked in the following order: Zn > Cu > Pb > As > Ni > Cr (VI) > Cd > Co. A total of 18% and 100% of the concentrations assessed in the samples for As and Cr, respectively, exceeded the corresponding threshold values of the Chinese Ambient Air Quality Standards and the WHO Global Air Quality Guidelines; the levels of As and Cr (VI) were also higher than those in some areas worldwide. Overall, higher $PM_{2.5}$ levels were found in this area, suffering from severe As and Cr (VI) contamination during winter haze periods.

Based on the PMF model, combustion was the largest contributor to $PM_{2.5}$-bound heavy metals (37.91%), followed by traffic emissions (32.19%) and industrial sources (29.9%). Finally, the potential health risks were estimated through exposure by the inhalation route. The non-carcinogenic risks of $PM_{2.5}$-bound heavy metals were negligible, but the carcinogenic risk values were all within the potential level (10^{-6}–10^{-4}), and both As and Cr (VI) were the main contributors. Therefore, more attention should be paid to the prevention and control of As and Cr (VI) pollution in the environment. By reducing coal burning and using clean energy in winter, we will continue to reduce industrial source pollution, thereby making a positive contribution to reduce the occurrence of winter haze periods.

Supplementary Materials: The following supporting information can be downloaded at: https://www.mdpi.com/article/10.3390/toxics10080467/s1. Text: Detailed description of chemical analyses; Table S1: Isotopes of heavy metals, corresponding internal standards, and detection limit for ICP-MS analysis; Table S2: $PM_{2.5}$ mass ($\mu g \cdot m^{-3}$) and its metals concentrations ($ng \cdot m^{-3}$) during winter in 2020 in the college town of Shanxi Province, China (n = 34); Table S3: The consumption, composition, and their proportion (%) of end-use energy (million ton) in recent years in Shanxi Province, China; Table S4: The days reached the standards of air quality in recent years and the annual mean concentrations of $PM_{2.5}$ ($\mu g \cdot m^{-3}$) in Jinzhong, Shanxi Province, China; Figure S1. The 72-h back trajectories on 25 November 2020 in Yuci College Town, Shanxi, China. References [21,62–69] are cited in the supplementary materials.

Author Contributions: Funding acquisition, L.L. and H.Q.; methodology, X.L.; visualization, X.L.; writing—original draft, L.L.; writing—review and editing, H.Q. All authors have read and agreed to the published version of the manuscript.

Funding: This work was supported by the Natural Science Foundation of Shanxi Province, China (201901D111299 and 201901D111301), and the Scientific and Technological Innovation Programs of Higher Education Institutions in Shanxi, China (2020L0584).

Institutional Review Board Statement: Not applicable.

Informed Consent Statement: Not applicable.

Data Availability Statement: Not applicable.

Conflicts of Interest: The authors declare no conflict of interest.

References

1. Cohen, A.J.; Brauer, M.; Burnett, R.; Anderson, H.R.; Frostad, J.; Estep, K.; Balakrishnan, K.; Brunekreef, B.; Dandona, L.; Dandona, R.; et al. Estimates and 25-year trends of the global burden of disease attributable to ambient air pollution: An analysis of data from the Global Burden of Diseases Study 2015. *Lancet* **2017**, *389*, 1907–1918. [CrossRef]
2. Turner, M.C.; Krewski, D.; Diver, W.R.; Pope, C.A.; Burnett, R.T.; Jerrett, M.; Marshall, J.D.; Gapstur, S.M. Ambient air pollution and cancer mortality in the cancer prevention study II. *Environ. Health Perspect.* **2017**, *125*, 087013. [CrossRef] [PubMed]
3. Landrigan, P.J.; Fuller, R.; Acosta, N.J.R.; Adeyi, O.; Arnold, R.; Basu, N.; Balde, A.B.; Bertollini, R.; Bose-O'Reilly, S.; Boufford, J.I.; et al. The Lancet Commission on pollution and health. *Lancet* **2018**, *391*, 462–512. [CrossRef]
4. WHO. *Air Quality Guidelines: Global Update 2005: Particulate Matter, Ozone, Nitrogen Dioxide, and Sulfur Dioxide*; World Health Organization: Geneva, Switzerland, 2005.
5. Soleimani, M.; Amini, N.; Sadeghian, B.; Wang, D.; Fang, L. Heavy metals and their source identification in particulate matter (PM$_{2.5}$) in Isfahan City, Iran. *J. Environ. Sci.* **2018**, *72*, 166–175. [CrossRef] [PubMed]
6. Abdulaziz, M.; Alshehri, A.; Yadav, I.C.; Badri, H. Pollution level and health risk assessment of heavy metals in ambient air and surface dust from Saudi Arabia: A systematic review and meta-analysis. *Air Qual. Atmos. Health* **2022**, *15*, 799–810. [CrossRef]
7. Zhang, X.; Eto, Y.; Aikawa, M. Risk assessment and management of PM$_{2.5}$-bound heavy metals in the urban area of Kitakyushu, Japan. *Sci. Total Environ.* **2021**, *795*, 148748. [CrossRef]
8. Li, Z.Y.; Ma, H.Q.; Fan, L.; Zhao, P.; Wang, L.; Jiang, Y.J.; An, C.X.; Liu, A.Q.; Hu, Z.S.; Jin, H. Size distribution of inorganic elements in bottom ashes from seven types of bio-fuels across Beijing-Tianjin-Hebei Region, China. *Aerosol Air Qual. Res.* **2017**, *17*, 2450–2462. [CrossRef]
9. Wu, T.; Liu, P.; He, X.; Xu, H.; Shen, Z. Bioavailability of heavy metals bounded to PM$_{2.5}$ in Xi'an, China: Seasonal variation and health risk assessment. *Environ. Sci. Pollut. Res.* **2021**, *28*, 35844–35853. [CrossRef] [PubMed]
10. Xie, J.; Jin, L.; Cui, J.; Luo, X.-S.; Li, J.; Zhang, G.; Li, X.-D. Health risk-oriented source apportionment of PM$_{2.5}$-associated trace metals. *Environ. Pollut.* **2020**, *262*, 114655. [CrossRef] [PubMed]
11. Liu, X.; Wang, Z.; Bai, H.; Zhang, S.; Mu, L.; Peng, L. Characteristics and health risk assessments of heavy metals in PM$_{2.5}$ in Taiyuan and Yuci College Town, China. *Air Qual. Atmos. Health* **2020**, *13*, 909–919. [CrossRef]
12. Farahani, V.J.; Soleimanian, E.; Pirhadi, M.; Sioutas, C. Long-term trends in concentrations and sources of PM$_{2.5}$-bound metals and elements in central Los Angeles. *Atmos. Environ.* **2021**, *253*, 118361. [CrossRef]
13. Yang, X.; Zheng, M.; Liu, Y.; Yan, C.; Liu, J.; Liu, J.; Cheng, Y. Exploring sources and health risks of metals in Beijing PM$_{2.5}$: Insights from long-term online measurements. *Sci. Total Environ.* **2022**, *814*, 151954. [CrossRef] [PubMed]
14. Sofowote, U.M.; Su, Y.; Dabek-Zlotorzynska, E.; Rastogi, A.K.; Brook, J.; Hopke, P.K. Sources and temporal variations of constrained PMF factors obtained from multiple-year receptor modeling of ambient PM$_{2.5}$ data from five speciation sites in Ontario, Canada. *Atmos. Environ.* **2015**, *108*, 140–150. [CrossRef]
15. Swietlik, R.; Trojanowska, M. Chemical fractionation in environmental studies of potentially toxic particulate-bound elements in urban air: A critical review. *Toxics* **2022**, *10*, 124. [CrossRef]
16. MEEC (Ministry of Ecology and Environment of China). *List of Toxic and Hazardous Air Pollutants (2018)*; Ministry of Ecology and Environment of China and National Health Commission of China: Beijing, China, 2019; (In Chinese). Available online: https://www.gov.cn/xinwen/2019-2002/2005/content_5364004.htm (accessed on 16 April 2022).
17. IARC (International Agency for Research on Cancer). Agents Classified by the IARC Monographs, Volumes 1–130. 2021. Available online: https://monographs.iarc.who.int/agents-classified-by-the-iarc/ (accessed on 18 April 2022).
18. Dong, Y.; Zhou, H.; Fu, Y.; Li, X.; Geng, H. Wavelet periodic and compositional characteristics of atmospheric PM$_{2.5}$ in a typical air pollution event at Jinzhong city, China. *Atmos. Pollut. Res.* **2021**, *12*, 245–254. [CrossRef]
19. Li, Z.; Ma, Z.; van der Kuijp, T.J.; Yuan, Z.; Huang, L. A review of soil heavy metal pollution from mines in China: Pollution and health risk assessment. *Sci. Total Environ.* **2014**, *468*, 843–853. [CrossRef]
20. MEEC (Ministry of Ecology and Environment of China). *Determination of Atmospheric Articles PM10 and PM2.5 in Ambient Air by Gravimetric Method*; China Environmental Press: Beijing, China, 2011. (In Chinese)
21. NEPSC (National Environmental Protection Standards of China). *Ambient Air and Stationary Sourc Emission: Determination of Metals in Ambient Particulate Matter–Inductively Coupled Plasma/Mass Spectrometry (ICP-MS)*; China Environmental Science Press NEPSC: Beijing, China, 2013. (In Chinese)

22. US EPA. Regional Screening Levels (RSLs)-Generic Tables as of November 2021. 2021. Available online: https://www.epa.gov/risk/regional-screening-levels-rsls-generic-tables (accessed on 16 April 2022).

23. Norris, G.; Duvall, R.; Brown, S.; Bai, S. *EPA Positive Matrix Factorization (PMF) 5.0 Fundamentals and User Guide*; EPA/600/R-14/108; U.S. Environmental Protection Agency: Washington, DC, USA, 2014; pp. 1–136.

24. US EPA. *Risk Assessment Guidance for Superfund: Volume I: Human Health Evaluation Manual (Part F, Supplemental Guidance for Inhalation Risk Assessment)*; U.S. Environmental Protection Agency, Office of Emergency and Remedial Response: Washington, DC, USA, 2009.

25. Oosthuizen, M.; Wright, C.Y.; Matooane, M.; Phala, N. Human health risk assessment of airborne metals to a potentially exposed community: A screening exercise. *Clean Air J.* **2015**, *25*, 51–57. [CrossRef]

26. State of Michigan, USA. Chemical Update Worksheet. Michigan Department of Environmental Quality. 2016. Available online: https://www.michigan.gov/documents/deq/deq-rrd-chem-LeadDatasheet_527864_527867.pdf (accessed on 16 April 2022).

27. NOAA (National Oceanic and Atmospheric Administration). Available online: https://www.ready.noaa.gov/HYSPLIT.php (accessed on 20 July 2022).

28. SSY (Shanxi Statistics Yearbook). (In Chinese). Available online: http://www.shanxi.gov.cn/sj/tjnj/ (accessed on 18 April 2022).

29. MEPC (Ministry of Environmental Protection of China). *Ambient Air Quality Standards (GB 3095-2012), National Standards of the People's Republic of China*; China Environmental Press: Beijing, China, 2012. (In Chinese)

30. WHO. *WHO Global Air Quality Guidelines: Particulate Matter (PM2.5 and PM10), Ozone, Nitrogen Dioxide, Sulfur Dioxide and Carbon Monoxide: Executive Summary*; WHO Regional Office for Europe: Copenhagen, Denmark, 2021. Available online: https://www.who.int/publications/i/item/9789240034433 (accessed on 18 April 2022).

31. Liu, J.; Cao, H.; Zhang, Y.; Chen, H. Potential years of life lost due to $PM_{2.5}$-bound toxic metal exposure: Spatial patterns across 60 cities in China. *Sci. Total Environ.* **2022**, *812*, 152593. [CrossRef]

32. Duan, X.; Yan, Y.; Li, R.; Deng, M.; Hu, D.; Peng, L. Seasonal variations, source apportionment, and health risk assessment of trace metals in $PM_{2.5}$ in the typical industrial city of Changzhi, China. *Atmos. Pollut. Res.* **2021**, *12*, 365–374. [CrossRef]

33. Zhao, S.; Tian, H.; Luo, L.; Liu, H.; Wu, B.; Liu, S.; Bai, X.; Liu, W.; Liu, X.; Wu, Y.; et al. Temporal variation characteristics and source apportionment of metal elements in $PM_{2.5}$ in urban Beijing during 2018-2019. *Environ. Pollut.* **2021**, *268*, 115856. [CrossRef]

34. Qu, Y.; Gao, T.; Yang, C. Elemental characterization and source identification of the near-road $PM_{2.5}$ using EDXRF in Chengdu, China. *X-ray Spectrom.* **2019**, *48*, 232–241. [CrossRef]

35. Li, Y.; Zhao, B.; Duan, K.; Cai, J.; Niu, W.; Dong, X. Contamination characteristics, mass concentration, and source analysis of metal elements in $PM_{2.5}$ in Lanzhou, China. *Elem. Sci. Anthr.* **2021**, *9*, 00125. [CrossRef]

36. Seibert, R.; Nikolova, I.; Volná, V.; Krejčí, B.; Hladký, D. Air pollution sources' contribution to $PM_{2.5}$ concentration in the northeastern part of the Czech Republic. *Atmosphere* **2020**, *11*, 522. [CrossRef]

37. Han, Y.; Wang, Z.; Zhou, J.; Che, H.; Tian, M.; Wang, H.; Shi, G.; Yang, F.; Zhang, S.; Chen, Y. $PM_{2.5}$-bound heavy metals in southwestern China: Characterization, sources, and health risks. *Atmosphere* **2021**, *12*, 929. [CrossRef]

38. Chen, R.; Jia, B.; Tian, Y.; Feng, Y. Source-specific health risk assessment of $PM_{2.5}$-bound heavy metals based on high time-resolved measurement in a Chinese megacity: Insights into seasonal and diurnal variations. *Ecotoxicol. Environ. Saf.* **2021**, *216*, 112167. [CrossRef] [PubMed]

39. Galon-Negru, A.G.; Olariu, R.I.; Arsene, C. Size-resolved measurements of PM2.5 water-soluble elements in Iasi, north-eastern Romania: Seasonality, source apportionment and potential implications for human health. *Sci. Total Environ.* **2019**, *695*, 133839. [CrossRef] [PubMed]

40. Kermani, M.; Jonidi Jafari, A.; Gholami, M.; Arfaeinia, H.; Shahsavani, A.; Fanaei, F. Characterization, possible sources and health risk assessment of $PM_{2.5}$-bound heavy metals in the most industrial city of Iran. *J. Environ. Health Sci. Eng.* **2021**, *19*, 151–163. [CrossRef]

41. Mitra, S.; Das, R. Health risk assessment of construction workers from trace metals in $PM_{2.5}$ from Kolkata, India. *Arch. Environ. Occup. Health* **2022**, *77*, 125–140. [CrossRef]

42. Zhong, S.; Zhang, L.; Jiang, X.; Gao, P. Comparison of chemical composition and airborne bacterial community structure in $PM_{2.5}$ during haze and non-haze days in the winter in Guilin, China. *Sci. Total Environ.* **2019**, *655*, 202–210. [CrossRef]

43. Cai, A.; Zhang, H.; Wang, L.; Wang, Q.; Wu, X. Source apportionment and health risk assessment of heavy metals in $PM_{2.5}$ in Handan: A typical heavily polluted city in North China. *Atmosphere* **2021**, *12*, 1232. [CrossRef]

44. Xue, H.; Liu, G.; Zhang, H.; Hu, R.; Wang, X. Similarities and differences in PM_{10} and $PM_{2.5}$ concentrations, chemical compositions and sources in Hefei City, China. *Chemosphere* **2019**, *220*, 760–765. [CrossRef] [PubMed]

45. Ledoux, F.; Kfoury, A.; Delmaire, G.; Roussel, G.; El Zein, A.; Courcot, D. Contributions of local and regional anthropogenic sources of metals in $PM_{2.5}$ at an urban site in northern France. *Chemosphere* **2017**, *181*, 713–724. [CrossRef] [PubMed]

46. Hulskotte, J.H.J.; Roskam, G.D.; Denier van der Gon, H.A.C. Elemental composition of current automotive braking materials and derived air emission factors. *Atmos. Environ.* **2014**, *99*, 436–445. [CrossRef]

47. Xu, M.; Yan, R.; Zheng, C.; Qiao, Y.; Han, J.; Sheng, C. Status of trace element emission in a coal combustion process: A review. *Fuel Process. Technol.* **2004**, *85*, 215–237. [CrossRef]

48. Cao, X. Policy and regulatory responses to coalmine closure and coal resources consolidation for sustainability in Shanxi, China. *J. Clean. Prod.* **2017**, *145*, 199–208. [CrossRef]

49. Liu, X.; Bai, Z.; Shi, H.; Zhou, W.; Liu, X. Heavy metal pollution of soils from coal mines in China. *Nat. Hazards* **2019**, *99*, 1163–1177. [CrossRef]
50. Lin, B.; Shi, L. Do environmental quality and policy changes affect the evolution of consumers' intentions to buy new energy vehicles. *Appl. Energy* **2022**, *310*, 118582. [CrossRef]
51. Zhang, Q.; Zheng, Y.; Tong, D.; Shao, M.; Wang, S.; Zhang, Y.; Xu, X.; Wang, J.; He, H.; Liu, W.; et al. Drivers of improved $PM_{2.5}$ air quality in China from 2013 to 2017. *Proc. Natl. Acad. Sci. USA* **2019**, *116*, 24463–24469. [CrossRef]
52. Tsai, P.J.; Young, L.H.; Hwang, B.F.; Lin, M.Y.; Chen, Y.C.; Hsu, H.T. Source and health risk apportionment for $PM_{2.5}$ collected in Sha-Lu area, Taiwan. *Atmos. Pollut. Res.* **2020**, *11*, 851–858. [CrossRef]
53. Yan, R.-H.; Peng, X.; Lin, W.; He, L.-Y.; Wei, F.-H.; Tang, M.-X.; Huang, X.-F. Trends and challenges regarding the source-specific health risk of $PM_{2.5}$-bound metals in a Chinese megacity from 2014 to 2020. *Environ. Sci. Technol.* **2022**, *56*, 6996–7005. [CrossRef] [PubMed]
54. Li, X.; Yan, C.; Wang, C.; Ma, J.; Li, W.; Liu, J.; Liu, Y. $PM_{2.5}$-bound elements in Hebei Province, China: Pollution levels, source apportionment and health risks. *Sci. Total Environ.* **2022**, *806*, 150440. [CrossRef]
55. Lin, X.; Gu, Y.; Zhou, Q.; Mao, G.; Zou, B.; Zhao, J. Combined toxicity of heavy metal mixtures in liver cells. *J. Appl. Toxicol.* **2016**, *36*, 1163–1172. [CrossRef] [PubMed]
56. Xu, X.; Li, Y.; Wang, Y.; Wang, Y. Assessment of toxic interactions of heavy metals in multi-component mixtures using sea urchin embryo-larval bioassay. *Toxicol. Vitr.* **2011**, *25*, 294–300. [CrossRef] [PubMed]
57. Alsop, D.; Wood, C.M. Metal and pharmaceutical mixtures: Is ion loss the mechanism underlying acute toxicity and widespread additive toxicity in zebrafish? *Aquat. Toxicol.* **2013**, *140*, 257–267. [CrossRef] [PubMed]
58. Gebara, R.C.; Alho, L.d.O.G.; Rocha, G.S.; da Silva Mansano, A.; Melão, M.d.G.G. Zinc and aluminum mixtures have synergic effects to the algae Raphidocelis subcapitata at environmental concentrations. *Chemosphere* **2020**, *242*, 125231. [CrossRef] [PubMed]
59. Zhu, B.; Wu, Z.-F.; Li, J.; Wang, G.-X. Single and joint action toxicity of heavy metals on early developmental stages of Chinese rare minnow (*Gobiocypris rarus*). *Ecotoxicol. Environ. Saf.* **2011**, *74*, 2193–2202. [CrossRef] [PubMed]
60. Mebane, C.A.; Dillon, F.S.; Hennessy, D.P. Acute toxicity of cadmium, lead, zinc, and their mixtures to stream-resident fish and invertebrates. *Environ. Toxicol. Chem.* **2012**, *31*, 1334–1348. [CrossRef] [PubMed]
61. Liu, Z.; Gao, W.; Yu, Y.; Hu, B.; Xin, J.; Sun, Y.; Wang, L.; Wang, G.; Bi, X.; Zhang, G. Characteristics of $PM_{2.5}$ mass concentrations and chemical species in urban and background areas of China: Emerging results from the CARE-China network. *Atmos. Chem. Phys.* **2018**, *18*, 8849–8871. [CrossRef]
62. SSY. Table 6–15: Consumption and Composition of End-Use Energy in 2013; Shanxi Statistics Yearbook. 2014. Available online: http://tjj.shanxi.gov.cn/tjsj/tjnj/nj2014/html/njcx.htm (accessed on 25 June 2022).
63. SSY. Table 6–13: Consumption and Composition of End-Use Energy in 2014; Shanxi Statistics Yearbook. 2015. Available online: http://tjj.shanxi.gov.cn/tjsj/tjnj/nj2015/indexch.htm (accessed on 25 June 2022).
64. SSY. Table 6–13: Consumption and Composition of End-Use Energy in 2015; Shanxi Statistics Yearbook. 2016. Available online: http://tjj.shanxi.gov.cn/tjsj/tjnj/nj2016/indexch.htm (accessed on 25 June 2022).
65. SSY. Table 6–13: Consumption and Composition of End-Use Energy in 2016; Shanxi Statistics Yearbook. 2017. Available online: http://tjj.shanxi.gov.cn/tjsj/tjnj/nj2017/indexch.htm (accessed on 25 June 2022).
66. SSY. Table 6–13: Consumption and Composition of End-Use Energy in 2017; Shanxi Statistics Yearbook. 2018. Available online: http://tjj.shanxi.gov.cn/tjsj/tjnj/nj2018/indexch.htm (accessed on 25 June 2022).
67. SSY. Table 6–13: Consumption and Composition of End-Use Energy in 2018; Shanxi Statistics Yearbook. 2020. Available online: http://tjj.shanxi.gov.cn/tjsj/tjnj/nj2019/zk/indexch.htm (accessed on 25 June 2022).
68. SSY. Table 6–14: Consumption and Composition of End-Use Energy in 2019; Shanxi Statistics Yearbook. 2020. Available online: http://tjj.shanxi.gov.cn/tjsj/tjnj/nj2020/zk/indexch.htm (accessed on 25 June 2022).
69. SSY. Table 6–13: Consumption and Composition of End-Use Energy in 2020; Shanxi Statistics Yearbook. 2021. Available online: http://tjj.shanxi.gov.cn/tjsj/tjnj/nj2021/zk/indexch.htm (accessed on 25 June 2022).

Article

Pollution Characteristics, Source Apportionment, and Health Risk Assessment of Potentially Toxic Elements (PTEs) in Road Dust Samples in Jiayuguan, Hexi Corridor, China

Kai Xiao [1], Xiaoqing Yao [1], Xi Zhang [1], Ning Fu [1,2], Qiuhong Shi [1], Xiaorui Meng [1] and Xuechang Ren [1,*]

[1]   College of Environmental and Municipal Engineering, Lanzhou Jiaotong University, Lanzhou 730070, China
[2]   Analysis and Testing Center, Gansu Province Environmental Monitoring Center, Lanzhou 730020, China
*   Correspondence: Rxchang1698@mail.lzjtu.cn

Abstract: The sources of potentially toxic elements (PTEs) in road dust are complex and potentially harmful to humans, especially in industrial cities. Jiayuguan is the largest steel-producing city in Northwest China, and this study was the first to conduct a related study on PTEs in road dust in this city, including the pollution characteristics, source apportionment, and health risk assessment of PTEs in road dust. The results showed that the highest concentration of PTEs in the local road dust samples were Mn, Ba, Zn, and Cr. The enrichment factor (EF) of Se was the highest, and it was "Very high enrichment" in areas other than the background area, indicating that the local Se was more affected by human activities. The geoaccumulation index (I_{geo}) of Se was also the highest, and the pollution level was 5 in all areas except the background area, indicating that the local Se was more polluted and related to coal combustion. The sources of PTEs in local road dust samples mainly included geogenic-industrial sources, coal combustion, traffic sources, and oil combustion. For the non-carcinogenic risk, the hazard index (HI) of each element of children was higher than that of adults, and the sum of the HI of each element was greater than 1, indicating that there was a non-carcinogenic risk under the combined influence of multiple elements, which was especially obvious in industrial areas. For the carcinogenic risk, the cancer risk (CR) of Cr at a certain point in the industrial area exceeded 10^{-4}, which was a carcinogenic risk, and the Cr in this area may be related to the topsoil of the local abandoned chromate plant.

Keywords: Jiayuguan; road dust; potentially toxic elements; pollution characteristics; source apportionment; health risk assessment

Citation: Xiao, K.; Yao, X.; Zhang, X.; Fu, N.; Shi, Q.; Meng, X.; Ren, X. Pollution Characteristics, Source Apportionment, and Health Risk Assessment of Potentially Toxic Elements (PTEs) in Road Dust Samples in Jiayuguan, Hexi Corridor, China. *Toxics* **2022**, *10*, 580. https://doi.org/10.3390/toxics10100580

Academic Editor: Ilaria Guagliardi

Received: 26 August 2022
Accepted: 26 September 2022
Published: 30 September 2022

Publisher's Note: MDPI stays neutral with regard to jurisdictional claims in published maps and institutional affiliations.

1. Introduction

Since human civilization entered the 21st century, with the activities and development of industry, the pollutants produced by human activities have increased day by day, which has brought certain challenges to the living environment and health of human beings. Inhalable particulate matter (PM_{10}) is particulate matter with a particle size of less than 10 μm in the air, and consists of organic and inorganic (trace metals, cations, and anions) species [1]. Long-term exposure to high concentrations of PM_{10} can cause certain harm to the human body, such as increasing the incidence of tuberculosis and causing circulatory diseases [2,3].

Road dust is composed of a mixture of particulate matter from various natural and man-made sources and contains certain harmful materials. It is one of the sources of PM_{10} and one of the important pollutants that causes urban environmental pollution [4–6]. The source of road dust is complex, and it is the carrier of various pollutants, and human activities have a great impact on the components of road dust. For example, under stable conditions, metallurgical dust, vehicle exhaust (non-exhaust particulate components), coal dust, and other particulate pollutants are mixed into road dust through gravity sedimentation and become a part of it, which leads to complex chemical components of road dust, including various ions, heavy metals and organics, etc. [7,8]. However, under the action of motor vehicles, pedestrians, and wind, low particle size road dust is usually suspended in the air again, causing the secondary pollution of particulate matter [9,10]. Therefore, the problem of road dust pollution has attracted public attention and has gradually become a research hotspot in urban pollution problems.

Potentially toxic elements (PTEs) have certain potential toxicity and can persist in dust, water, soil, and other media. Although some PTEs are beneficial to the organism at low concentrations and are essential trace elements, they are toxic when the concentration exceeds the tolerated dose of the organism, such as Cu, Zn, and Se [11,12]. Relevant studies have shown that PTEs in road dust have negative effects on human health, such as causing poisoning, cancer, respiratory diseases, and even psychological diseases [13–15]. In the road dust in different areas of the city, the content of PTEs usually shows a certain spatial difference due to the geographical location [16,17], which makes the impact of urban road dust on human health particularly complex. In the past few years, many scholars have done a lot of related research on PTEs in road dust, such as source apportionment, influencing factors, health risk assessment, and component characteristics [18–20]. The research sites of these studies were mostly in major cities in the world, but there were a few relevant research cases in the desert areas of northwest China, especially for industrial cities in the desert areas of northwest China, and a few reports have been reported in recent years.

Jiayuguan (98°17′ E, 39°47′ N) is located in the Hexi Corridor of Gansu Province, China, and is the largest steel production base in Northwest China. The local industry is dominated by metallurgical industry and material processing industry, supplemented by tourism. As one of the important industrial cities in the northwest China, road dust in this region is greatly affected by industrial emissions. According to the results of field investigation in this area, there were many chemical enterprises in Jiayuguan a few years ago, such as chromate plants and cement plants. Although these factories have been shut down one after another in recent years, the topsoil of the waste factories still contains a large amount of PTEs [21], which undoubtedly has a certain impact on the concentration of PTEs in the local road dust, and poses potential harm to the health of local residents. However, there are no relevant research reports on road dust in Jiayuguan, especially for PTEs in road dust in this area. For industrial cities, research on PTEs in road dust is of great significance. Such research can not only provide a scientific basis for local pollution prevention and control, but also provide a reference for local health departments to regulate relevant policies to ensure public health. Therefore, based on the above situation, the purpose of this study is as follows: (1) To collect road dust samples (16 samples in total) in each functional area and measure the concentration of PTEs to analyze the pollution characteristics of PTEs in road dust; (2) To determine the pollution levels and identifying sources of PTEs in road dust; and (3) Assess the health risks of PTEs in road dust.

2. Materials and Methods

2.1. Collection of Road Dust Samples and Chemical Analysis of PTEs

Related studies in recent years showed that the urbanization rate of Jiayuguan was extremely high, exceeding 90% [22]. Therefore, this study focused on road dust in the urban area of Jiayuguan. In this study, the urban area was divided into industrial area, commercial area, residential area, and background area. Among the four functional areas, four sampling points were set up near the main road in each area, and the distribution of sampling points is shown in Figure 1. For the industrial area, the vehicles in this area were mainly large diesel vehicles, and the sampling point was close to Jiabei Industrial Park and "Jiuquan Iron & Steel (Group) Co., Ltd. (JISCO)" (Gansu, China), the largest iron and steel producer in the northwest China. For the commercial area, the vehicles in this area were mainly private cars, taxis, and buses, with a high traffic flow, and there were many shopping malls and various stores near the sampling point. For the residential area, the vehicles in this area were mainly private cars and buses, and there were many apartments and residential houses near the sampling point. For the background area, the area was far from the city center and was less affected by human activities.

In the winter of 2020, road dust samples were collected on the main roads around each sampling point. During sampling, brushes and dustpans were used to collect road dust samples, and 4–6 samples were collected around each sampling point and mixed. The mass of each sample was greater than 100 g. After the sample collection was completed, the samples were sealed with polyethylene bags, and the samples were promptly transferred to the laboratory to dry naturally. Then, 20-mesh (0.85-mm) and 100-mesh (0.15-mm) nylon sieves were used in sequence to remove the broken leaves, hair, stones, and other parts that are difficult to resuspend in the road dust. After the sample preparation was completed, the samples were stored under 4 °C and protected from light before chemical analysis.

Referring to the research results of other scholars [17,23], this study selected 12 PTEs (V, Cr, Mn, Co, Ni, Cu, Zn, As, Se, Cd, Ba, and Pb) with high pollution potential for chemical analysis. The sample digestion method is as follows: (1) Put 0.1 g of the sieved sample into a 50 mL Teflon crucible (Dechuang, Huizhou, China), add 5 mL of HNO_3 (Aladdin, Shanghai, China), and digest on an electric hot plate (Keheng, Shanghai, China) at 120 °C for 5 min. (2) Add 2 mL H_2O_2 (Damao, Tianjin, China) to the crucible, raise the temperature to 160 °C for digestion for 5 min. (3) Add 2 mL of HF (Aladdin, Shanghai, China) to the crucible, raise the temperature to 180 °C for digestion for 5 min. (4) Remove the Teflon crucible from the electric hot plate, add 5 mL HNO_3, 3 mL HF, 3 mL $HClO_4$ (Xinyuan, Tianjin, China) in turn after cooling, cover the crucible, and digest on the electric hot plate at 180 °C for 1 h. (5) Remove the lid, wait for the solution to evaporate to nearly dryness, dilute it with ultrapure water in a 50 mL volumetric flask. After the samples were digested, the concentrations of 12 PTEs were measured by inductively coupled plasma mass spectrometer (ICP-MS) (7900, Agilent, Santa Clara, CA, USA). Strict quality control was implemented during the experiment. Blank samples, reference material of soil (GBW07446), and duplicate samples were used for quality assurance, and the pretreatment method was the same as that of road dust samples. In this study, no metal was measured in the blank sample, and the relative standard deviation of the duplicate samples was less than 2%. Information on the limits of detection and quantification is shown in Table S1.

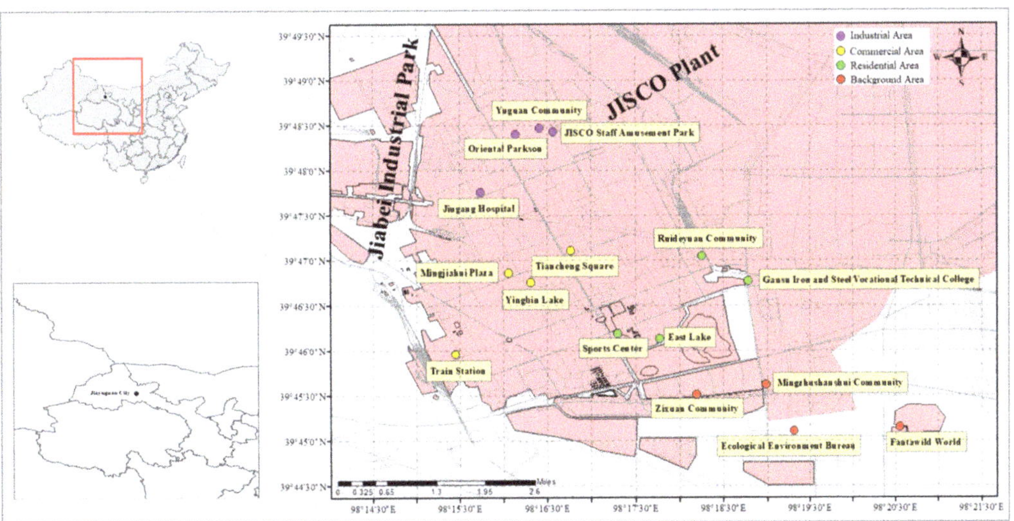

Figure 1. Distribution of sampling points.

2.2. Methods of Data Analysis

2.2.1. Enrichment Factor (EF)

In the related studies of PTEs, Enrichment Factor (EF) is usually used to analyze the enrichment of PTEs and evaluate the impact of human activities on PTEs [24], and was calculated as follows:

$$EF_i = \frac{(C_i/C_n)_{road\ dust}}{(C_i/C_n)_{soil}} \tag{1}$$

In the formula, EF_i is the enrichment factor of element i; C_i is the concentration of element i (mg·kg^{-1}); C_n is the concentration of reference element n (mg·kg^{-1}). For the selection of reference elements, elements that have high concentration values in the environment and do not have any antagonistic and synergistic effects on the measured elements are usually selected, such as Mn, Al, Fe, and Sc [25,26]. Considering that the production process of the local metallurgical industry may contributed a certain amount of Fe and Al to the environment, Mn was selected as the reference element in this study. The background value of soil elements in Gansu Province [27] was selected as the concentration value of each element in the soil, and detailed information is shown in Table S2. The value of EF corresponds to the five pollution categories [28,29], as shown in Table 1.

Table 1. The value of EF and the pollution category.

Value of EF	Pollution Category
(0, 2)	Deficiency to minimal enrichment
[2, 5)	Moderate enrichment
[5, 20)	Significant enrichment
[20, 40)	Very high enrichment
[40, +∞)	Extremely high enrichment

2.2.2. Index of Geoaccumulation (I_{geo})

An index of geoaccumulation is usually used to evaluate the pollution level of PTEs in an area [30]. It has been widely used in the study of sediment, soil, particulate matter, river, etc. [31–33]. The calculation formula is as follows:

$$I_{geo} = \log_2\left(\frac{C_n}{k \times B_n}\right) \tag{2}$$

In the formula, C_n is the concentration of element n in the sample (mg·kg^{-1}); B_n is the background value of element n in the soil (mg·kg^{-1}); k is the correction coefficient, usually 1.5. The value of the geoaccumulation index corresponds to the six pollution categories [28], as shown in Table 2.

Table 2. The value of I_{geo} and the pollution category.

Value of I_{geo}	Level	Pollution Category
[−∞, 0)	0	Unpolluted
[0, 1)	1	Unpolluted to moderately polluted
[1, 2)	2	Moderately polluted
[2, 3)	3	Moderately to strongly polluted
[3, 4)	4	Strongly polluted
[4, 5)	5	Strongly to extremely polluted
[5, +∞)	6	Extremely polluted

2.2.3. Principal Component Analysis (PCA)

Principal component analysis (PCA) can reduce the dimensionality of the data, the principle of this method is to try to recombine the original variables into a new set of several comprehensive variables that are independent of each other, and according to actual needs, less comprehensive variables can be taken out to reflect as much information of the original variables as possible. At present, PCA is widely used in source apportionment studies of pollutants in various media, such as road dust, particulate matter, and soil [34,35]. In this study, PCA was used to determine the source of PTEs in road dust, and the SPSS (Statistical Product Service Solutions) software was used for the calculation of PCA.

2.2.4. Health Risk Assessment

In this study, the EPA health risk assessment model was used to assess the health risk of PTEs in road dust [36]. The model can separately calculate the health risks of PTEs entering the human body through ingestion, inhalation, and dermal contact [37,38]. The average daily dose (ADD) for the three exposure pathways is calculated as follows:

$$ADD_{ing} = \frac{c \times IngR \times EF \times ED \times CF}{BW \times AT} \tag{3}$$

$$ADD_{inh} = \frac{c \times InhR \times EF \times ED}{PEF \times BW \times AT} \tag{4}$$

$$ADD_{dermal} = \frac{c \times SA \times AF \times ABF \times EF \times ED \times CF}{BW \times AT} \tag{5}$$

In this formula, ADD_{ing}, ADD_{inh}, and ADD_{dermal} are the average daily doses (mg·kg^{-1}·d^{-1}) of PTEs entering the human body through ingestion, inhalation, and dermal contact, respectively. The information of other parameters is shown in Table 3.

For the calculation of non-carcinogenic risk and carcinogenic risk, the calculation formula is as follows:

$$HQ_{ij} = \frac{ADD_{ij}}{RfD_i} \tag{6}$$

$$HI_i = \sum HQ_{ij} \tag{7}$$

$$CR_{ij} = ADD_{ij} \times SF_i \tag{8}$$

$$CR_i = \sum CR_{ij} \tag{9}$$

In these formulas, i is the PTE; j is the pathway of entry into the human body (ingestion, inhalation, dermal contact); HQ is the hazard quotient; HI is the hazard index; CR is the cancer risk; RfD is the reference dose of the PTE; SF is the slope factor. The information of RfD and SF is shown in Table 4. If $HI<1$, there is no obvious non-carcinogenic risk, and if $HI > 1$, there is a non-carcinogenic risk. If $CR < 10^{-6}$, the carcinogenic risk is negligible, if the CR is between 10^{-6} and 10^{-4}, the carcinogenic risk is at an acceptable level, and if $CR > 10^{-4}$, there is a carcinogenic risk [39].

Table 3. Exposure parameters for health risk assessment model.

Parameter	Meaning	Unit	Children	Adult	Source
IngR	Ingestion rate	mg·d^{-1}	200	100	
InhR	Inhalation rate	mg·d^{-1}	5	20	
EF	Exposure frequency	d·a^{-1}	180	180	
ED	Exposure duration	a	6	24	
BW	Body weight	kg	15	70	
PEF	Particle emission factor	m^3·kg^{-1}	1.32×10^9	1.32×10^9	[25,40–44]
AT (carcinogen)	Average time	d	365·70	365·70	
AT (non-carcinogen)	Average time	d	365·ED	365·ED	
SA	Skin surface area	cm^2	2800	5700	
AF	Skin adherence factor	mg·(cm^2·d)$^{-1}$	0.2	0.7	
ABF	Absorption factor	/	0.001	0.001	
CF	Conversion factor	kg/mg	1×10^{-6}	1×10^{-6}	

Table 4. RfD and SF value of different elements for different exposure routes.

PTEs	RfDing	RfDinh	RfDdermal	SF$_{ing}$	SF$_{inh}$	SF$_{dermal}$
V	7.00×10^{-3}	1.40×10^{-5}	7.00×10^{-5}	-	-	-
Mn	4.66×10^{-2}	1.43×10^{-5}	1.84×10^{-3}	-	-	-
Pb	3.50×10^{-3}	3.50×10^{-3}	5.25×10^{-4}	-	-	-
Cu	3.70×10^{-2}	4.00×10^{-2}	1.20×10^{-2}	-	-	-
Zn	3.00×10^{-1}	3.00×10^{-1}	6.00×10^{-2}	-	-	-
Se	5.00×10^{-3}	5.70×10^{-5}	2.20×10^{-3}	-	-	-
Ba	2.00×10^{-1}	1.40×10^{-2}	1.43×10^{-4}			
Cd	1.00×10^{-3}	1.00×10^{-3}	1.00×10^{-5}	0.38	6.30	3.00
Co	3.00×10^{-3}	6.00×10^{-5}	2.80×10^{-5}	-	9.80	-
As	3.00×10^{-4}	1.23×10^{-4}	1.23×10^{-4}	1.50	15.10	3.66
Cr	3.00×10^{-3}	2.85×10^{-5}	6.00×10^{-5}	0.50	42.00	20.00
Ni	2.00×10^{-2}	2.06×10^{-2}	5.40×10^{-3}	0.84	1.70	42.50
Source			[25,44,45]			

3. Results and Discussion

3.1. Concentration Characteristics of PTEs in Road Dust

The concentration of PTEs in road dust samples in each area is shown in Figure 2. The mean total metal concentrations in the road dust samples decreased in the order: Mn > Ba > Zn > Cr > Pb > Cu > Ni > V > Co > As > Se > Cd. Mn is one of the elements widely distributed in the crust, and Cr Zn are mainly related to industrial sources. According to related research, the average content of Ba in Chinese coal (159 mg·kg^{-1}) exceeded the world average (150 mg·kg^{-1}), and the content of Ba in coal ash was about 4.2% [46], so coal combustion is one of the potential sources of Ba in the environment.

From the perspective of different areas, the concentration of PTEs in the industrial area was generally higher than that in the other three areas. The reason for this phenomenon was that there were a large number of metallurgical plants and chemical enterprises in the industrial area. These enterprises emitted a large amount of flue gas and particulate matter during the production process, and part of the flue gas and particulate matter settled to the surface under the action of gravity and mixed into the road dust, thereby increasing the concentration of PTEs in road dust in industrial areas, such as As and Se in coal combustion [47,48], and Cr, Cd, Cu, and Zn in metallurgy [49,50]. For the commercial area, the concentrations of Pb and Ni in the road dust in the commercial area were the highest among the four areas, and vehicle exhaust emissions were one of the potential sources of Pb and Ni [50,51]. Therefore, the phenomenon was caused by the emission of a large number of private cars and taxis in the commercial area. The concentrations of PTEs in road dust in residential area and background area were generally lower than those in industrial areas and commercial areas. Compared with the industrial area and the commercial area, the residential area and the background area are farther away from the industrial enterprises, with less traffic flow and less pollution.

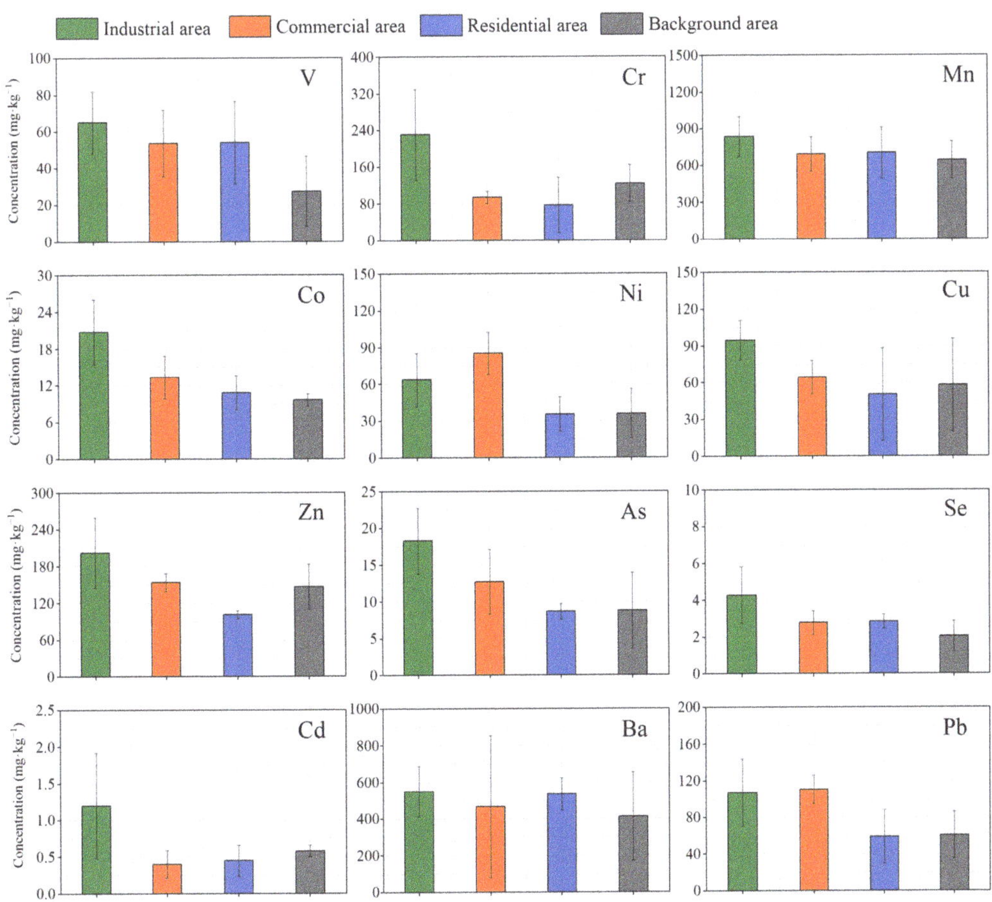

* The data in the figure represent the "Mean ± SD"

Figure 2. Concentration of PTEs in each area.

3.2. Enrichment Characteristics of PTEs in Road Dust

The Enrichment Factor (EF) of PTEs in road dust samples is shown in Figure 3. From the EF value, the EF of Se was the highest, and it was more than 20 in all areas except the background area, which was "Very high enrichment". The EF of Cd in the industrial area was greater than 5, which was "Significant enrichment", and the EF in the other three areas was between 2 and 5, which was "Moderate enrichment". The EF of Pb in the commercial area was greater than 5, which was "Significant enrichment", and the EF in the other three areas was between 2 and 5, which was "Moderate enrichment". Except for Se, Cd, and Pb, the EF of other PTEs in all areas were lower than 5, which were "Deficiency to minimal enrichment" or "Moderate enrichment" and were less affected by human activities.

From the perspective of various areas, the EF of Cr, Co, Cu, Zn, As, Se, and Cd in industrial areas were higher than those in other areas. The EF of Ni and Pb in commercial area were the highest, and the EF of these two elements were generally lower in other areas. According to the actual situation of each area, most of the industrial areas were metallurgical plants and chemical enterprises, and the commercial areas had high traffic flow. Therefore, compared to residential area and background area, industrial area and commercial area were more affected by human activities, especially vehicle exhaust emissions and the particulate matter emissions during industrial production and coal burning, which greatly increased the content of PTEs in road dust and increased the EF [52,53].

The EF of Se in road dust in Jiayuguan was the highest. Compared with Huainan [18] and Chengdu [54], which are both Chinese cities, the concentration of Se in road dust in Jiayuguan was higher than that in Huainan (1.200 mg·kg^{-1}) and Chengdu (0.373 mg·kg^{-1}), and the EF was much higher than that of Huainan (0.5), which is also an industrial city. Due to the large scale of steel production in Jiayuguan and the huge consumption of coal, a large amount of particulate matter was emitted in the process of burning coal, and part of it became part of the road dust through gravity sedimentation. In addition, unlike Huainan and Chengdu, Jiayuguan is located in northern China and is a typical northern city with a heating period. Therefore, in winter, the central heating of Jiayuguan increased the consumption of coal, and due to the limitation of the heating system, some villages and towns around the urban area could not obtain municipal heating supply. The residents of these villages and towns chose to burn coal for heating, which also increased the pollution caused by coal combustion and the EF of Se.

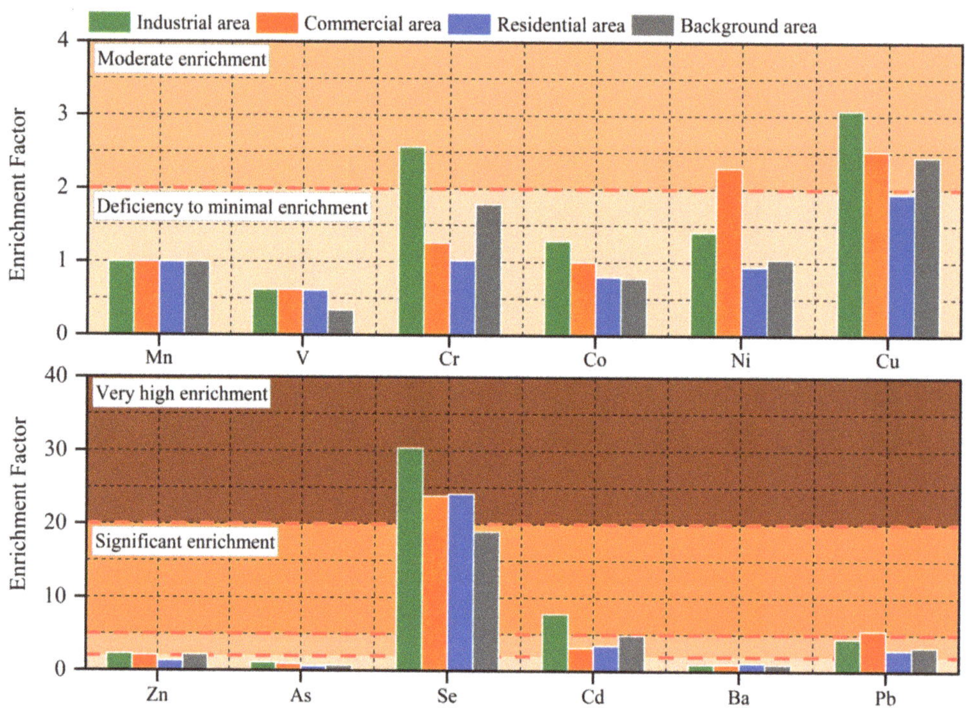

Figure 3. EF of PTEs in road dust in various areas.

3.3. The Pollution Level of Each Area

The geoaccumulation index (I_{geo}) of PTEs in road dust samples in various areas is shown in Figure 4. Among all PTEs, Se had the highest I_{geo} and was greater than 4 in all areas except the background area, and the pollution level was 5, which was "strong to extremely polluted". The I_{geo} of Cd in the industrial area was between 2 and 3, and the pollution level was 3, which was "moderately to strongly pollution". Except for Se and Cd, the I_{geo} of other PTEs in each area was less than 2, and the pollution level was below level 2. It is worth noting that the I_{geo} of Mn, V, As, and Ba were all less than 0, unpolluted. Similar to the enrichment factor, except for Ni and Pb, the I_{geo} of other PTEs in the industrial area was the highest in all areas, which indicated that the pollution in the industrial area was the most serious. For the commercial area, the I_{geo} of Ni and Pb in the commercial area was the highest, Ni and Pb mainly came from the exhaust emissions of motor vehicles [55,56], and could be mixed with the particulate matter into the road dust, so the result was in line with the characteristics of the large traffic flow in the commercial area.

According to the actual situation of industrial production in Jiayuguan, a large amount of coal was used in the production process of the local metallurgical industry, and the large amount of coal consumption increased the emission of elements such as Se and As [57,58]. Therefore, a large amount of particulate matter was emitted during the burning of coal, and this particulate matter contained a lot of Se. Additionally, the sedimentation to the surface increased the concentration of Se on the surface and increased the pollution of road dust. In addition, due to the limitation of the heating system, the residents of villages and towns around the Jiayuguan urban area used coal for heating in the winter, which also increased the emission of Se in the environment.

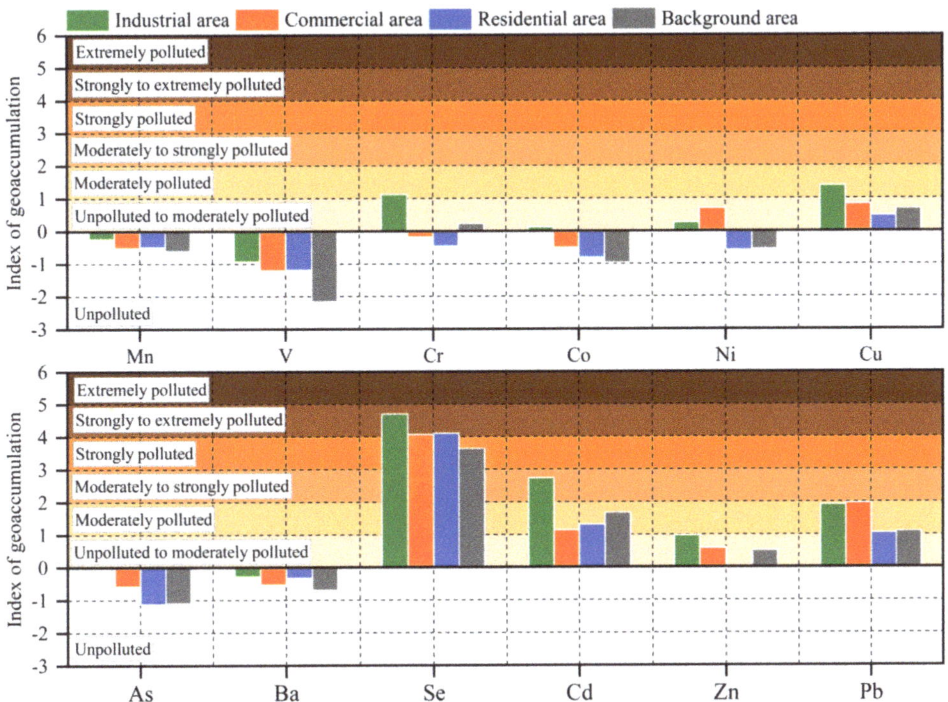

Figure 4. I_{geo} of PTEs in road dust in various areas.

3.4. Source Apportionment of PTEs

Considering the limited number of samples in each area, in the process of principal component analysis, this study combined the sample data of four areas to determine the source of PTEs in road dust samples, and finally reflected the overall situation of road dust in Jiayuguan. The result is shown in Table 5. According to the result of PCA, a total of four components with eigenvalue greater than 1 were identified, and the cumulative variance contribution rate reached 74.67%.

PC1 explained 28.88% of the total variance, with higher loading for Cr, Co, and Cd. Although related research showed that Cr and Cd are related to emissions in metallurgical industrial production activities [17], some studies have shown that Co comes from iron ore and can be used as a tracer of particulate matter emitted by steel production [51]. However, according to the analysis results of the enrichment factor and geoaccumulation index in this study, the EF of Cr in all areas except industrial areas was less than 2, which was "Deficiency to minimal enrichment", and only showed "Moderately polluted" in industrial areas. The EF of Co in all areas was less than 2, which was "Deficiency to minimal enrichment", and studies have shown that Co is mainly from natural sources [59]. After the above analysis, the main source of Cr and Co in PC1 was considered to be a geogenic source. Therefore, the main source of PC1 was the geogenic source, followed by the industrial source.

PC2 explained 20.86% of the total variance, with higher loading for Mn, Ni, As, Se, and Pb. Among the elements with high loading, As and Se are representative elements of coal combustion [57,58]. Coal combustion is one of the sources of Mn in the environment. In addition to being related to vehicle exhaust emissions [55,56], Pb and Ni also come from the process of coal combustion [53,60]. Therefore, PC2 was identified as a coal combustion.

PC3 explained 13.35% of the total variance, with higher loading for Zn and As. Zn mainly comes from the wear of motor vehicle parts and exhaust emissions [1,53,61]. Usually, As is considered to be the representative element of coal combustion, but some studies have shown that As also comes from vehicle exhaust emissions [62]. Therefore, PC3 was identified as the traffic source.

PC4 explained 11.59% of the total variance, with higher loading for V and Ba. V is recognized as a tracer for oil combustion [63], and Ba is associated with diesel consumption emissions from diesel vehicles [64]. Therefore, PC4 was identified as oil combustion.

PTEs in road dust samples in Jiayuguan mainly came from geogenic-industrial sources, coal combustion, traffic sources, and oil combustion. The sources of road dust were complex, which was a collection of various particulate pollutants. According to the results of source apportionment, PTEs in road dust in Jiayuguan were mainly related to human activities. As the largest steel manufacturing city in Northwest China, the industrial production process of Jiayuguan was accompanied by huge consumption of coal and oil. Therefore, a large amount of particulate matter was emitted during industrial production and motor vehicle transportation. These particles contained a large amount of PTEs, and finally settled to the surface by gravity and mixed into the road dust.

Table 5. Principal component analysis of overall road dust in Jiayuguan.

Elements	Principal Components			
	PC1	PC2	PC3	PC4
V	0.433	0.415	0.004	0.532
Cr	0.893	−0.076	0.197	0.154
Mn	0.300	0.714	−0.271	0.293
Co	0.765	0.344	0.185	−0.294
Ni	−0.147	0.850	0.182	−0.048
Cu	0.450	0.012	0.279	−0.764
Zn	0.057	−0.048	0.929	0.012
As	0.242	0.463	0.613	0.084
Se	0.694	0.558	−0.026	−0.170
Cd	0.903	0.124	−0.056	−0.142
Ba	−0.058	−0.041	0.302	0.512
Pb	0.469	0.657	0.098	−0.081
Eigenvalue	3.47	2.50	1.60	1.39
Variance contribution (%)	28.88	20.86	13.35	11.59
Cumulative variance contribution (%)	28.88	49.74	63.09	74.67

3.5. Health Risk Assessment of PTEs in Road Dust

3.5.1. Non-Carcinogenic Risk Assessment

The hazard quotient (HQ) for each category of ingestion is shown in Table 6. From the perspective of different groups, the HQ_{ing} and HQ_{inh} of children were higher than that of adults, and the HQ_{dermal} of adults was higher than that of children. This phenomenon is similar to the research results of some scholars [18,23,53]. Compared with adults, children have a weaker constitution, so they have lower resistance to PTEs pollution in the environment, especially for inhalation and ingestion. For the dermal contact route, the skin surface area (SA) and skin adherence factor (AF) of adult are higher than those of children. Due to the large difference in these parameters, and AF having a large influence on the intake of PTEs [65,66], the calculation result of HQ usually presents as "HQ_{dermal} (adult) > HQ_{dermal} (children)". For each element, the HQ of each element for various exposure routes was less than 1, and there was no obvious non-carcinogenic risk for a single exposure route corresponding to a single element.

Figure 5 reflects the HI_i of each sampling point and the sum of the HI_i ($\sum HI_i$) of each area. For children, the non-carcinogenic risk of Cr and As was more obvious, the HI of Cr and As in industrial area was close to 1, and the non-carcinogenic risk was relatively high. The HI of each element for adults was lower than 1, and each element had no obvious non-carcinogenic risk. From the perspective of different areas, the HI of each element in the industrial area was relatively higher than other areas, which indirectly reflected the actual situation of serious pollution in the industrial area. The $\sum HI_i$ of each element was calculated to reflect the comprehensive impact of all elements on the non-carcinogenic risk of each group. The $\sum HI_i$ of adults in the four areas was less than 1, indicating no non-carcinogenic risk. For children, the $\sum HI_i$ of all points in the industrial area was greater than 1, the $\sum HI_i$ of three points in the commercial area was greater than 1, the $\sum HIi$ of all points in the residential area was less than 1, and the $\sum Hii$ of two points in the background area was greater than 1.

To sum up, the non-carcinogenic risk of PTEs in road dust in Jiayuguan was higher for children than for adults, and except for residential areas, PTEs, at some points in the rest of the area, present non-carcinogenic risks to local children. Therefore, it is suggested that the local environmental protection and health departments pay attention to the pollution of PTEs in road dust, especially the emission of PTEs in industrial areas, which should receive strengthened supervision.

Table 6. HQ_{ij} for different intake route.

Elements	Children			Adult		
	HQ_{ing}	HQ_{inh}	HQ_{dermal}	HQ_{ing}	HQ_{inh}	HQ_{dermal}
V	4.70×10^{-2}	4.45×10^{-4}	1.31×10^{-2}	5.03×10^{-3}	3.81×10^{-4}	2.01×10^{-2}
Cr	2.88×10^{-1}	5.73×10^{-4}	4.03×10^{-2}	3.08×10^{-2}	4.92×10^{-4}	6.15×10^{-2}
Mn	1.01×10^{-1}	6.25×10^{-3}	7.19×10^{-3}	1.09×10^{-2}	5.36×10^{-3}	1.10×10^{-2}
Co	2.99×10^{-2}	2.83×10^{-5}	8.97×10^{-3}	3.20×10^{-3}	2.43×10^{-5}	1.37×10^{-2}
Ni	1.81×10^{-2}	3.33×10^{-7}	1.88×10^{-4}	1.94×10^{-3}	2.86×10^{-7}	2.87×10^{-4}
Cu	1.19×10^{-2}	2.08×10^{-7}	1.03×10^{-4}	1.27×10^{-3}	1.78×10^{-7}	1.57×10^{-4}
Zn	3.31×10^{-3}	6.26×10^{-8}	4.63×10^{-5}	3.54×10^{-4}	5.37×10^{-8}	7.07×10^{-5}
As	2.65×10^{-1}	1.23×10^{-5}	1.81×10^{-3}	2.84×10^{-2}	1.05×10^{-5}	2.77×10^{-3}
Se	3.93×10^{-3}	6.53×10^{-6}	2.50×10^{-5}	4.21×10^{-4}	5.59×10^{-6}	3.82×10^{-5}
Cd	4.32×10^{-3}	8.17×10^{-8}	1.21×10^{-3}	4.62×10^{-4}	7.00×10^{-8}	1.84×10^{-3}
Ba	1.62×10^{-2}	4.37×10^{-6}	6.33×10^{-2}	1.73×10^{-3}	3.75×10^{-6}	9.66×10^{-2}
Pb	1.58×10^{-1}	2.99×10^{-6}	2.95×10^{-3}	1.69×10^{-2}	2.57×10^{-6}	4.50×10^{-3}

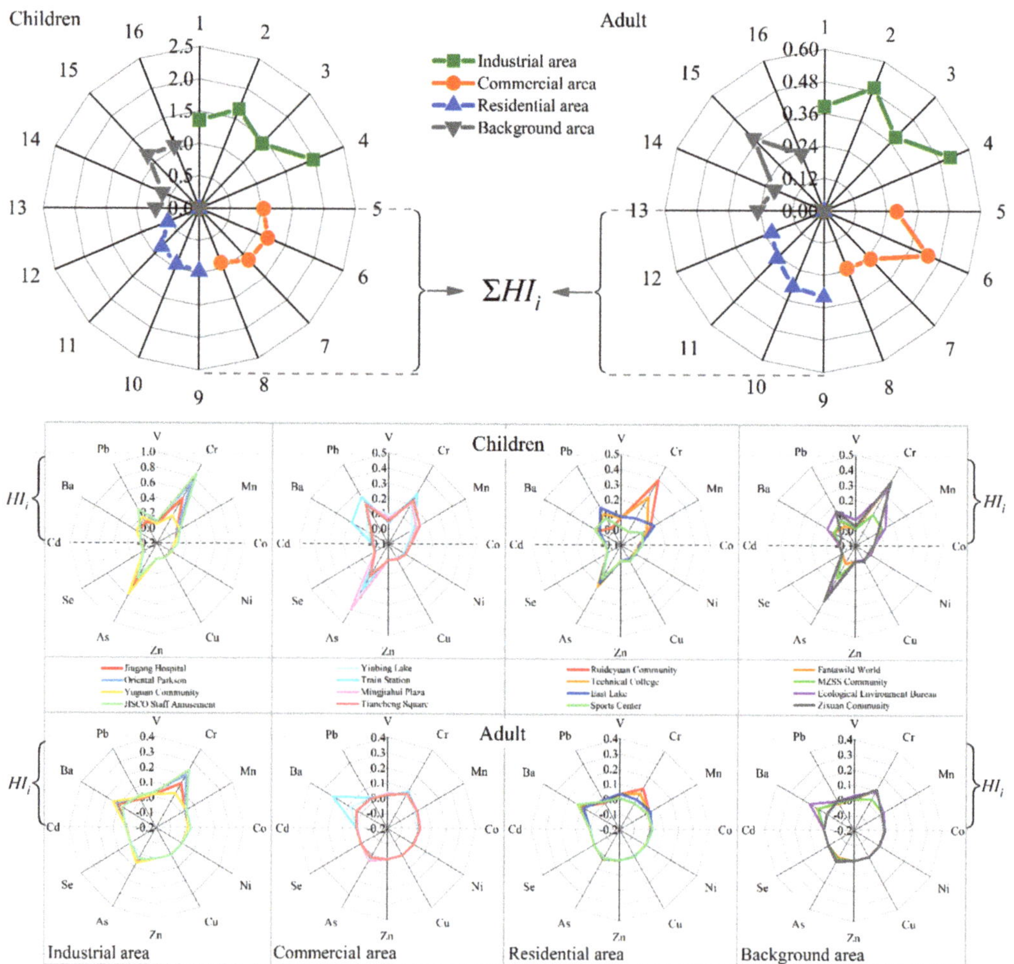

Figure 5. Characterization results for non-carcinogenic risk.

3.5.2. Carcinogenic Risk Assessment

The cancer risk (CR) for each category of intake is shown in Table 7. From the perspective of different groups, the CR_{ing} of children was greater than that of adults, and the CR_{inh} and CR_{dermal} of children were smaller than that of adults, which indicates that children are more susceptible to ingestion pathways than adult. The mean values of CR_{ij} in the four areas were all between 10^{-4} and 10^{-6}, which indicated that the carcinogenic risk of each element under a single intake route was at an acceptable level.

The calculation results and related information of CR_i for children and adults in each area are shown in Table 8. For each element, the mean values of CR of Co and Cd in each area were lower than 10^{-6}, and its carcinogenic risk was negligible. The mean values of CR of Cr, Ni, and As were between 10^{-4} and 10^{-6}, and the carcinogenic risk was at an acceptable level. It is worth noting that in the industrial area, the maximum CR of Cr to children and adults reached 1.09×10^{-4} and 1.10×10^{-4}, respectively, which exceeded 1×10^{-4}, and there was a risk of cancer. Cr can accumulate in the human body and cause a series of gastrointestinal reactions before the lesions, and when too much Cr accumulates in the human body, it may cause lung cancer and gastric cancer [67]. According to the local field investigation results, there were some chromate plants in the "Jiabei industrial park" in the north of the city, and these chromate plants were discontinued a few years ago. The Cr in the topsoil of these historically abandoned factories seriously exceeded the standard, which presents potential harm to the health of the local population. At present, Cr in the topsoil of these abandoned factories has attracted the attention of local environmental protection departments, and soil remediation work has been carried out since 2020. In future research on the local environment, this work deserves continuous attention.

The mean CR of PTEs in Jiayuguan road dust samples was lower than 10^{-4}, and the carcinogenic risk was at an acceptable level or low risk. However, the CR of Cr at individual point in the industrial area was higher than 10^{-4}, and there was still a carcinogenic risk. A potential source of Cr in industrial area was topsoil from abandoned chromate plants. Therefore, it is recommended that the local environmental protection and health departments continue to pay attention to the soil remediation work in this area.

Table 7. CR_{ij} for different intake route.

Elements	Children			Adult		
	CR_{ing}	CR_{inh}	CR_{dermal}	CR_{ing}	CR_{inh}	CR_{dermal}
Cr	3.70×10^{-5}	5.88×10^{-8}	4.14×10^{-6}	1.59×10^{-5}	2.02×10^{-7}	2.53×10^{-5}
Co	-	1.43×10^{-9}	-	-	4.89×10^{-9}	-
Ni	2.61×10^{-5}	1.00×10^{-9}	3.70×10^{-6}	1.12×10^{-5}	3.43×10^{-9}	2.26×10^{-5}
As	1.02×10^{-5}	1.95×10^{-9}	7.00×10^{-8}	4.39×10^{-6}	6.69×10^{-9}	4.27×10^{-7}
Cd	1.41×10^{-7}	4.41×10^{-11}	3.11×10^{-9}	6.02×10^{-8}	1.51×10^{-10}	1.90×10^{-8}

Table 8. CR_i for children and adults in various areas.

Area		Children					Adult				
		Cr	Co	Ni	As	Cd	Cr	Co	Ni	As	Cd
Industrial area	Max	1.09×10^{-4}	3.07×10^{-9}	5.09×10^{-5}	2.17×10^{-5}	5.26×10^{-7}	1.10×10^{-4}	1.05×10^{-8}	5.77×10^{-5}	1.02×10^{-5}	2.90×10^{-7}
	Min	2.76×10^{-5}	1.58×10^{-9}	1.90×10^{-5}	1.20×10^{-5}	1.09×10^{-7}	2.77×10^{-5}	5.42×10^{-9}	2.16×10^{-5}	5.61×10^{-6}	6.05×10^{-8}
	Mean	7.23×10^{-5}	2.17×10^{-9}	3.44×10^{-5}	1.56×10^{-5}	2.63×10^{-7}	7.26×10^{-5}	7.44×10^{-9}	3.90×10^{-5}	7.28×10^{-6}	1.45×10^{-7}
Commercial area	Max	3.62×10^{-5}	1.88×10^{-9}	5.54×10^{-5}	1.53×10^{-5}	1.31×10^{-7}	3.64×10^{-5}	6.46×10^{-9}	6.28×10^{-5}	7.17×10^{-6}	7.26×10^{-8}
	Min	2.53×10^{-5}	8.79×10^{-10}	3.51×10^{-5}	5.45×10^{-6}	2.19×10^{-8}	2.54×10^{-5}	3.01×10^{-9}	3.98×10^{-5}	2.55×10^{-6}	1.21×10^{-8}
	Mean	2.94×10^{-5}	1.40×10^{-9}	4.62×10^{-5}	1.08×10^{-5}	8.76×10^{-8}	2.96×10^{-5}	4.79×10^{-9}	5.24×10^{-5}	5.06×10^{-6}	4.84×10^{-8}
Residential area	Max	4.97×10^{-5}	1.61×10^{-9}	2.93×10^{-5}	8.26×10^{-6}	1.31×10^{-7}	4.99×10^{-5}	5.52×10^{-9}	3.32×10^{-5}	3.86×10^{-6}	7.26×10^{-8}
	Min	4.71×10^{-7}	8.26×10^{-10}	8.43×10^{-6}	6.04×10^{-6}	2.19×10^{-8}	4.73×10^{-7}	2.83×10^{-9}	9.56×10^{-6}	2.83×10^{-6}	1.21×10^{-8}
	Mean	2.41×10^{-5}	1.13×10^{-9}	1.91×10^{-5}	7.39×10^{-6}	9.85×10^{-8}	2.42×10^{-5}	3.88×10^{-9}	2.17×10^{-5}	3.45×10^{-6}	5.44×10^{-8}
Background area	Max	4.86×10^{-5}	1.17×10^{-9}	2.78×10^{-5}	1.24×10^{-5}	1.53×10^{-7}	4.88×10^{-5}	4.02×10^{-9}	3.15×10^{-5}	5.81×10^{-6}	8.47×10^{-8}
	Min	1.72×10^{-5}	8.89×10^{-10}	8.11×10^{-7}	1.28×10^{-6}	1.09×10^{-7}	1.73×10^{-5}	3.05×10^{-9}	9.19×10^{-7}	5.97×10^{-7}	6.05×10^{-8}
	Mean	3.88×10^{-5}	1.01×10^{-9}	1.94×10^{-5}	7.49×10^{-6}	1.26×10^{-7}	3.90×10^{-5}	3.45×10^{-9}	2.20×10^{-5}	3.50×10^{-6}	6.95×10^{-8}

4. Conclusions

Based on the data of PTEs in Jiayuguan road dust samples, this study calculated the enrichment factor and geoaccumulation index of PTEs in Jiayuguan road dust, identified the source of PTEs by principal component analysis, and the health risk of PTEs to different local groups was assessed by the EPA health risk assessment model. The conclusions are as follows.

(1) Among the 12 PTEs (V, Cr, Mn, Co, Ni, Cu, Zn, As, Se, Cd, Ba, and Pb) in road dust samples, the highest concentrations were Mn, Ba, Zn, and Cr, and the concentration of PTEs in industrial areas was higher than the other three functional areas as a whole.

(2) For the EF of PTEs in road dust samples, the EF of Se was the highest, and the EF of Se in industrial area, commercial area, and residential area all exceeded 20, which was "very high enrichment". The EF of Cd in the industrial area exceeded 5, which was "Significant enrichment". The EF of Pb in the commercial area exceeded 5, which was "significant enrichment". The EF of the remaining elements were low, with "Moderate enrichment" or "Deficiency to minimal enrichment " in each area.

(3) The calculation of the I_{geo} showed that for most elements, the pollution level of the industrial area was higher than that of the rest area. For each element, the Igeo of Se was the highest, which was higher than 4 in all areas except the background area, and the pollution

level was 5, which was "Strongly to extremely polluted". The pollution level of Cd in the industrial area was 3, which was "Moderately to strongly polluted". The pollution level of other elements in each area did not exceed level 2.

(4) According to the source apportionment results, the PTEs in Jiayuguan road dust mainly came from geogenic-industrial sources, coal combustion, traffic sources, and oil combustion.

(5) For non-carcinogenic risk, the HI of each element for children was higher than that for adults, and the sum of the HI of each element to children at some points in the industrial area, commercial area and background area exceeds 1, indicating a non-carcinogenic risk. From the perspective of different areas, the HI of the points in the industrial area was higher than that of the other areas, indicating that the non-carcinogenic risk of the industrial area was higher than that of the other areas.

(6) For the carcinogenic risk, the mean CR of each element in each region was less than 10^{-4}, indicating that the carcinogenic risk was at an acceptable level or low risk. However, the CR at a certain point in the industrial area exceeded 10^{-4}, indicating a carcinogenic risk, and the source of Cr in this area may be related to the topsoil of the local abandoned chromate plant.

According to the above conclusions, it is suggested that the local government should pay attention to coal-fired emissions and introduce relevant policies and regulations to reduce the pollution of Se to the environment, and focus on the health of children, especially the health effects of PTEs in industrial areas on children. In addition, the local government should continue to pay attention to the topsoil pollution of the abandoned chromate plant and take relevant measures to reduce the Cr pollution of the topsoil in the area.

Supplementary Materials: The following are available online at https://www.mdpi.com/article/10.3390/toxics10100580/s1, Table S1: Detection limit and quantification limit, Table S2: Background values of soil elements in Gansu Province, China.

Author Contributions: Conception and design of study: K.X., X.R. and X.Y.; Acquisition of data: K.X., X.Z., Q.S. and X.M.; Analysis and interpretation of data: K.X., N.F. and X.Y.; Drafting the manuscript: K.X.; Revising the manuscript critically for important intellectual content: X.R. All authors have read and agreed to the published version of the manuscript.

Funding: This work was supported by the project of Air Pollution Cause Analysis and Countermeasure Research of Jiayuguan Ecological Environment Bureau (JYGZCDL2019041G). Author Xuechang Ren has received research support from Jiayuguan Ecological Environment Bureau.

Institutional Review Board Statement: Not applicable.

Informed Consent Statement: Not applicable.

Data Availability Statement: All data included in this study are available upon request by contact with the corresponding author.

Conflicts of Interest: The authors declare that they have no competing interests.

References

1. Pant, P.; Harrison, R.M. Critical Review of Receptor Modelling for Particulate Matter: A Case Study of India. *Atmos. Environ.* **2012**, *49*, 1–12. [CrossRef]
2. Zhang, Y.; Ma, Y.; Feng, F.; Cheng, B.; Wang, H.; Shen, J.; Jiao, H. Association between PM10 and Specific Circulatory System Diseases in China. *Sci. Rep.* **2021**, *11*, 12129. [CrossRef] [PubMed]
3. Xiang, K.; Xu, Z.; Hu, Y.Q.; He, Y.S.; Dan, Y.L.; Wu, Q.; Fang, X.H.; Pan, H.F. Association between Ambient Air Pollution and Tuberculosis Risk: A Systematic Review and Meta-Analysis. *Chemosphere* **2021**, *277*, 130342. [CrossRef] [PubMed]
4. Jayarathne, A.; Egodawatta, P.; Ayoko, G.A.; Goonetilleke, A. Assessment of Ecological and Human Health Risks of Metals in Urban Road Dust Based on Geochemical Fractionation and Potential Bioavailability. *Sci. Total Environ.* **2018**, *635*, 1609–1619. [CrossRef]
5. Wei, B.; Yang, L. A Review of Heavy Metal Contaminations in Urban Soils, Urban Road Dusts and Agricultural Soils from China. *Microchem. J.* **2010**, *94*, 99–107. [CrossRef]
6. Sutherland, R.A.; Tolosa, C.A. Multi-Element Analysis of Road-Deposited Sediment in an Urban Drainage Basin, Honolulu, Hawaii. *Environ. Pollut.* **2000**, *110*, 483–495. [CrossRef]

7. Jose, J.; Srimuruganandam, B. Source Apportionment of Urban Road Dust Using Four Multivariate Receptor Models. *Environ. Earth Sci.* **2021**, *80*. [CrossRef]
8. Shi, D.; Lu, X. Accumulation Degree and Source Apportionment of Trace Metals in Smaller than 63 Mm Road Dust from the Areas with Different Land Uses: A Case Study of Xi'an, China. *Sci. Total Environ.* **2018**, *636*, 1211–1218. [CrossRef]
9. Alshetty, D.; Shiva Nagendra, S.M. Urban Characteristics and Its Influence on Resuspension of Road Dust, Air Quality and Exposure. *Air Qual. Atmos. Health* **2022**, *15*, 273–287. [CrossRef]
10. Beauchemin, S.; Levesque, C.; Wiseman, C.L.S.; Rasmussen, P.E. Quantification and Characterization of Metals in Ultrafine Road Dust Particles. *Atmosphere* **2021**, *12*, 1564. [CrossRef]
11. Adrees, M.; Ali, S.; Rizwan, M.; Ibrahim, M.; Abbas, F.; Farid, M.; Zia-ur-Rehman, M.; Irshad, M.K.; Bharwana, S.A. The Effect of Excess Copper on Growth and Physiology of Important Food Crops: A Review. *Environ. Sci. Pollut. Res.* **2015**, *22*, 8148–8162. [CrossRef] [PubMed]
12. Liu, P.; Zhang, Y.; Feng, N.; Zhu, M.; Tian, J. Potentially Toxic Element (PTE) Levels in Maize, Soil, and Irrigation Water and Health Risks through Maize Consumption in Northern Ningxia, China. *BMC Public Health* **2020**, *20*, 1729. [CrossRef] [PubMed]
13. Wang, X.; Liu, E.; Lin, Q.; Liu, L.; Yuan, H.; Li, Z. Occurrence, Sources and Health Risks of Toxic Metal(Loid)s in Road Dust from a Mega City (Nanjing) in China. *Environ. Pollut.* **2020**, *263*, 114518. [CrossRef] [PubMed]
14. Kim, J.A.; Park, J.H.; Hwang, W.J. Heavy Metal Distribution in Street Dust from Traditional Markets and the Human Health Implications. *Int. J. Environ. Res. Public Health* **2016**, *13*, 820. [CrossRef] [PubMed]
15. Veremchuk, L.V.; Tsarouhas, K.; Vitkina, T.I.; Mineeva, E.E.; Gvozdenko, T.A.; Antonyuk, M.V.; Rakitskii, V.N.; Sidletskaya, K.A.; Tsatsakis, A.M.; Golokhvast, K.S. Impact Evaluation of Environmental Factors on Respiratory Function of Asthma Patients Living in Urban Territory. *Environ. Pollut.* **2018**, *235*, 489–496. [CrossRef] [PubMed]
16. Alves, C.A.; Evtyugina, M.; Vicente, A.M.P.; Vicente, E.D.; Nunes, T.V.; Silva, P.M.A.; Duarte, M.A.C.; Pio, C.A.; Amato, F.; Querol, X. Chemical Profiling of PM10 from Urban Road Dust. *Sci. Total Environ.* **2018**, *634*, 41–51. [CrossRef]
17. Krupnova, T.G.; Rakova, O.V.; Gavrilkina, S.V.; Antoshkina, E.G.; Baranov, E.O.; Yakimova, O.N. Road Dust Trace Elements Contamination, Sources, Dispersed Composition, and Human Health Risk in Chelyabinsk, Russia. *Chemosphere* **2020**, *261*, 127799. [CrossRef]
18. Liu, Y.; Liu, G.; Yousaf, B.; Zhou, C.; Shen, X. Identification of the Featured-Element in Fine Road Dust of Cities with Coal Contamination by Geochemical Investigation and Isotopic Monitoring. *Environ. Int.* **2021**, *152*, 106499. [CrossRef]
19. Chen, S.; Zhang, X.; Lin, J.; Huang, J.; Zhao, D.; Yuan, T.; Huang, K.; Luo, Y.; Jia, Z.; Zang, Z.; et al. Fugitive Road Dust PM2.5 Emissions and Their Potential Health Impacts. *Environ. Sci. Technol.* **2019**, *53*, 8455–8465. [CrossRef]
20. Wang, S.; Wang, L.; Huan, Y.; Wang, R.; Liang, T. Concentrations, Spatial Distribution, Sources and Environmental Health Risks of Potentially Toxic Elements in Urban Road Dust across China. *Sci. Total Environ.* **2022**, *805*, 150266. [CrossRef]
21. Xin-chuan, L.; Jian-guo, Y.; Chang-liang, Y.; Fang, Z.; Sheng-yin, Z.; Yu, Z. Distribution and Assessment of Soil Heavy Metals in Northern Jiayuguan City, China. *J. Lanzhou Univ. Sci.* **2019**, *55*. [CrossRef]
22. Hou, K.; Wen, J. Quantitative Analysis of the Relationship between Land Use and Urbanization Development in Typical Arid Areas. *Environ. Sci. Pollut. Res.* **2020**, *27*, 38758–38768. [CrossRef] [PubMed]
23. Shahab, A.; Zhang, H.; Ullah, H.; Rashid, A.; Rad, S.; Li, J.; Xiao, H. Pollution Characteristics and Toxicity of Potentially Toxic Elements in Road Dust of a Tourist City, Guilin, China: Ecological and Health Risk Assessment. *Environ. Pollut.* **2020**, *266*, 115419. [CrossRef] [PubMed]
24. Buck, C.S.; Aguilar-Islas, A.; Marsay, C.; Kadko, D.; Landing, W.M. Trace Element Concentrations, Elemental Ratios, and Enrichment Factors Observed in Aerosol Samples Collected during the US GEOTRACES Eastern Pacific Ocean Transect (GP16). *Chem. Geol.* **2019**, *511*, 212–224. [CrossRef]
25. Ahamad, A.; Janardhana Raju, N.; Madhav, S.; Gossel, W.; Ram, P.; Wycisk, P. Potentially Toxic Elements in Soil and Road Dust around Sonbhadra Industrial Region, Uttar Pradesh, India: Source Apportionment and Health Risk Assessment. *Environ. Res.* **2021**, *202*, 111685. [CrossRef]
26. González-Macías, C.; Schifter, I.; Lluch-Cota, D.B.; Méndez-Rodríguez, L.; Hernández-Vázquez, S. Distribution, Enrichment and Accumulation of Heavy Metals in Coastal Sediments of Salina Cruz Bay, México. *Environ. Monit. Assess.* **2006**, *118*, 211–230. [CrossRef]
27. China National Environmental Monitoring Centre. *Background Values of Soil Elements in China*; China Environmental Press: Beijing, China, 1990; ISBN 978780007726.
28. Loska, K.; Wiechuła, D.; Korus, I. Metal Contamination of Farming Soils Affected by Industry. *Environ. Int.* **2004**, *30*, 159–165. [CrossRef]
29. Sutherland, R.A. Bed Sediment-Associated Trace Metals in an Urban Stream, Oahu, Hawaii. *Environ. Geol.* **2000**, *39*, 611–627. [CrossRef]
30. Müller, G. Index of Geoaccumulation in Sediments of the Rhine River. *Geol. J.* **1969**, *2*, 108–118.
31. Chen, H.; Teng, Y.; Lu, S.; Wang, Y.; Wang, J. Contamination Features and Health Risk of Soil Heavy Metals in China. *Sci. Total Environ.* **2015**, *512–513*, 143–153. [CrossRef]
32. Ackah, M. Soil Elemental Concentrations, Geoaccumulation Index, Non-Carcinogenic and Carcinogenic Risks in Functional Areas of an Informal e-Waste Recycling Area in Accra, Ghana. *Chemosphere* **2019**, *235*, 908–917. [CrossRef] [PubMed]

33. Looi, L.J.; Aris, A.Z.; Yusoff, F.M.; Isa, N.M.; Haris, H. Application of Enrichment Factor, Geoaccumulation Index, and Ecological Risk Index in Assessing the Elemental Pollution Status of Surface Sediments. *Environ. Geochem. Health* **2019**, *41*, 27–42. [CrossRef]

34. Jeong, H.; Choi, J.Y.; Lim, J.; Ra, K. Pollution Caused by Potentially Toxic Elements Present in Road Dust from Industrial Areas in Korea. *Atmosphere* **2020**, *11*, 1366. [CrossRef]

35. Anaman, R.; Peng, C.; Jiang, Z.; Liu, X.; Zhou, Z.; Guo, Z.; Xiao, X. Identifying Sources and Transport Routes of Heavy Metals in Soil with Different Land Uses around a Smelting Site by GIS Based PCA and PMF. *Sci. Total Environ.* **2022**, *823*, 153759. [CrossRef]

36. Mean, B. US EPA Risk Assessment Guidance for Superfund Volume I: Human Health Evaluation Manual (Part F, Supplemental Guidance for Inhalation Risk Assessment). *Off. Superfund Remediat. Technol. Innov. Environ. Prot. Agency* **2009**, *I*, 1–68.

37. Nilkarnjanakul, W.; Watchalayann, P.; Chotpantarat, S. Spatial Distribution and Health Risk Assessment of As and Pb Contamination in the Groundwater of Rayong Province, Thailand. *Environ. Res.* **2022**, *204*, 111838. [CrossRef]

38. Men, C.; Liu, R.; Wang, Q.; Miao, Y.; Wang, Y.; Jiao, L.; Li, L.; Cao, L.; Shen, Z.; Li, Y.; et al. Spatial-Temporal Characteristics, Source-Specific Variation and Uncertainty Analysis of Health Risks Associated with Heavy Metals in Road Dust in Beijing, China. *Environ. Pollut.* **2021**, *278*, 116866. [CrossRef]

39. U.S. EPA. *Office of Emergency and Remedial Response. Risk Assessment Guidance for Superfund. Volume 1 Human Health Evaluation Manual (Part A). Interim Final*; U.S. EPA: Washington, DC, USA, 1989.

40. U.S. EPA. Risk Assessment Guidance for Superfund Volume I: Human Health Evaluation Manual (Part E, Supplemental Guidance for Dermal Risk Assessment), Chapter 3: Exposure Assessment. *Off. Superfund Remediat. Technol. Innov.* **2004**, *II*, 1–89.

41. U.S. EPA. Supplemental Guidance for Developing Soil Screening Levels for Superfund Sites. *United States Environ. Prot. Agency* **2002**, *12*, 1–187.

42. Zheng, N.; Liu, J.; Wang, Q.; Liang, Z. Health Risk Assessment of Heavy Metal Exposure to Street Dust in the Zinc Smelting District, Northeast of China. *Sci. Total Environ.* **2010**, *408*, 726–733. [CrossRef]

43. Rinklebe, J.; Antoniadis, V.; Shaheen, S.M.; Rosche, O.; Altermann, M. Health Risk Assessment of Potentially Toxic Elements in Soils along the Central Elbe River, Germany. *Environ. Int.* **2019**, *126*, 76–88. [CrossRef]

44. Safiur Rahman, M.; Khan, M.D.H.; Jolly, Y.N.; Kabir, J.; Akter, S.; Salam, A. Assessing Risk to Human Health for Heavy Metal Contamination through Street Dust in the Southeast Asian Megacity: Dhaka, Bangladesh. *Sci. Total Environ.* **2019**, *660*, 1610–1622. [CrossRef]

45. Bhutiani, R.; Kulkarni, D.B.; Khanna, D.R.; Gautam, A. Water Quality, Pollution Source Apportionment and Health Risk Assessment of Heavy Metals in Groundwater of an Industrial Area in North India. *Expo. Health* **2016**, *8*, 3–18. [CrossRef]

46. Hao, H.; Zhang, M.; Wang, J.; Fu, Z.; Balaji, P.; Jiang, S. Barium in Coal and Coal Combustion Products: Distribution, Enrichment and Migration. *Energy Explor. Exploit.* **2022**, *40*, 889–907. [CrossRef]

47. Han, J.; Lee, S.; Mammadov, Z.; Kim, M.; Mammadov, G.; Ro, H.M. Source Apportionment and Human Health Risk Assessment of Trace Metals and Metalloids in Surface Soils of the Mugan Plain, the Republic of Azerbaijan. *Environ. Pollut.* **2021**, *290*, 118058. [CrossRef] [PubMed]

48. Fan, M.Y.; Zhang, Y.L.; Lin, Y.C.; Cao, F.; Sun, Y.; Qiu, Y.; Xing, G.; Dao, X.; Fu, P. Specific Sources of Health Risks Induced by Metallic Elements in PM2.5 during the Wintertime in Beijing, China. *Atmos. Environ.* **2021**, *246*, 118112. [CrossRef]

49. Zhang, R.; Jing, J.; Tao, J.; Hsu, S.C.; Wang, G.; Cao, J.; Lee, C.S.L.; Zhu, L.; Chen, Z.; Zhao, Y.; et al. Chemical Characterization and Source Apportionment of PM2.5 in Beijing: Seasonal Perspective. *Atmos. Chem. Phys.* **2013**, *13*, 7053–7074. [CrossRef]

50. Zhang, N.; Cao, J.; Xu, H.; Zhu, C. Elemental Compositions of PM2.5 and TSP in Lijiang, Southeastern Edge of Tibetan Plateau during Pre-Monsoon Period. *Particuology* **2013**, *11*, 63–69. [CrossRef]

51. Hsu, C.Y.; Chiang, H.C.; Chen, M.J.; Chuang, C.Y.; Tsen, C.M.; Fang, G.C.; Tsai, Y.I.; Chen, N.T.; Lin, T.Y.; Lin, S.L. Ambient PM2.5 in the Residential Area near Industrial Complexes: Spatiotemporal Variation, Source Apportionment, and Health Impact. *Sci. Total Environ.* **2017**, *590*, 204–214. [CrossRef]

52. Pey, J.; Querol, X.; Alastuey, A. Discriminating the Regional and Urban Contributions in the North-Western Mediterranean: PM Levels and Composition. *Atmos. Environ.* **2010**, *44*, 1587–1596. [CrossRef]

53. Men, C.; Liu, R.; Xu, F.; Wang, Q.; Guo, L.; Shen, Z. Pollution Characteristics, Risk Assessment, and Source Apportionment of Heavy Metals in Road Dust in Beijing, China. *Sci. Total Environ.* **2018**, *612*, 138–147. [CrossRef] [PubMed]

54. WANG, X.; LI, X.; GAO, L.; GAO, Y.; LANG, C. Speciation and Ecological Risk of Selenium in Street Dust in Downtown Chengdu. *Environ. Chem.* **2022**, *41*, 2014–2021. [CrossRef]

55. Zhou, S.; Yuan, Q.; Li, W.; Lu, Y.; Zhang, Y.; Wang, W. Trace Metals in Atmospheric Fine Particles in One Industrial Urban City: Spatial Variations, Sources, and Health Implications. *J. Environ. Sci.* **2014**, *26*, 205–213. [CrossRef]

56. Zhang, Y.; Cao, S.; Xu, X.; Qiu, J.; Chen, M.; Wang, D.; Guan, D.; Wang, C.; Wang, X.; Dong, B.; et al. Metals Compositions of Indoor PM2.5, Health Risk Assessment, and Birth Outcomes in Lanzhou, China. *Environ. Monit. Assess.* **2016**, *188*, 325. [CrossRef] [PubMed]

57. Hu, J.; Sun, Q.; He, H. Thermal Effects from the Release of Selenium from a Coal Combustion during High-Temperature Processing: A Review. *Environ. Sci. Pollut. Res.* **2018**, *25*, 13470–13478. [CrossRef] [PubMed]

58. Wang, J.; Li, S.; Cui, X.; Li, H.; Qian, X.; Wang, C.; Sun, Y. Bioaccessibility, Sources and Health Risk Assessment of Trace Metals in Urban Park Dust in Nanjing, Southeast China. *Ecotoxicol. Environ. Saf.* **2016**, *128*, 161–170. [CrossRef] [PubMed]

59. Jin, Y.; O'Connor, D.; Ok, Y.S.; Tsang, D.C.W.; Liu, A.; Hou, D. Assessment of Sources of Heavy Metals in Soil and Dust at Children's Playgrounds in Beijing Using GIS and Multivariate Statistical Analysis. *Environ. Int.* **2019**, *124*, 320–328. [CrossRef]

60. Zhang, Y.; Wang, X.; Chen, H.; Yang, X.; Chen, J.; Allen, J.O. Source Apportionment of Lead-Containing Aerosol Particles in Shanghai Using Single Particle Mass Spectrometry. *Chemosphere* **2009**, *74*, 501–507. [CrossRef]
61. Hsu, C.Y.; Chiang, H.C.; Lin, S.L.; Chen, M.J.; Lin, T.Y.; Chen, Y.C. Elemental Characterization and Source Apportionment of PM10 and PM2.5 in the Western Coastal Area of Central Taiwan. *Sci. Total Environ.* **2016**, *541*, 1139. [CrossRef]
62. Luilo, G.B.; Othman, O.C.; Mrutu, A. Arsenic: A Toxic Trace Element of Public Health Concern in Urban Roadside Soils in Dar Es Salaam City. *J. Mater. Environ. Sci.* **2014**, *5*, 1742–1749.
63. Shafer, M.M.; Toner, B.M.; Overdier, J.T.; Schauer, J.J.; Fakra, S.C.; Hu, S.; Herner, J.D.; Ayala, A. Chemical Speciation of Vanadium in Particulate Matter Emitted from Diesel Vehicles and Urban Atmospheric Aerosols. *Environ. Sci. Technol.* **2012**, *46*, 189–195. [CrossRef] [PubMed]
64. Jandacka, D.; Durcanska, D.; Bujdos, M. The Contribution of Road Traffic to Particulate Matter and Metals in Air Pollution in the Vicinity of an Urban Road. *Transp. Res. Part D Transp. Environ.* **2017**, *50*, 397–408. [CrossRef]
65. Wang, Y.L.; Tsou, M.C.M.; Pan, K.H.; Özkaynak, H.; Dang, W.; Hsi, H.C.; Chien, L.C. Estimation of Soil and Dust Ingestion Rates from the Stochastic Human Exposure and Dose Simulation Soil and Dust Model for Children in Taiwan. *Environ. Sci. Technol.* **2021**, *55*, 11805–11813. [CrossRef]
66. Tarafdar, A.; Sinha, A. Health Risk Assessment and Source Study of PAHs from Roadside Soil Dust of a Heavy Mining Area in India. *Arch. Environ. Occup. Health* **2019**, *74*, 252–262. [CrossRef] [PubMed]
67. Shi, G.; Chen, Z.; Bi, C.; Wang, L.; Teng, J.; Li, Y.; Xu, S. A Comparative Study of Health Risk of Potentially Toxic Metals in Urban and Suburban Road Dust in the Most Populated City of China. *Atmos. Environ.* **2011**, *45*, 764–771. [CrossRef]

Article

Pollution and Health Risk Assessments of Potentially Toxic Elements in the Fine-Grained Particles (10–63 μm and <10 μm) in Road Dust from Apia City, Samoa

Hyeryeong Jeong [1,2] and Kongtae Ra [2,3,*]

1   Ifremer, Département Ressources Biologiques et Environnement (RBE), Unité Contamination Chimique des Ecosystèmes Marins (CCEM), F-44000 Nantes, France
2   Marine Environmental Research Center, Korea Institute of Ocean Science and Technology (KIOST), Busan 49111, Korea
3   Department of Ocean Science (Oceanography), KIOST School, University of Science and Technology (UST), Daejeon 34113, Korea
*   Correspondence: ktra@kiost.ac.kr

Abstract: Fine road dust is a major source of potentially toxic elements (PTEs) pollution in urban environments, which adversely affects the atmospheric environment and public health. Two different sizes (10–63 and <10 μm) were separated from road dust collected from Apia City, Samoa, and 10 PTEs were analyzed using inductively coupled plasma mass spectrometry (ICP-MS). Fine road dust (<10 μm) had 1.2–2.3 times higher levels of copper (Cu), zinc (Zn), arsenic (As), cadmium (Cd), antimony (Sb), lead (Pb), and mercury (Hg) than 10–63 μm particles. The enrichment factor (EF) value of Sb was the highest among PTEs, and reflected significant contamination. Cu, Zn, and Pb in road dust were also present at moderate to significant levels. Chromium (Cr), cobalt (Co), and nickel (Ni) in road dust were mainly of natural origins, while Cu, Zn, Sb, and Pb were due to traffic activity. The levels of PTEs in road dust in Samoa are lower than in highly urbanized cities, and the exposure of residents in Samoa to PTEs in road dust does not pose a noncarcinogenic health risk. Further studies of the effects of PTEs contamination in road dust on the atmosphere and living organisms are needed.

Keywords: potentially toxic elements; source identification; risk assessment; Samoa

Citation: Jeong, H.; Ra, K. Pollution and Health Risk Assessments of Potentially Toxic Elements in the Fine-Grained Particles (10–63 μm and <10 μm) in Road Dust from Apia City, Samoa. *Toxics* **2022**, *10*, 683. https://doi.org/10.3390/toxics10110683

Academic Editor: Ilaria Guagliardi

Received: 5 October 2022
Accepted: 8 November 2022
Published: 11 November 2022

Publisher's Note: MDPI stays neutral with regard to jurisdictional claims in published maps and institutional affiliations.

1. Introduction

There have been many studies on metal contamination in road dust in association with increasing transportation activities worldwide, due to urbanization and industrialization [1–5]. In urban environments, road dust is highly contaminated with various potentially toxic elements (PTEs) of natural origin, as well as PTEs derived from human activities, including industrial and traffic emissions [6,7]. These PTEs do not decompose in the environment and accumulate in living organisms. Some metals in road dust, including chromium (Cr), copper (Cu), zinc (Zn), antimony (Sb), cadmium (Cd), and lead (Pb), are derived from the corrosion of plated parts of vehicles and wear of brake pads, discs, tires, and asphalt [8–11]. Although some metals, such as Cu and Zn, are classified as essential, high concentrations in the environment have toxic effects on living organisms [12,13]. Some metals (Cr, Cd, and Pb) are classified as toxic elements and can present a carcinogenic risk to humans [14–16].

Fine road dust particles generally have relatively high PTEs concentrations compared to coarse particles due to their large surface area, and PTEs concentrations tend to increase as the particle size becomes smaller [17–19]. Fine particles of road dust are resuspended by wind and rainfall runoff, transported into the atmosphere and surrounding aquatic environments, and eventually deposited in coastal areas [20–24]. Therefore, road dust is a major source of PTEs contamination in urban environments, as well as the atmosphere and coastal environments. PTEs are of interest due to concerns about the detrimental effects of contaminated road dust on the atmosphere, living organisms, environment, and human health. Many countries are working to reduce the harmful effects of road and street dust on the environment and public health by regularly cleaning road surfaces using sweeping machines and vehicles to manage atmospheric pollution and stormwater quality [25–29]. These efforts have reduced the amount of road dust and metal contamination, but are limited to urban areas with large traffic volumes and high population density.

The fine particles in road dust are not removed efficiently, so they accumulate on the road surface over long periods. PTEs concentrations in road dust have been reported to be relatively high in areas of heavy traffic, including industrial, residential, and commercial areas [30–32]. Relatively high concentrations of PTEs were found in river sediments near urban and commercial regions, and PTEs-contaminated road dust is considered as a cause of contamination to river and coastal environments [24,33]. As more than half of the world's population lives in urban cities, residents living in areas with heavy traffic and high population densities are at higher risk of exposure to PTEs present in road dust [34,35]. Especially, road dust with a diameter <10 µm can enter the human body through inhalation, ingestion, and dermal contact [36–38]. Long-term exposure adversely affects human health, and may even have carcinogenic effects.

Studies on PTEs in road dust are concentrated in large cities and industrial regions, and the human health risk of residents and employees are evaluated. In Samoa, the tourism industry is making an important economic contribution. Traffic activities related to tourism is one major cause of PTE contamination in road dust and can have a detrimental effect on residents as well as tourists. However, research on PTE contamination and risk assessments in road dust from Pacific Island countries, including Samoa, is very limited. Therefore, this study was performed to evaluate PTE contamination and health risks posed by fine road dust (10–63 µm, <10 µm) collected in Apia City, Samoa.

2. Materials and Methods

2.1. Study Area

Samoa is located in the Polynesian region of south-central Pacific Ocean and consists of the two large island of Upolo and Savai'i and eight small islands. Our study area, Upolo island, is a volcanic island formed by a massive basaltic shield volcano rising from the seafloor of the Pacific Ocean. The volcanic rocks of Upolo Island include basaltic lavas, scoria cones, tuffs, and pyroclastic [39,40]. There is mountainous interior covered with dense rain forests with a height about 1100 m of volcanic origin. The length and area of Upolo Island are 75 km and 1125 km^2, respectively, and Apia, the capital of Samoa, is located in the northern part of the Upolu Island. The climate of study area is oceanic tropical characterized by high humidity and temperature.

2.2. Sampling and PTE Analysis

Eighteen road dust samples were collected using a vacuum cleaner (DC35; Dyson, Wiltshire, UK) from Apia City, Samoa, in July 2014 (Figure 1). To prevent cross-contamination during road dust sampling, the vacuum cleaner was disassembled and cleaned after sampling.

Four surface soil samples were also collected near the R2, R11, R12, and R16 sites to provide background values for pollution assessment. In the laboratory, particles of two size classes (10–63 and <10 µm) were separated using a vibratory sieve shaker (Analysette 3 Pro; Fritsch Co., Idar-Oberstein, Germany) with a nylon sieve. For PTE analysis, all samples were ground and homogenized with a planetary mono mill (Pulverisette 6; Fritsch Co.).

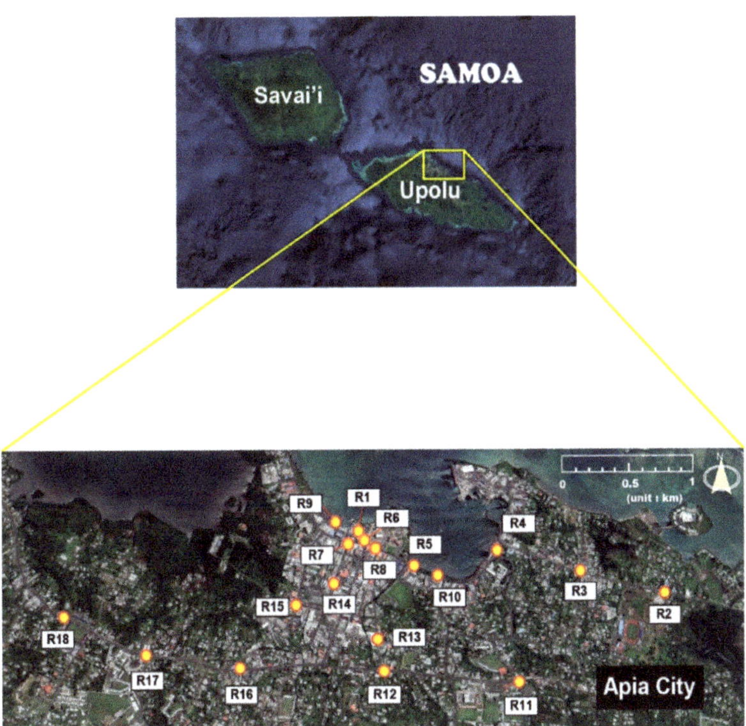

Figure 1. Map of sampling sites for road dust from Apia city of Samoa (base map from Google Earth).

After weighing about 100 mg of sample in a Teflon vessel, high-purity acid mixture (HF, HNO_3, $HClO_4$ = 4:3:1 v/v) was added and heated at 180 °C for 24 h on a hot plate for complete digestion. The decomposed samples were evaporated to dryness and dissolved in 2% HNO_3. PTEs other than Hg were analyzed by inductively coupled plasma mass spectrometry (ICP-MS; i-CAPQ; Thermo Fisher Scientific, Dreieich, Germany) at the Korea Institute of Ocean Science and Technology (KIOST). Hg concentrations were determined using a direct mercury analyzer (DMA-80; Milestone Inc., Sorisole, Italy).

Two certified reference materials (CRMs), MESS4 and BCR723, were decomposed together with the samples and analyzed to verify the reliability of PTE analysis. The recovery of CRMs was 97.6–103.3% for MESS-4 ($n = 6$) and 96.2–106.1% for BCR723 ($n = 6$), which were highly consistent with the certified values.

2.3. Pollution Assessment

Many pollution indices are widely used to evaluate the contamination levels of single or multiple elements. In this study, the enrichment factor (EF) and pollution load index (PLI) were used to evaluate PTE contamination in road dust.

EF was calculated by the following equation, and aluminum (Al) was used as a reference element.

$$EF| = |\frac{(metal/Al)_{sample}}{(metal/Al)_{background}}$$

where $(metal/Al)_{sample}$ and $(metal/Al)_{background}$ are the ratios of PTE and Al concentrations in two different size classes of road dust and background soil, respectively. The concentrations of Cr, Co, Ni, Cu, Zn, As, Cd, Sb, Pb, and Hg in surface soil, as background values, were 519, 96.3, 346, 37.9, 109, 7.8, 0.33, 0.50, 13.1, and 0.05 mg/kg, respectively (Table 1).

The calculated EF values were classified into five categories as follows [41]: EF < 2 (depletion to minimum enrichment), 2 < EF < 5 (moderate enrichment), 5 < EF < 20 (significant enrichment), 20 < EF < 40 (very high enrichment), EF > 40 (extremely high enrichment).

PLI was evaluated using the following equation, reflecting the overall contamination level of 10 PTEs in road dust analyzed in this study [42].

$$PLI| = |\left(\frac{C_{i1}}{C_{b1}}| \times |\frac{C_{i2}}{C_{b2}}| \times |\frac{C_{i3}}{C_{b3}}| \times | \cdots \frac{C_{in}}{C_{bn}}\right)^{\frac{1}{n}}$$

where C_i and C_b are the concentrations of element *i* in road dust and background soil, respectively.

The calculated PLI values were divided into four categories, as follows [43,44]: PLI < 1 (unpolluted), 1 < PLI < 2 (low polluted), 2 < PLI < 3 (moderately polluted), and PLI > 3 (heavily polluted).

2.4. Health Risk Assessment

The noncarcinogenic health risks to adults and children were evaluated considering three exposure pathways, i.e., ingestion (hand to mouth), inhalation (mouth and nose), and dermal contact. The level of PTE exposure by fine road dust particles (<10 μm) was calculated by the following equation using the chronic daily intake [45].

$$ADD_{ing}| = |C_i| \times |\left(\frac{IngR| \times |EF| \times |ED}{BW| \times |AT}\right)| \times |10^{-6}$$

$$ADD_{inh}| = |C_i| \times |\left(\frac{InhR| \times |EF| \times |ED}{PEF| \times |BW| \times |AT}\right)$$

$$ADD_{derm}| = |C_i| \times |\left(\frac{SL| \times |SA| \times |ABS| \times |EF| \times |ED}{BW| \times |AT}\right)| \times |10^{-6}$$

Here, C_i is the concentration of PTE i in road dust (<10 μm).

The noncarcinogenic health risk for each PTE was calculated using the following equation.

$$HQ_{ing}| = |\frac{ADD_{ing}}{RfD_{ing}}; HQ_{inh}| = |\frac{ADD_{inh}}{RfD_{inh}}; HQ_{derm}| = |\frac{ADD_{derm}}{RfD_{derm}}$$

The hazard index (HI) was calculated as the sum of the HQ values for the three exposure pathways.

HI values > 1 indicate potential chronic effects, and HI < 1 indicates no risk of adverse health effects [46].

RfD is the reference dose for ingestion, inhalation, and dermal contact, and is an estimate of daily exposure in residents [47,48]. The values for each parameter in the ADD calculation were obtained from Adamiec and Jarosz-Krzeminska [49] and are presented by Jeong et al. [4].

2.5. Statistical Analysis

The mean, minimum, maximum, and coefficient of variation (CV) values were carried out using Microsoft Excel program based on the PTE concentration. The Pearson's correlation analysis was applied to understand the potential source of each PTE in road dust (<10 μm) through the correlation between elements and was performed using PASW Statistics 18.0.

3. Results and Discussion

3.1. PTE Concentration

The minimum, maximum, mean, and coefficient of variation (CV) PTE levels in road dust are presented in Table 1. Figure 2 shows 10 PTE concentrations for particle sizes in road dust of 10–63 and <10 μm. In road dust with particle size 10–63 μm, Cr had the highest mean concentration and the PTE concentrations decreased in the order Cr (739 mg/kg) > Zn (351 mg/kg) > Ni (272 mg/kg) > Cu (82.8 mg/kg) > Co (66.7 mg/kg) > Pb (35.3 mg/kg) > As (4.5 mg/kg) > Sb (2.1 mg/kg) > Cd (0.31 mg/kg) > Hg (0.03 mg/kg) (Table 1) [47–49]. The CV represents the spatial variation according to sampling site for the PTE concentrations analyzed in this study. A high CV reflects heterogeneity of the PTE concentration among environmental samples, indicating the presence of hotspots of PTE contamination from specific anthropogenic sources [50,51]. The CVs for Cr, Ni, and Zn were <20%, indicating low regional variability, while the other PTEs showed moderate regional variability (20% < CV < 50%).

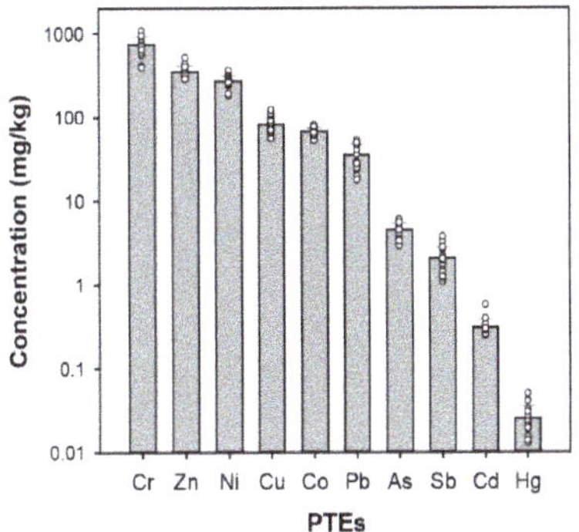
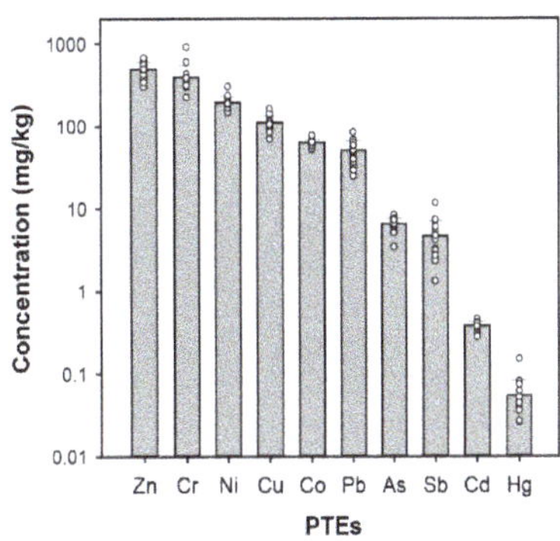

Figure 2. Comparison of PTEs concentrations with particle size of 10–63 μm (**left**) and <10 μm (**right**) in road dust.

In road dust with particle size <10 μm, Zn showed the highest mean concentration, at 494 mg/kg (Figure 2). The mean concentrations of Cr, Co, and Ni in road dust with particle size 10–63 μm were higher than those in fine road dust (<10 μm). However, the fine road dust had 1.2–2.3 times higher levels of Cu, Zn, As, Cd, Sb, Pb, and Hg than road dust with particle size 10–63 μm.

Table 1. Comparison of mean values (minimum and maximum value in parenthesis) and coefficient of variation (CV; %) for Al and PTEs concentration in the fine particles of road dust from Samoa and those in the other reported data (mean value).

Size		Al	Cr	Co	Ni	Cu	Zn	As	Cd	Sb	Pb	Hg	References
		%	mg/kg	mg/kg	mg/kg	mg/kg	mg/kg	mg/kg	mg/kg	mg/kg	mg/kg	mg/kg	
	mean	5.8	739	66.7	272	82.8	351	4.49	0.31	2.08	35.3	0.03	
10–63 μm	min.	4.9	395	53.0	184	55.5	288	2.92	0.25	1.10	17.7	0.01	This study
	max.	6.6	1078	81.6	364	123	520	5.99	0.58	3.76	52.6	0.05	
	CV (%)	8	26	11	16	21	18	21	25	36	33	40	
	mean	7.0	395	62.9	195	111	494	6.5	0.38	4.68	50.5	0.05	
<10 μm	min.	5.8	226	52.3	149	70.0	301	3.5	0.28	1.34	24.8	0.03	This study
	max.	8.5	913	78.0	304	163	677	8.4	0.46	11.7	85.0	0.15	
	CV (%)	9	38	9	18	21	21	18	12	53	31	18	
Soil	mean	8.5	519	96.3	346	37.9	109	7.8	0.33	0.50	13.1	0.05	This study
<10 μm	mean	-	300	19.8	75.6	513	3007	17.0	5.6	61.6	480	0.6	Korea [52]
<10 μm	mean	-	53.8	9.3	34.2	100	302	17.0	0.6	-	48.8	-	China [53]
<10 μm	mean	-	68	15	46	184	1026	-	0.8	12	91	-	Russia [54]

The highest concentrations of Cu (163 mg/kg), Zn (677 mg/kg), and Sb (11.69 mg/kg) were observed at site R17, which had a high traffic volume. Pb (85.0 mg/kg) and Hg (0.15 mg/kg) concentrations were highest at site R5, which is near the intersection in the downtown direction of Apia City. PTE concentrations tended to be relatively high at sampling sites with high traffic volume. The CV of Sb was 53%, which was higher than those of other PTEs. Most PTEs showed low CVs and the spatial distributions in fine road dust were not different. These results suggest that PTEs in road dust may be affected by human activities, such as transportation of residents and tourists, rather than by specific contamination sources, such as industrial emissions.

3.2. Pollution Assessment of PTEs in Road Dust

The mean EF values for 10 PTEs are shown in Figure 3. Sb had the highest mean EF value among the PTEs examined in both fractions of road dust. The mean EF values of PTEs in road dust with a particle size of 10–63 µm decreased in the order Sb > Zn > Pb > Cu > Cr > Cd > Ni > Co > As > Hg, whereas those in fine road dust (<10 µm) decreased in the order Sb > Zn > Pb > Cu > Cd > Hg > As > Cr > Co > Ni; the EF values for Cr, Co, and Ni, which are of natural origin, were relatively low. For 10–63-µm road dust particles, the mean EF value of Sb was 6.1, corresponding to significant contamination. Cu, Zn, and Pb had mean EF values between 2 and 5, indicating moderate contamination. For Zn, the EF values exceeded 5 at five sampling sites with high traffic volume.

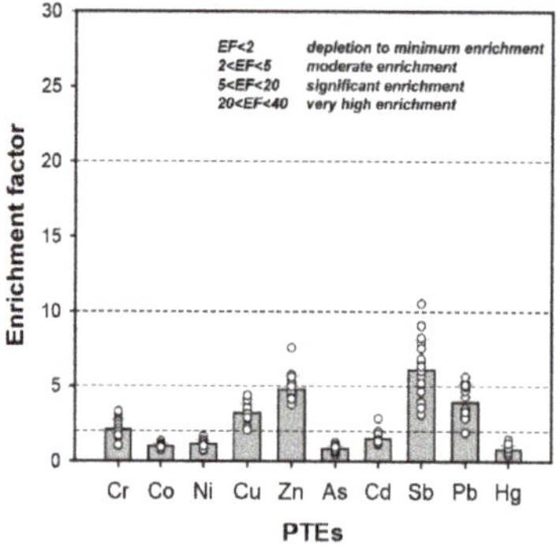
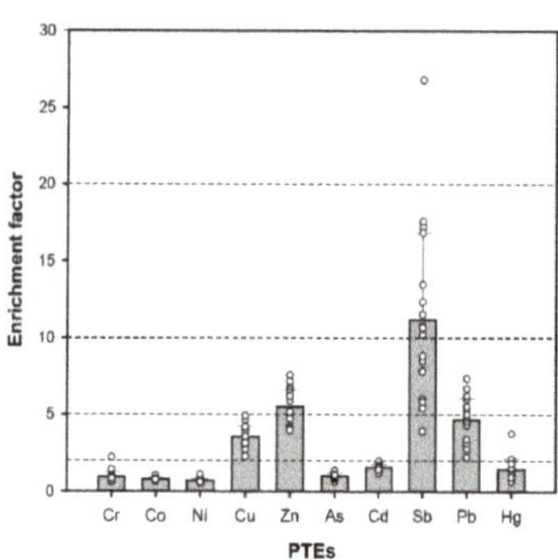

Figure 3. Comparison of enrichment factor (EF) for PTEs with particle size 10–63 µm (**left**) and <10 µm (**right**) in road dust.

For fine road dust (<10 µm), the mean EF of Sb was 11.2, corresponding to significant contamination. At sampling site R17, which had the widest road and dense traffic, the EF value of Sb was 26.8, indicating very high contamination. The mean EF of Zn, contamination with which is mainly due to tire wear in urban areas, indicates significant contamination. The contamination level of road dust in Samoa was lower than in large cities with high population density and traffic activity, but high levels of Cu, Zn, Sb, and Pb contamination were observed in relation to traffic activity (Table 1). Notably, Sb, which is present in high levels in brake pads [9,55], was a significant contaminant.

The PLI value of road dust with particle size of 10–63 µm ranged from 0.9 to 1.6 and (mean value = 1.3), indicating low pollution levels for 10 PTEs (Figure 4). The mean PLI value of fine road dust (<10 µm) was 1.6, corresponding to low pollution. However, two sampling sites (R5 and R17) had higher PLI values (exceeding 2) than the other sampling sites 2, indicating that that fine road dust in Samoa was moderately polluted with PTEs.

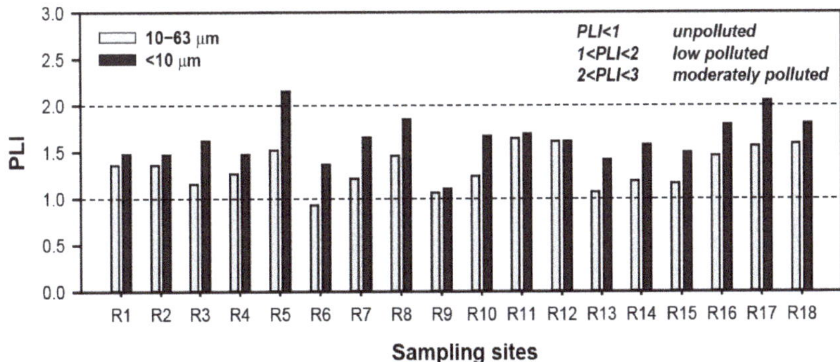

Figure 4. Spatial distribution of pollution load index (PLI) in road dust at sampling site.

Table 2 shows the relationship between the concentration of PTEs in <10 μm of road dust of this study. The concentrations of Cr, Ni, and Co in road dust were closely related. The mean concentrations of Cr, Ni, and Co in the surface soil were 519, 96.3, and 346 mg/kg, respectively, which were very high compared to road dust in this study and the upper continental crust [56]. In a study of heavy metals in soils of a volcanic island (Réunion) located in the western part of the Indian Ocean, Dœlsch et al. [57] reported that the Cr and Ni contents in soil were closely related to volcanic rocks. In this study, no clear correlations were found between Cr, Ni, or Co, and other PTEs (Cu, Zn, Sn, and Pb), mainly arising from nonexhaust emission sources, such as the wear of tires, brake pads, and asphalt due to traffic activity. However, there were significant correlations among Cu, Zn, Sb, and Pb. Therefore, the high Cr, Ni, and Co concentrations in road dust in Samoa, and their high correlations, were strongly influenced by natural processes, such as the weathering of volcanic parent rocks.

Table 2. Results of Pearson's correlation matrix between PTEs in <10 μm of road dust. Bold indicates that correlation is significant at the 0.05 level (2-tailed).

	Cr	Ni	Co	Cu	Zn	As	Cd	Sb	Pb	Hg
Cr	1									
Ni	**0.72**	1								
Co	**0.95**	**0.82**	1							
Cu	−0.13	0.15	−0.13	1						
Zn	−0.36	−0.12	−0.40	**0.63**	1					
As	−0.26	−0.01	−0.22	0.29	0.45	1				
Cd	**−0.64**	−0.32	**−0.62**	0.47	**0.67**	0.51	1			
Sb	−0.16	−0.03	−0.17	**0.87**	**0.68**	0.26	**0.50**	1		
Pb	0.08	0.30	0.13	**0.63**	**0.57**	0.34	0.44	**0.51**	1	
Hg	0.11	0.30	0.23	0.22	0.25	0.30	0.10	0.16	**0.59**	1

3.3. Non-Carcinogenic Risk of PTEs in Fine Road Dust (10 μm)

Table 3 presents the mean HQ and HI values for adults and children, and three exposure pathways (in descending order): $HQ_{ing} > HQ_{derm} > HQ_{inh}$. These observations indicated that the pathway of ingestion from hand to mouth was the main exposure route, having a detrimental effect on human health compared to other processes. The contribution of ingestion was 76% for adults and 79% for children, but the sum of the HQ values for PTEs was 0.62 for children, which was higher than that for adults (0.07). HQ and HI values for all PTEs were <1 at all sampling sites, indicating no noncarcinogenic health risk for adults or children. The mean HI values for adults and children were 0.09 and 0.78, respectively. Therefore, the noncarcinogenic health risk of exposure to PTEs in road dust was about nine times higher in children than adults.

Table 3. The median values of hazard quotient (HQ) and hazard index (HI) of non-carcinogenic hazards for PTEs with particle size of less than 10 μm in this study.

	Adult				Children			
	HQ_{ing}	HQ_{inh}	HQ_{derm}	HI	HQ_{ing}	HQ_{inh}	HQ_{derm}	HI
Cr	4.3×10^{-2}	3.5×10^{-3}	4.0×10^{-2}	8.7×10^{-2}	4.1×10^{-1}	3.2×10^{-2}	9.1×10^{-2}	5.3×10^{-1}
Co	1.1×10^{-3}	3.0×10^{-3}	3.2×10^{-5}	4.1×10^{-3}	1.0×10^{-2}	2.8×10^{-2}	7.3×10^{-5}	3.8×10^{-2}
Ni	3.3×10^{-3}	2.5×10^{-3}	2.8×10^{-4}	3.6×10^{-3}	3.1×10^{-2}	2.4×10^{-5}	6.4×10^{-4}	3.2×10^{-2}
Cu	9.5×10^{-4}	7.3×10^{-7}	7.3×10^{-5}	1.0×10^{-3}	8.9×10^{-3}	6.8×10^{-6}	1.7×10^{-4}	9.1×10^{-3}
Zn	5.9×10^{-4}	4.5×10^{-7}	6.7×10^{-5}	6.6×10^{-4}	5.5×10^{-3}	4.2×10^{-6}	1.5×10^{-4}	5.7×10^{-3}
As	7.7×10^{-3}	5.9×10^{-6}	4.3×10^{-4}	8.2×10^{-3}	7.2×10^{-2}	5.5×10^{-5}	9.9×10^{-4}	7.3×10^{-2}
Cd	1.4×10^{-4}	1.0×10^{-7}	3.1×10^{-4}	4.5×10^{-4}	1.3×10^{-3}	9.7×10^{-7}	7.2×10^{-4}	2.0×10^{-3}
Sb	3.8×10^{-3}	2.9×10^{-6}	4.4×10^{-3}	8.2×10^{-3}	3.6×10^{-2}	2.7×10^{-5}	1.0×10^{-2}	4.6×10^{-2}
Pb	5.0×10^{-3}	3.8×10^{-6}	7.7×10^{-4}	5.8×10^{-3}	4.7×10^{-2}	3.6×10^{-5}	1.8×10^{-3}	4.9×10^{-2}
Hg	5.4×10^{-5}	4.1×10^{-8}	1.7×10^{-5}	7.1×10^{-5}	5.0×10^{-4}	3.8×10^{-7}	4.0×10^{-5}	5.4×10^{-4}

The HI value of Cr was higher than that of other PTEs, and the order of HI values was Cr > As > Sb > Pb > Ni > Co > Cu > Zn > Cd > Hg in adults. These results suggest that Cr and As may be major contributors to noncarcinogenic risk among PTEs in road dust accumulated on road surfaces in Samoa.

Cr is widely used in many industries, including metallurgy, steel manufacturing, automobile industry, electroplating, and wood preservation, because of its anticorrosive properties [58–61]. Cr is highly mobile and has extremely toxic effects even at very low concentrations; it is classified as a Group A carcinogen element in humans [62]. Cr is used for the wheels of vehicles, and many parts of vehicles are plated with Cr as it has an aesthetic appearance and is not easily corroded. Many studies have reported that Cr contamination in road dust and soil is due to traffic and industrial activities [63–66].

Although the noncarcinogenic risk of Cr is higher than those of other PTEs, the mean EF value of Cr in road dust (<10 μm) was 0.9 (range 0.6–2.2) in this study, and most of the sampling sites showed depletion to minimal enrichment. The contamination level of Cr in road dust was lower than those of Cu, Zn, Sb, and Pb, which are mainly derived from traffic activity. High concentrations of Cr and Ni have been reported in the soil of islands formed by volcanic activity [57,67].

The mean concentrations of Cr in road dust (<10 μm) and surface soil were 395 and 519 mg/kg, respectively, in this study, which were higher than those in the upper continental crust (92.0 mg/kg) [55]. Our results indicated that Cr concentration in road dust in Samoa is derived from natural sources, mainly due to the weathering of volcanic parent materials. Therefore, the noncarcinogenic risk of Cr in road dust on Samoa seemed to be much lower than in highly urbanized cities.

PTEs such as Cu, Zn, Sb, and Pb have a negligible health risk on the human body, but road dust is contaminated with these elements which come from traffic activities. Pacific Island countries, including Samoa, import and use used cars for the tourisms industry. The activation of the tourism industry and the environmental factors (high humidity and heavy precipitation) can increase the contamination levels of PTEs in road dust, as these accelerate the corrosion of vehicles. In addition, road dust contaminated with PTEs is easily resuspended via stormwater runoff, transport to rivers and streams, and is finally deposited in the coastal environments. Road dust in urban environments is one of the major sources of PTEs contamination that contaminates marine sediments and has a detrimental effect on marine organisms. Efficient management by road cleaning is necessary to prevent the impact on the environments and ecosystem caused by road dust contaminated with PTEs.

4. Conclusions

In this study, road dust was collected in Apia City, Samoa, and divided into two fractions (10–63 and <10 μm). Analyzing road dust is important for public health, and for understanding environmental PTE contamination. Pollution and health risk assessments were performed for 10 PTEs in road dust. Cr, Co, and Ni concentrations in road dust with a particle size of 10–63 µm were higher than in fine road dust (<10 µm), while Cu, Zn, As, Cd, Sb, Pb, and Hg were present at relatively high concentrations in fine road dust. The PTE concentration in road dust was lower than in large cities with high levels of human activity. The CVs of PTEs in road dust suggested that anthropogenic sources related to traffic activity rather than industrial emissions affect road dust in Apia City, Samoa. Both particle sizes of road dust had significant levels of Sb, and the levels of Cu, Zn, and Pb, which are derived from nonexhaust emission arising from traffic activity, were also moderately high. The results of correlation analyses and pollution assessments showed that Cr, Co, and Ni are greatly affected by natural sources due to the weathering of volcanic parent rocks. There were significant correlations among Cu, Zn, Sb, and Pb. In addition, relatively high concentrations were observed in sampling sites with high traffic volume, suggesting that contamination with these PTEs (Cu, Zn, Sb, and Pb) was mainly due to traffic activity. The ingestion exposure route posed a greater health risk than inhalation and dermal contact. The mean HI values for adults and children were 0.09 and 0.78, respectively, indicating that the noncarcinogenic health risk in children was about nine times higher than that in adults. However, HI values for adults and children were <1 at all sampling sites, suggesting no noncarcinogenic risk from exposure to PTEs in road dust. Among the PTEs, Cr had a higher HQ value than the other elements, and posed a major noncarcinogenic health risk. The Cr in road dust in Samoa was mainly derived from natural sources, i.e., the weathering of volcanic parent rocks rather than anthropogenic sources (traffic activity). Therefore, our results suggest that the health risks posed by PTEs in road dust in Samoa are negligible compared to highly urbanized cities. However, further studies regarding the effects of PTE-contaminated road dust on the atmosphere and living organisms, including humans, are needed.

Author Contributions: Conceptualization, H.J. and K.R.; investigation, K.R.; methodology, H.J. and K.R.; formal analysis, H.J. and K.R.; data curation, H.J.; visualization, K.R.; writing—original draft preparation, H.J.; writing—review and editing, K.R.; funding acquisition, K.R. All authors have read and agreed to the published version of the manuscript.

Funding: This study was supported by grants (PEA0012) from the Korea Institute of Ocean Science and Technology (KIOST).

Institutional Review Board Statement: Not applicable.

Informed Consent Statement: Not applicable.

Data Availability Statement: All data for this study are available from the first or corresponding authors upon request.

Acknowledgments: We gratefully appreciate the assistance of Ministry of Natural Resources and Environment of Samoa for helping with this research.

Conflicts of Interest: The authors declare no conflict of interest.

References

1. Alghamdi, A.G.; El-Saeid, M.H.; Alzahrani, A.J.; Ibrahim, H.M. Heavy metal pollution and associated health risk assessment of urban dust in Riyadh, Saudi Arabia. *PLoS ONE* **2022**, *17*, e0261957. [CrossRef] [PubMed]
2. Al-Shidi, H.K.; Al-Reasi, H.A.; Sulaiman, H. Heavy metals levels in road dust from Muscat, Oman: Relationship with traffic volumes, and ecological and health risk assessments. *Int. J. Environ. Health Res.* **2022**, *32*, 264–276. [CrossRef] [PubMed]
3. Huang, F.; Liu, B.; Yu, Y.; Lv, L.; Luo, L.; Yin, F. Heavy metals in road dust across China: Occurrence, sources and health risk assessment. *Bull. Environ. Contam. Toxicol.* **2022**, *109*, 323–331. [CrossRef] [PubMed]
4. Jeong, H.; Ra, K. Source apportionment and health risk assessment for potentially toxic elements in size-fractionated road dust in Busan Metropolitan City, Korea. *Environ. Monit. Assess.* **2022**, *194*, 350. [CrossRef] [PubMed]
5. Zglobicki, W.; Telecka, M.; Skupinski, S. Assessment of short-term changes in street dust pollution with heavy metals in Lublin (E Poland)—Levels, sources and risks. *Environ. Sci. Pollut. Res.* **2019**, *26*, 35049–35060. [CrossRef] [PubMed]

6. Cai, Y.; Li, F.; Zhang, J.; Zhu, X.; Li, Y.; Fu, J.; Chen, X.; Liu, C. Toxic metals in size-fractionated road dust from typical industrial district: Seasonal distribution, bioaccessibility and stochastic-fuzzy health risk management. *Environ. Technol. Innov.* **2021**, *23*, 101643. [CrossRef]

7. Jeong, H.; Choi, J.Y.; Ra, K. Potentially toxic elements pollution in road deposited sediments around the active smelting industry of Korea. *Sci. Rep.* **2021**, *11*, 7238. [CrossRef] [PubMed]

8. Aquilera, A.; Bautista, F.; Gutierrez-Ruiz, M.; Ceniceros-Gomez, A.E.; Cejudo, R.; Goguitchaichvili, A. Heavy metal pollution of street dust in the largest city of Mexico, sources and health risk assessment. *Environ. Monit. Assess.* **2021**, *193*, 193. [CrossRef]

9. Jeong, H.; Ryu, J.S.; Ra, K. Characteristics of potentially toxic elements and multi-isotope signatures (Cu, Zn, Pb) in non-exhaust traffic emission sources. *Environ. Pollut.* **2022**, *292*, 118339. [CrossRef]

10. Wiseman, C.L.S.; Levesque, C.; Rasmussen, P.E. Characterizing the sources, concentrations and resuspension potential of metals and metalloids in the thoracic fraction of urban road dust. *Sci. Total Environ.* **2021**, *786*, 147467. [CrossRef]

11. Zhao, L.; Hu, G.; Yan, Y.; Yu, R.; Cui, J.; Wang, X.; Yan, Y. Source apportionment of heavy metals in urban road dust in a continental city of eastern China: Using Pb and Sr isotopes combined with multivariate statistical analysis. *Atmos. Environ.* **2019**, *201*, 201–211. [CrossRef]

12. Festa, R.A.; Thiele, D.J. Copper: An essential metal in biology. *Curr. Biol.* **2011**, *21*, R877–R883. [CrossRef]

13. Arif, N.; Yadav, V.; Singh, S.; Singh, S.; Ahmad, P.; Mishra, R.K.; Sharma, S.; Tripathi, D.K.; Dubey, N.K.; Chauhan, D.K. Influence of high and low Levels of plant-beneficial heavy metal ions on plant growth and development. *Front. Environ. Sci.* **2016**, *4*, 69. [CrossRef]

14. Anedo, P.; Netvichian, R.; Thiendedsakui, P.; Khaodhiar, S.; Tulayakui, P. Carcinogenic Risk of Pb, Cd, Ni, and Cr and Critical Ecological Risk of Cd and Cu in Soil and Groundwater around the Municipal Solid Waste Open Dump in Central Thailand. *J. Environ. Public Health* **2022**, *2022*, 3062215. [CrossRef] [PubMed]

15. Faisal, M.; Wu, Z.; Wang, H.; Hussain, Z.; Azam, M.I. Human health risk assessment of heavy metals in the urban road dust of Zhengzhou Metropolis, China. *Atmosphere* **2021**, *12*, 1213. [CrossRef]

16. Irigoyen-Arredondo, M.S.; Moreno-Sanchez, X.G.; Escobar-Sanchez, O.; Soto-Jimenez, M.F.; Marin-Enriquez, E.; Abitia-Cardenas, L.A. Essential (Cu, Zn) and nonessential (Pb, Cd) metals in the muscle of leopard groupers (*Mycteroperca rosacea*) from a mining port in the Gulf of California, Mexico: Human health risk assessment. *Environ. Sci. Pollut. Res.* **2022**, *29*, 35001–35011. [CrossRef] [PubMed]

17. Jeong, H.; Choi, J.Y.; Lee, J.; Lim, J.; Ra, K. Heavy metal pollution by road-deposited sediments and its contribution to total suspended solids in rainfall runoff from intensive industrial areas. *Environ. Pollut.* **2020**, *265*, 115028. [CrossRef]

18. Jeong, H.; Ra, K. Multi-isotope signatures (Cu, Zn, Pb) of different particle sizes in road-deposited sediments: A case study from industrial area. *J. Anal. Sci. Technol.* **2021**, *12*, 39. [CrossRef]

19. Logiewa, A.; Miazgowicz, A.; Krennhuber, K.; Lanzerstofer, C. Variation in the concentration of metals in road dust size fractions between 2 μm and 2 mm: Results from three metallurgical centres in Poland. *Arch. Environ. Contam. Toxicol.* **2020**, *78*, 46–59. [CrossRef] [PubMed]

20. Amato, F.; Alastuey, A.; de la Rosa, J.; Castanedo, Y.G.; de la Campa, A.M.S.; Pandolfi, M.; Lozano, A.; Gonzalez, J.C.; Querol, X. Trends of road dust emissions contributions on ambient air particulate levels at rural, urban and industrial sites in southern Spain. *Atmos. Chem. Phys.* **2014**, *14*, 3533–3544. [CrossRef]

21. Charters, F.J.; Cochrane, T.A.; O'Sullivan, A.D. Untreated runoff quality from roof and road surfaces in a low intensity rainfall climate. *Sci. Total Environ.* **2016**, *550*, 265–272. [CrossRef] [PubMed]

22. Gao, S.; Wang, X.; Li, H.; Kong, Y.; Chen, J.; Chen, Z. Heavy metals in road-deposited sediment and runoff in urban and intercity expressways. *Transp. Saf. Environ.* **2022**, *4*, tdab030. [CrossRef]

23. Faisal, M.; Wu, Z.; Wang, H.; Lin, X.; Hussain, Z.; Azam, M.I. Potential Heavy Metals Pollution Contribution from Wash-Off of Urban Road-Dust. *Toxics* **2022**, *10*, 397. [CrossRef] [PubMed]

24. Jeong, H.; Choi, J.Y.; Lim, J.; Shim, W.J.; Kim, Y.O.; Ra, K. Characterization of the contribution of road deposited sediments to the contamination of the close marine environment with trace metals: Case of the port city of Busan (South Korea). *Mar. Pollut. Bull.* **2021**, *161*, 111717. [CrossRef]

25. Amato, F.; Querol, X.; Johansson, C.; Nagl, C.; Alastuey, A. A review on the effectiveness of street sweeping, washing and dust suppressants as urban PM control methods. *Sci. Total Environ.* **2010**, *408*, 3070–3084. [CrossRef]

26. Calvillo, S.; Williams, E.S.; Brooks, B.W. Street dust: Implications for stormwater and air quality, and environmental through street sweeping. *Rev. Environ. Contam. Toxicol.* **2015**, *233*, 71–128. [CrossRef]

27. Hanfi, M.Y.; Mostafa, M.Y.A.; Zhukovsky, M.V. Heavy metal contamination in urban surface sediments: Sources, distribution, contamination control, and remediation. *Environ. Monit. Assess.* **2020**, *192*, 32. [CrossRef]

28. Kim, D.G.; Jeong, K.; Ko, S.O. Removal of road deposited sediments by sweeping and its contribution to highway runoff quality in Korea. *Environ. Technol.* **2014**, *35*, 2546–2555. [CrossRef]

29. Selbig, W.R.; Bannerman, R.T. Evaluation of Street Sweeping as a Stormwater-Quality-Management Tool in Three Residential Basins in Madison, Wisconsin. Scientific Investigation Report 2007-5156. 2002. Available online: https://pubs.usgs.gov/sir/2007/5156/ (accessed on 1 October 2022).

30. Apeagyei, E.; Bank, M.S.; Spengler, J.D. Distribution of heavy metals in road dust along an urban-rural gradient in Massachusetts. *Atmos. Environ.* **2011**, *45*, 2310–2323. [CrossRef]

31. Miazgowicz, A.; Krennhuber, K.; Lanzerstofer, C. Metals concentrations in road dust from high traffic and low traffic area: A size dependent comparison. *Int. J. Environ. Sci.* **2020**, *17*, 3365–3372. [CrossRef]

32. Trujillo-Gonzalez, J.M.; Torres-Mora, M.A.; Keesstra, S.; Brevik, E.C.; Jimenez-Ballesta, R. Heavy metal accumulation related to population density in road dust samples taken from urban sites under different land uses. *Sci. Total Environ.* **2016**, *553*, 636–642. [CrossRef] [PubMed]

33. Harris, A.; Xanthos, S.J.; Galiotos, J.K.; Douvris, C. Investigation of the metal content of sediments around the historically polluted Potomac River basin in Washington D.C., United States by inductively coupled plasma-optical emission spectroscopy (ICP-OES). *Microchem. J.* **2018**, *142*, 140–143. [CrossRef]

34. Kaonga, C.C.; Kosamu, I.B.M.; Utembe, W.R. A review of metal levels in urban dust, their methods of determination, and risk assessment. *Atmosphere* **2021**, *12*, 891. [CrossRef]

35. Roy, S.; Gupta, S.K.; Prakash, J.; Habib, G.; Kumar, P. A global perspective of the current state of heavy metal contamination in road dust. *Environ. Sci. Pollut. Res.* **2022**, *29*, 33230–33251. [CrossRef] [PubMed]

36. Kastury, F.; Smith, E.; Juhasz, A.L. A critical review of approaches and limitations of inhalation bioavailability and bioaccessibility of metal(loid)s from ambient particulate matter or dust. *Sci. Total Environ.* **2017**, *574*, 1054–1074. [CrossRef] [PubMed]

37. Levesque, C.; Wiseman, C.L.S.; Beauchemin, S.; Rasmussen, P.E. Thoracic fraction (PM_{10}) of resuspended urban dust: Geochemistry, particle size distribution and lung bioaccessibility. *Geosciences* **2021**, *11*, 87. [CrossRef]

38. Suvetha, M.; Charles, P.E.; Vinothkannan, A.; Rajaram, R.; Paray, B.A.; Ali, S. Are we at risk because of road dust? An ecological and health risk assessment of heavy metals in a rapid growing city in South India. *Environ. Adv.* **2022**, *7*, 100165. [CrossRef]

39. Richmond, B.M. Coastal Geology of Upolu, Western Samoa. In *Geology and Mineral Resources of the Central Pacific Basin, Circum-Pacific Council of Energy and Mineral Resources Earth Science Series 14*; Keating, B.H., Bolto, B.R., Eds.; Springer: New York, NY, USA, 1992; pp. 101–125.

40. Goodwin, I.D.; Grossman, E.E. Middle to late Holocene coastal evolution along the south coast of Upolu Island, Samoa. *Mar. Geol.* **2003**, *202*, 1–16. [CrossRef]

41. Sutherland, R.A. Bed sediment-associated trace metals in an urban stream, Oahu, Hawaii. *Environ. Geol.* **2000**, *39*, 611–627. [CrossRef]

42. Tomlinson, D.L.; Wilson, J.G.; Harris, C.R.; Jeffrey, D.W. Problems in the assessment of heavy-metal levels in estuaries and the formation of a pollution index. *Helgoländer Meeresunters.* **1980**, *33*, 566–575. [CrossRef]

43. Hayrat, A.; Eziz, M. Identification of the spatial distributions, pollution levels, sources, and health risk of heavy metals in surface dusts from Korla, NW China. *Open Geosci.* **2020**, *12*, 1338–1349. [CrossRef]

44. Nguyen, V.T.; Dat, N.D.; Vo, T.D.H.; Nguyen, D.H.; Nguyen, T.B.; Nguyen, L.S.P.; Nguyen, X.C.; Dinh, V.C.; Nguyen, T.H.H.; Huynh, T.M.T.; et al. Characteristics and risk assessment of 16 metals in street dust collected from a highway in a densely populated Metropolitan Area of Vietnam. *Atmosphere* **2021**, *12*, 1548. [CrossRef]

45. USEPA (U.S. Environmental Protection Agency). *Exposure Factors Handbook (Final)*; EPA/600/R-09/052F; USEPA: Washington, DC, USA, 2011.

46. USEPA (U.S. Environmental Protection Agency). Risk assessment guidance for superfund. In *Part A, Process for Conducting Probabilistic Risk Assessment*; EPA 540-R-02–002; Office of Emergency and Remedial Response: Washington, DC, USA, 2001; Volume III.

47. Ferreira-Baptista, L.; de Miguel, E. Geochemistry and risk assessment of street dust in Luanda, Angola: A tropical urban environment. *Atmos. Environ.* **2005**, *39*, 4501–4512. [CrossRef]

48. Ma, L.; Aduduwaili, J.; Liu, W. Spatial distribution and health risk assessment of potentially toxic elements in surface soils of Bosten Lake basin, Central Asia. *Int. J. Environ. Health Res.* **2019**, *16*, 3741. [CrossRef]

49. Adamiec, E.; Jarosz-Kreminska, E. Human health risk assessment associated with contaminants in the finest fraction of sidewalk dust collected in proximity to trafficked roads. *Sci. Rep.* **2019**, *9*, 16364. [CrossRef] [PubMed]

50. Phil-Eze, P.O. Variability of soil properties related to vegetation cover in a tropical rainforest landscape. *J. Geogr. Reg. Plann.* **2010**, *3*, 177–184.

51. Pan, H.; Lu, X.; Lei, K. A comprehensive analysis of heavy metals in urban road dust of Xi'an, China: Contamination, source apportionment and spatial distribution. *Sci. Total Environ.* **2017**, *609*, 1361–1369. [CrossRef]

52. Jeong, H.; Ra, K. Characteristics of potentially toxic elements, risk assessments, and isotopic compositions (Cu-Zn-Pb) in the PM_{10} fraction of road dust in Busan, South Korea. *Atmosphere* **2021**, *12*, 1229. [CrossRef]

53. Kong, S.; Lu, B.; Ji, Y.; Zhao, X.; Bai, Z.; Xu, Y.; Liu, Y.; Jiang, H. Risk assessment of heavy metals in road and soil dusts within $PM_{2.5}$, PM_{10} and PM_{100} fractions in Dongying city, Shandong Province, China. *J. Environ. Monit.* **2012**, *14*, 791–803. [CrossRef]

54. Vlasov, D.; Kosheleva, N.; Kasimov, N. Spatial distribution and sources of potentially toxic elements in road dust and its PM_{10} fraction of Moscow megacity. *Sci. Total Environ.* **2021**, *761*, 143267. [CrossRef]

55. Iijima, A.; Sato, K.; Yano, K.; Tago, H.; Kato, M.; Kimura, H.; Furuta, N. Particle size and composition distribution analysis of automotive brake abrasion dust for the evaluation of antimony sources of airborne particulate matter. *Atmos. Environ.* **2007**, *41*, 4908–4919. [CrossRef]

56. Rudnick, R.L.; Gao, S. Composition of the continental crust. In *Treatise Geochem*; Hollad, D.H., Turekian, K.K., Eds.; Elsevier: Oxford, UK, 2003; Volume 3, pp. 1–64. [CrossRef]

57.   Doelsch, E.; de Kerchove, V.V.; Macary, H.S. Heavy metal content in soils of Réunion (Indian Ocean). *Geoderma* **2006**, *134*, 119–134. [CrossRef]

58.   Gao, Y.; Xia, J. Chromium contamination accident in China: Viewing environment policy of China. *Environ. Sci. Technol.* **2011**, *45*, 8605–8606. [CrossRef] [PubMed]

59.   Shahid, M.; Shamshad, S.; Rafiq, M.; Khalid, S.; Bibi, I.; Niazi, N.K.; Dumat, C.; Imtiaz, R. Chromium speciation, bioavailability, uptake, toxicity and detoxification in soil-plant system: A review. *Chemosphere* **2017**, *178*, 513–533. [CrossRef] [PubMed]

60.   del Real, A.E.P.; Perez-Sanz, A.; Garcia-Gonzalo, R.; Castillo-Michel, H.; Gismera, M.J.; Lobo, M.C. Evaluating Cr behaviour in two different polluted soils: Mechanisms and implications for soil functionality. *J. Environ. Manag.* **2020**, *276*, 111073. [CrossRef]

61.   Jeong, H.; Ra, K. Investigations of metal pollution in road dust of steel industrial area and application of magnetic separation. *Sustainability* **2022**, *14*, 919. [CrossRef]

62.   USEPA (U.S. Environmental Protection Agency). *Toxicological Review of Hexavalent Chromium*; CAS No. 18540-29-9; USEPA: Washington, DC, USA, 1998.

63.   Rajaram, B.S.; Suryawanshi, P.V.; Bhanarkar, A.D.; Rao, C.V.C. Heavy metals contamination in road dust in Delhi city, India. *Environ. Earth Sci.* **2014**, *72*, 3929–3938. [CrossRef]

64.   Rabak, J.; Wrobel, M.; Bihalowicz, J.S.; Rogula-Kozlowska, W. Selected metals in urban road dust: Upper and lower Silesia case study. *Atmosphere* **2020**, *11*, 290. [CrossRef]

65.   Ahmad, T.; Ahmad, K.; Khan, Z.I.; Munir, Z.; Khalofah, A.; Al-Qthanin, R.N.; Alsubeie, M.S.; Alamri, S.; Hashem, M.; Farooq, S.; et al. Chromium accumulation in soil, water and forage samples in automobile emission area. *Saudi J. Biol. Sci.* **2021**, *28*, 3517–3522. [CrossRef]

66.   Denny, M.; Baskaran, M.; Curdick, S.; Tummala, C.; Dittrich, T. Investigation of pollutant metals in road dust in a post-industrial city: Case study from Detroit, Michigan. *Front. Environ. Sci.* **2022**, *10*, 974237. [CrossRef]

67.   Ahn, J.S.; Chon, C.M. Geochemical distributions of heavy metals and Cr behavior in natural and cultivated soils of volcanic Jeju Island, Korea. *Geosystem Eng.* **2010**, *13*, 9–20. [CrossRef]